ESTIMATING FOR LANDSCAPE & IRRIGATION CONTRACTORS

by

James R. Huston

Smith Huston, Inc.

P. O. Box 1244
Englewood, CO 80150-1244
PHONE 800-451-5588
FAX 800-451-5494
E-mail: shi@smith-huston.com

Avis is a trademark of the Avis Rent-a-Car System Corporation. Dataquest is a company of the Dun & Bradsheet Corporation. *Dodge Reports* is a trademark of MCGraw-Hill Corporation. The Gallup Organization is a trademark of the Gallup Organization Corporation. Excel is a trademark of the Microsoft Corporation. Hertz is a trademark of the Hertz Corporation. Home Depot is a trademark of the Home Depot Corporation. Lotus 1-2-3 is a trademark of the Lotus Development Corporation. McDonald's Restaurants is a trademark of the McDonald's Corporation. *Money* magazine is a copyright of Time, Incorporated. Price Club is a trademark of the Price Corporation. Rain Bird is a trademark of the Rain Bird National Sales Corporation. *Sesame Street* is a trademark of the Children's Television Workshop. Toro is a trademark of the Toro Company.

Excerpts from *Strategic Planning for Landscape & Irrigation Contractors* by James R. Huston. Copyright 1990 by Smith Huston, Inc. Used by permission.

Excerpts from *Preparing for & Responding to a Down Economy* by James R. Huston. Copyright 1991 by Smith Huston, Inc. Used by permission.

Library of Congress Catalog Card Number: 93-091781

ISBN 0-9628521-2-0   **$75.00 Soft cover**

Library of Congress Cataloging in Publication Data

Huston, James Ray.
   Estimating for landscape & irrigation contractors / James Ray Huston
      p.cm.
      Includes index.
      ISBN 0-9628521-2-0
      1. Landscaping industry—Estimates. 2. Irrigation—Estimates.
         I. Title. II. Title: Estimating for landscape and irrigation contractors.

SB472.565.H87 1993

                                    712'.068'1
                                    QBI93-22113

DEDICATION

The contracting business is complex. And being a small business contractor generally means facing monumental challenges with very limited resources. It is the small business entrepreneur who, in spite of all of the headaches, bureaucracy, and hassles, provides the majority of the jobs in North America, mine included.

It is to the small business landscape and irrigation contractor that this book is dedicated.

James R. Huston
February, 1994

ACKNOWLEDGEMENTS

*The following individuals have contributed
substantially to this work; without their
feedback and assistance its completion
would have been impossible:*

Susan Burns: word processing.

Emmett Childress: word processing and draft editing.

Uriah J. Huston: illustrations.

Sandi Lord and Jerry Mills, The Printing Business:
book and cover design, layout, and final editing.

Susan Titus Osborn: copy editing.

Steve Smith: conceptual design and consultation.

Contributing Contractors/Estimators/Vendors

Mark Christl, office manager/estimator,
The Jonathan Company, Tarzana CA

Robert Dobson, irrigation contractor,
Middletown Sprinkler Company, Inc., Port Monmouth NJ

Marc Dutton, irrigation contractor,
Marc Dutton Irrigation, Inc., Waterford MI

Michael Flowers, construction & maintenance contractor,
Flowers Landscape Development, Inc., Harwinton CT

Dan Foley, construction & maintenance contractor,
D. Foley Landscape, Inc., Walpole MA

Jan Huston, process quality engineer,
Storage Technology Corporation, Louisville CO

Dennis Kendrick, estimator/field supervisor,
Lowries Landscape, Inc., Clarkston MI

Mark Pendergast, nurseryman & construction contractor,
Salmon Falls Nursery, Inc., Berwick ME

Jay Tripathi and Peter Estournes,
construction & maintenance contractors,
Gardenworks, Inc., Healdsburg CA

Robert Vaughan, construction contractor,
Westview Landscape, San Diego CA

Todd Williams, maintenance contractor,
RBI Maintenance, Inc., Littleton CO

John Witonis, wholesale & retail distributor,
Sherwin Williams Company, Inc., Portland ME

CONTENTS

APPENDICES

PREFACE

Do you remember "The Count" on *Sesame Street?* He loves to count. Numbers drive him wild.

Some people actually do love numbers, mathematics, and arithmetic. They love numbers so much that they often think that they can reduce the entire universe down to laws of physics, figures, symbols and numerals. Others are intimidated by mathematics and arithmetic. Myself, I look at numbers as being something like cars: I want them to get me to where I want to go without a whole lot of hassle.

Numbers are necessary to all of us in this world; they can even be a friend. I am amazed at how numbers can provide significant benefit and insight into a construction and maintenance company—if they are properly used.

Most contractors do not realize how important and beneficial the proper formatting and utilization of numbers can be to their company. I like what one large and very successful contractor said: he runs his entire company right from the bid sheet.

Bidding (the science of determining the costs in a job) and estimating (the art of adding markups to costs in order to determine a price for a job) are a vital part of any business. In fact, I would go so far as to say that the way you price work is the foundation (the cornerstone) of your entire business—assuming, of course, that you are in a nation or culture that supports the free enterprise process.

If you cannot price a job correctly, you don't want the job. If you cannot consistently put the right price on a job, you and your company either are in trouble or soon will be.

I have seen many landscape and irrigation companies around the country with fatal flaws in their estimating systems; for some, this has cost them their company. Others have identified the problem(s), made the necessary corrections, and moved ahead.

As I said earlier, estimating is foundational to your entire company. Just about any problem you are experiencing, in one way or another, can be traced to the estimating process. Job-costing/accounting, field production and scheduling, marketing, or strategic planning—all are reflected in estimating.

We will cover the five most common methods used in the market today to estimate and price work. The pros and cons of each will be discussed and you will see the advantage of understanding each method when negotiating work at the bid table or when discussing pricing with a prospective customer.

A good estimating system does not operate in a "vacuum". It should undergird and permeate your whole operation. This is why I say that an estimating system should provide you with a **PLAN** and a **PROCESS** as well as with an accurate **PRICE.** These three items provide the skeleton of the "systems" in your company upon which all else hangs.

And systems are the essential practical ingredient—the manifestation of the philosophies, methodologies, and strategies which encompass the entire company. I refer to this as **PMSS (Philosophies, Methodologies, Systems and Strategies).** You take care of the systems, and the systems will take care of you. You take care of the business, and the business will take care of you.

The **purpose of this book** is to

1. **Teach you the estimating process,** step-by-step, so that you can develop an accurate budget (a plan written in numbers) for your company and your jobs.

And to

2. **Show you how to run your company** with confidence right from the bid sheet using the methods, systems, and strategies explained in this book.

You do not have to love numbers (like "The Count") in order to do so. You MUST, however, develop an appreciation for what they can do for you and for your business.

Let's get started.

INTRODUCTION

Planning is essential for any business person. It is especially valuable and necessary for the landscape and irrigation contractor.

I like what Phillip Crosby says in his book, *Quality Is Free*:

> *"Good things happen only when planned.*
> *Bad things happen on their own."*[1]

Business is a controlled science. It is the process of establishing goals, directing many diverse events toward those goals, collecting data to measure what is happening, analyzing the data, and then adjusting the whole process as necessary to achieve the goals. To accomplish this, business people employ the same methodology used by scientists to conduct research and development (experiments). *The Webster's Encyclopedic Unabridged Dictionary* defines the scientific method as follows:

> *"A method of research in which a problem is identified, relevant data are gathered, a*
> *hypothesis is formulated from these data, and the hypothesis is empirically tested."*

When applying this definition to the landscape and irrigation business process, the project to be bid would be the *problem*. The *relevant data gathered* would be the plans and specifications for the project (or job). The estimate or budget for the job (containing objective measurable standards for materials, labor, labor burden, equipment, etc.) would be the *hypothesis*. The *empirical testing of the hypothesis* would be accomplished by job costing, where you would *test* or compare your budget to actual performance in the field as you performed the project.

Great strides and advancements have been made in the area of medicine over the last fifty years because of the use of the analytical methodology of science in that field. Landscape and irrigation contractors can achieve similar results if a similar methodology is used. Unfortunately, the methods of most contractors resemble "witch doctor" medicine and "voodoo" economics more than they do modern medicine and business.

It is this very process that allows you to control your business and to grow in the same way that scientists control their environments and make progress. The business process provides services and products and produces wealth, while the scientific process creates products which lead to progress in the areas of medicine, business, agriculture, etc. The ends may be different, but both can and usually do supplement one another as they employ the same methodology to achieve the results desired. Business and science control this process by means of **objective, analytical data.** It is this data and this methodology which provide the foundation upon which to build your business.

BUDGETS AND GOALS

Business needs two primary types of plans to be run effectively. The first is the **STRATEGIC PLAN.** Basically, this is a plan put into words. For this, I wrote my first book, *Strategic Planning for Landscape & Irrigation Contractors.*

The second item that you need is a **BUDGET.** A budget is a plan that is reduced to (rendered into) numbers. These numbers help you to organize, to control, and to direct the company. They also provide the vital feedback necessary for analyzing the results and effectiveness of your organizational ability.

These two documents are what are needed to **drive, direct, change,** and **control** your company.

The budgeting process produces key standards (or targets) that will help you to run jobs, as well as the overall business. Diagram (2-1) provides a model for estimating. Diagram (I-1) provides such a model for your entire company.

As you can see, estimating is one portion of the four main functional areas in your business. I would say, however, that it is the most important because everything else, including **total quality management** (TQM), is built upon it.

The budgeting process produces cost standards and targets for:

- Sales (by division)
- Direct costs (by division)
- Gross profit margin (by division)
- Overhead (not necessarily by division)
- Net profit margin (not necessarily by division)
- Labor burden for office and field personnel
- Field-labor hours
- Equipment
- Average wage

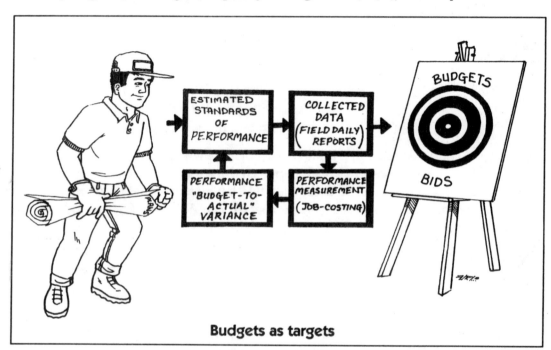

Budgets as targets

These cost standards are then combined with other job-related information in order to determine the cost of the project. Once we identify direct costs on an individual job, we then add markups (profit, overhead, labor burden, sales tax, etc.). The break-even point **(BEP)** for the project is identified once labor burden, sales tax, and overhead are added to the direct costs identified in Phases I and II. A contingency factor and net profit are added to the BEP in order to arrive at the **FINAL PRICE** for the project. It is these two numbers, the BEP and the **FINAL PRICE,** that you are constantly attempting to identify during the estimating process. They answer two critical questions and identify a bidding "envelope" or range within which you need to price your jobs. The BEP answers the question, "**HOW LOW** can you go to get the job and still cover all costs?". The BEP is a compilation of costs for a job and is therefore fairly analytical in nature. However, the FINAL PRICE is subjective and addresses the question, "**HOW HIGH** can you go and still get the job?". What pricing structure will the market bear? Can you add 40% gross profit margin (GPM) to the project and still get the job? Or will this market and this particular job realistically only allow you to add 15-18% GPM above direct costs in order to have a real chance of winning the bid?

We constantly want to identify and monitor the BEP and GPM on jobs being bid. We do so in order to measure, control, and allocate the overhead dollars in our overhead budget. Because overhead (G&A) costs are **indirect** costs, there is no right or wrong amount of overhead costs to allocate to a job being bid. Subsequently, the five various methods of pricing jobs that we will cover are not right or wrong concerning overhead recovery. There simply is **no right or wrong way** to determine the amount of overhead costs to put in a particular job being bid.

How high? How low?

The question to ask is not "How much overhead should you add to a job and by which method?". Rather, it is, "If you sell a particular job for a certain price and perform it on-budget, will it help you achieve your budgeted gross and net profit dollar goals for the year?".

The goal now is to measure, to allocate, and to control overhead costs throughout the budget year, not to argue over which method is correct—there is none!

The process, however, does not stop at estimating. You need to constantly monitor the standards or targets identified above—for the company, as well as for individual jobs.

In order to run your jobs effectively, you job cost them. To run a company effectively, you monitor the profit and loss (P&L) financial statement, as well as the balance sheet. These are, if you will, your report cards for the company or division as a whole.

1. Individual jobs need to be monitored by job costing.
2. We monitor our entire company as though it were **one big job.** What difference would it make if individual jobs were profitable, if you do not have enough of them to support the overhead structure and the costs for the company as a whole?

   Major items to monitor are as follows:

   - Overhead costs
   - Field-labor hours and costs
   - Equipment costs

These are the items you **need to be checking constantly** in order to run your entire company—and not just your individual jobs.

We will cover (in detail) the methods and reports used to monitor these two key areas of your business in a later section.

The budgeting and controlling process of a landscape and irrigation contracting company is a never-ending series of events that occur simultaneously. You are constantly setting goals (budgeting targets), taking action, monitoring progress, adjusting, and fine-tuning. The goal of this process is to have your company operate as smoothly as a Swiss watch amidst the constant tumult.

Proper budgeting and estimating will not solve **all** of your problems, but it will address and provide the tools for solving the majority of those problems—certainly, the most important ones.

THREE KINDS OF BUDGETS

In a business, you need three types of budgets in order to plan and to control the business effectively:

1. The **ESTIMATING BUDGET** is designed to help you bid your work and to recover ALL costs accurately.
2. The **ACCOUNTING** (or tax) **BUDGET** is required in order to help you meet government/tax requirements and regulations.
3. The **CASH FLOW BUDGET** is designed to help you:
   a. Predict your cash flow.
   b. Identify those periods where it may be reduced, so that you can take the necessary steps to compensate (e.g., by means of a loan or an increased line of credit).

I. MANAGEMENT/ADMINISTRATION
(Directing the Company)

1. Strategic planning
2. Marketing strategy
3. Estimating strategy
4. Training/safety
5. Networking/support groups
6. Finance
7. Quality control
8. Production control
9. Computer systems
10. Information management
11. Equipment planning
12. Budgeting/goal setting
13. Tax planning
14. Board of directors/advisors
15. Customer service
16. _____

II. MARKETING:
SALES/ESTIMATING/
CONTRACT ADMINISTRATION
(Getting the work)

1. Marketing strategy
2. Bidding strategy
3. Estimating budget
4. Field-labor hour budget
5. Bidding goals
6. Sales goals
7. Contract administration
8. Customer service
9. _____
10. _____
11. _____
12. _____

III. FIELD PRODUCTION
(Producing the work)

1. Quality control
2. Pre-construction meetings
3. Job startup/coordination
4. Post-construction meetings
5. Daily field reports
6. Crew/job scheduling
7. Equipment scheduling
8. Training/safety
9. Customer service
10. Purchasing materials
11. Inventory control
12. _____
13. _____
14. _____
15. _____
16. _____

IV. FINANCE/ACCOUNTING
(Keeping score)

1. Job costing
2. Purchase order control
3. Information management/systems
4. Accounts payable
5. Accounts receivable
6. Cash flow budget
7. Accounting system/software
8. Financial statements
9. Charts/graphs/reports
10. Bonus systems
11. _____
12. _____
13. _____
14. _____

DIAGRAM I-1. COMPANY OVERVIEW

In this book, we will cover estimating budgets. You should consult with your accountant or certified public accountant (CPA) to develop the other two.

BUDGET PERIODS

We develop an estimating budget in twelve-month increments, not necessarily a fiscal or calendar year. It can be either, but we can also develop a budget for the immediate next twelve months, starting at any time during the year. It is easier to prepare an estimating budget parallel to a company's fiscal year (which may be the calendar year), but it is not necessary to do so.

I tell seasonal contractors that it takes two to three seasons to really get your operation under control, to get it running smoothly (like that Swiss watch). Year-round businesses (such as those in Southern California or Florida) usually take about the same amount of time (two to three years), as well to achieve the same results. This may seem like a long time, but remember, what you are trying to do is like **trying to fix a flat tire on a car that is still moving down the road.** There will be lots of other things to do while you are trying to implement the systems, strategies, and methodologies that are outlined in this book. This is why I tell contractors to think long-term: *the process takes* **time** *and* **persistence.**

Do not misunderstand. Significant progress is expected immediately. You will see a real, initial impact on your business as a result of implementing the methodologies, systems, and strategies contained in this book. However, there is **no magic wand** to wave or fairy dust to throw at your company. Plan on spending the next couple of years focusing and fine-tuning this entire process throughout your business.

Fixing a tire on a moving car...

HISTORICAL DATA: THE FOUNDATION

We start by reviewing historical data: financial statements, balance sheets, insurance rates, etc. We build from there, based on assumptions tailored to your company.

We hope that you have good historical data; data that is accurate, properly formatted, and consistent (categories are consistently defined from one year to the next). If we do not have good historical data—if there is no rhyme or reason to it; you cannot identify trends; the data is either inconsistent, not of the proper kind, or is not formatted correctly—then we have what I call **HYSTERICAL** (not "historical") data: it makes no sense.

START-UP COMPANIES

We can easily budget for start-up situations. Even though we do not have historical data, using estimates about our costs, we can still make basic assumptions, go through the planning process, and put together a fairly accurate estimating budget.

INCREMENTAL GOAL TIME PERIODS

Year One: The first year of this two to three year process is a year of setting goals, making major changes in the estimating process, and collecting and monitoring data in specific formats.

Year Two: The second year is, generally, a period of re-evaluation:
* Did we meet last year's goals?
* Did we budget and plan accurately and effectively?
* Did we make the appropriate subsequent adjustments when we repeated the budgeting process?

Year Three: The third year is more of the same: further evaluation and fine-tuning of the organization and the processes used to control and direct it.

QUALITY CONTROL AND TOTAL QUALITY MANAGEMENT (TQM)

Once you develop your estimating budgets and begin to estimate properly (which you can begin to do after one or two days of training), you can then decide how to format and monitor key types of data throughout your business by using the objective, analytical methodology of science described earlier.

Comparing objective, quantifiable (or measurable) plans (budgets, goals, targets, and standards) to actual performance is a measuring process that is at the heart of the TOTAL QUALITY CONTROL process for your company.

If you cannot budget and estimate properly (produce an accurate price) on a job, you can never achieve significant quality control, let alone "total" quality management of your company. Your estimating systems should be the foundation upon which you develop TQM throughout your entire organization.

SUMMARY

Even though your estimating system is only one segment of your business, it is the brain and nerve center (the command post) of the entire operation. It should not only provide you with a *PRICE,* but it should also provide a communicable *PLAN* and a quality control *PROCESS* that facilitates total quality control throughout your company. This process is based upon the objective, analytical, and quantifiable methods of science which allows you and your staff to direct and to control the company to predetermined goals and profitability.

Section I

CONTRACTING MANAGEMENT
FOR THE YEAR 2000

CHAPTER 1
THE NEXT TEN YEARS

While the basic free enterprise philosophy may be the same, business methodologies, systems, and strategies are changing at an amazingly accelerating rate. Technological advances, personal computers and all facets of business are developing so quickly, product and service obsolescence often occur within months of their introduction to the marketplace.

Further compounding this problem for the landscape and irrigation industry is the fact that, as an industry, construction and maintenance contracting lags behind the general business community in addressing and implementing these technological advances.

TRENDS

There are two very important trends to be aware of in the landscape and irrigation industry.

1. SURVIVAL OF THE "FATTEST"

It is easier to stay "king of the hill" than it is to get on top of that hill in the first place.

As noted in my first book, *Strategic Planning for Landscape & Irrigation Contractors:*

> "Landscape Management, *in an article entitled 'Survival of the Fattest' (February, 1989), hit on something very big when it noted an important, yet often overlooked, trend in construction today. It said, 'Among those companies reporting (1988) gross sales above $3 million, the average increase in sales was $830,000. Smaller companies, on average, lost $55,000 in gross sales.' Look at it this way: when you are playing 'king of the hill', it is always easier to stay on top of the hill than it is to get there. Getting there the first time is the real battle.*

> "Construction contracting appears to be going through today what the farm industry went through years ago. At one time, there were hundreds of thousands of small farmers, each doing a relatively small amount of business. Today, there are far fewer farmers, but—individually—they are producing a far greater amount of gross sales.*

> "The Landscape Management *article talked about the 'staying power' of the big boys. I choose to call it 'buying power,' and the main commodity being purchased is strategic 'people power.' If you grow and are profitable in the process, you can afford to replace yourself in many key positions within the daily operation of your business. This allows you to research and develop key areas of the business*

(marketing, estimating, computers, standardizing field operations, finances, etc.) and to develop more effective systems and strategies for running it. As the saying goes, 'It is impossible to drain the swamp when you are up to your (waist) in alligators.' The big boys, if they are profitable and smart, buy themselves the opportunity to drain not only today's 'swamp' but tomorrow's 'swamp' as well."[2]

2. THE TURNKEY (OR FRANCHISE) APPROACH TO LANDSCAPE AND IRRIGATION CONTRACTING.

While not entirely valid, this approach does have its applications.

McDonald's Restaurants, for example, developed and popularized the turnkey—or franchise—approach to business. Essentially, it is the philosophy of placing proven methods, systems, and strategies in place before you unlock the front door and open for business. Success follows almost immediately *if* you have done your homework and have planned properly.

Although the turnkey approach has more validity in landscape maintenance and irrigation service businesses, the concept does have applications in the construction business, as well.

I have seen many young, aggressive contractors implement the right philosophies, methodologies, systems, and strategies and take off. Not just a few have successfully weathered the inevitable storms of adversity and recessions because they understood their businesses—and their methodologies, systems, and strategies—well enough to keep them on track.

On the other hand, I have seen some of the oldest and biggest landscaping contractors self-destruct from within, because they ignored essential elements of their organization. They did not have a "wholistic" approach to their business.

Keep in mind that if you only have three of the four tires necessary on a race car, you really do not have a race car. You just have an expensive piece of metal alloy, rubber, plastic, and steel. Similarly, if your landscape and irrigation contracting company lacks an essential component of the systems necessary to run it effectively, all you have is a disaster looking for a place to happen.

The landscape and irrigation contracting companies that will still be thriving in the year 2000 will be doing so because of proper planning. These firms will have met the tumult and pressures of the marketplace and will have learned to capitalize upon and advance through adversity. They will have done so because they implemented proven methodologies, systems, and strategies.

Essentially, they will have heeded Tom Peters's challenge of "thriving on chaos" outlined in his book of the same name.

CONCLUSION

There is a train that is leaving for the year 2000. That train is revolutionizing contracting management in construction and maintenance services businesses. Smart contractors (large and small) are boarding it more and more. It won't necessarily be an easy trip, but it should be fun, and it should help us to make money.

The choice is up to you. Will you board that train and join a growing breed of savvy market- and business-wise contractors? Or will you be left standing at the station?

CHAPTER 2
ESTIMATING SYSTEM OVERVIEW

PURPOSE

To provide an overview of an estimating system

OBJECTIVES OF AN ESTIMATING SYSTEM

An estimating system should provide:

1. An accurate **PRICE** on a job for your company

2. A communicable **PLAN** that the field can use to run the job

3. An objective quality-control **PROCESS** that accounting (or your accountants) can use to job cost the project. One that your estimators can use to become better estimators and that will train new estimators.

It is this **PROCESS** that allows us to address the most common complaint that I hear from landscape and irrigation contractors: *"I do not know if I am making money on this project."*

Most estimating systems produce a questionable price for a job and little else. The system that is taught in this book will do all three of the above. It will provide a **PRICE,** a **PLAN,** and a **PROCESS**—a quality-control process for controlling the entire spectrum of events that occur within a company.

We're going to cover a lot of ground in this book. In order to illustrate our points, we're going to use two models or business paradigms. The first is the "Company Overview" in Diagram (I-1), and the second is the "OPPH Bidding Method Overview" in Diagram (2-1).

Diagram (2-2), the "Traditional or MORS Bidding Method Overview," is included for comparison purposes which will be discussed in detail in subsequent chapters.

Keep referring back to these models, especially if you get confused. Once you put a bid together, use the models to help you to review a bid in order to see if you have forgotten anything.

Also, be on the constant alert for **LEAKS.** For every check you write from your checkbook, you should ask: *"Where is this expense included in the model?".*

In other words, you should be able to find that check (or see where that check is recovered back to you) in these estimating models. Also, you want to look for areas where you may be **"DOUBLE-DIPPING"**—that is, where you are charging twice for the same item.

Let me guide you through the models and the three phases found in an estimate.

THREE PHASES OF AN ESTIMATE

1. THE FINISHED PRODUCT OR SERVICE

This stage of the bidding process includes costs for materials physically included in the finished product or service provided and the labor, equipment, and subcontractor costs associated with it. If it's part of the finished product or service provided, the cost is included here.

A. **MATERIAL** is included AT COST.

B. **LABOR** is calculated in FIELD-LABOR HOURS multiplied by either a crew average wage (CAW) or specific wage rates for differing classes of labor.

C. **EQUIPMENT** is included by multiplying hours by the cost per hour (CPH) for each piece of equipment.

D. **SUBCONTRACTORS** are included AT COST.

2. GENERAL CONDITIONS

Included here are costs for those items required to produce the finished product or service (material, labor, equipment, and subcontractors as above), but which are not directly required to produce the end product or service. For instance, portable fences, toilets, or traffic control may be required by or "tied to" a specific job being bid, but they are not directly required to produce (mow) the mown lawn, plant or trim the trees, fertilize the site, mulch the planting areas, etc. Although there is sometimes a fine line between Phase I and II items, the detailed discussion and list of general condition items in Chapter 12 should clarify any confusion concerning them. It should also be noted that distinguishing whether a particular item should be in Phase I or II is less critical than ensuring that it is accurately included in one of them.

General condition items include (but are not limited to) the following:

- Crew pickup trucks
- Supervision time
- Crew drive time
- Mobilization of equipment to and from the job site
- Material hauling labor and equipment costs
- Portable toilets
- Portable fences
- Permits, dump fees, licenses, soil tests, as-builts, etc.

3. MARKUPS

A. Sales tax

Appropriate sales tax markups are applied to materials and other items as required by state and local laws. In some states, tax is now applied to the final price of the job, not just to the materials.

B. Labor burden

Labor is marked up by a predetermined percentage in order to recover FICA, FUTA, SUTA, general liability insurance, field-crew vacations, holidays, sick days, workers' compensation insurance, medical insurance, etc.

C. Overhead

A number of methods are commonly used by contractors to recover overhead. Usually, either a percentage markup is used on materials, labor (including labor burden), equipment, and subcontractors (or just one of these items); or an overhead cost per field-labor hour is used.

D. Net Profit

When added together, the items identified up to now produce your break-even point (BEP). NET PROFIT is then ADDED to your BEP.

I. PRODUCTION OF FINISHED PRODUCT/SERVICE:

MATERIAL	LABOR	EQUIPMENT	SUBCONTRACTORS
(at wholesale cost)	(labor hours multiplied by crew average wage or labor rate)	(equipment hours on job multiplied by predetermined cost per hour)	(at cost)

II. GENERAL CONDITIONS:

MATERIAL	LABOR	EQUIPMENT	SUBCONTRACTORS
(same)	(same)	(same)	(same)

III. MARKUPS:

MATERIAL	LABOR	EQUIPMENT	SUBCONTRACTORS

A. Sales tax on materials:
B. Labor burden on labor:
C. Overhead recovery (total number of field-labor hours in sections I and II above multiplied by company overhead cost per field-labor hour [OPH]):
D. Contingency factor (if applicable):
E. Net profit (a straight percentage markup on total of all costs for the job or desired company average profit per hour [PPH] multiplied by total number of field-labor hours in sections I and II above):

DIAGRAM 2-1. BIDDING OVERVIEW (THE OPPH METHOD)

I. PRODUCTION OF FINISHED PRODUCT/SERVICE:

MATERIAL	LABOR	EQUIPMENT	SUBCONTRACTORS
(at wholesale cost)	(labor hours multiplied by crew average wage or labor rate)	(equipment hours on job multiplied by predetermined cost per hour)	(at cost)

II. GENERAL CONDITIONS:

MATERIAL	LABOR	EQUIPMENT	SUBCONTRACTORS
(same)	(same)	(same)	(same)

III. MARKUPS:

MATERIAL	LABOR	EQUIPMENT	SUBCONTRACTORS

A. Sales tax on materials:
B. Labor burden on labor:
C. Overhead recovery:

MATERIAL	LABOR	EQUIPMENT	SUBCONTRACTORS
10%	variable %	25%	5%

D. Contingency factor (if applicable):
E. Net profit (a straight percentage markup on total of all costs for the job):

DIAGRAM 2-2. BIDDING OVERVIEW (THE MORS OR "TRADITIONAL" METHOD)

The usual NET PROFIT MARKUP range for construction projects is:

(1). For highly competitive commercial jobs: 6% to 10%

(2). For residential projects: 15% to 20%

Your choice of a specific percent to use on a particular job is usually determined by:

(1). How badly you need the job

(2). Its size

(3). The risk involved

(4). What the market will bear—or what you think you can get away with

(5). Your ability to negotiate and to win the confidence or trust of the client

E. Contingency Factor

Contractors will sometimes add a fifth markup category that I often refer to as the **aggravation factor.** Basically, this is additional money tacked on to the price of a job that will make it worth it to you to incur additional risk or to put up with certain situations or individuals (architects, owners, developers, homeowners, etc.) with whom you would rather not deal.

Remember that except for the contingency factor, you are constantly attempting to differentiate between and to determine two numbers in the estimating process: total costs (which includes direct and overhead costs) or the BEP, and the maximum price that the market will bear.

COMPLETING THE BID

We can now equate or define **bidding** as determining those costs involved with Phases I and II in Diagrams (2-1) and (2-2) plus sales tax and labor burden.

The **estimating** process is completed when we add markups in Phase III to the first two phases and arrive at the **final price** for a particular job.

One last note: direct costs consist of all costs in Phases I and II plus sales tax on material and labor burden added to labor. Indirect (general and administrative) or overhead costs are all other costs that cannot be directly identified and tied to a particular job.

Section II

ESTABLISHING BUDGETS:
Determining Costs and Costing Standards

PURPOSE

To teach you how to develop and determine cost data and costing standards for your company or division for the budget items listed below. These will then be used by you to bid your jobs.

1. Sales projections
2. Direct costs for materials and subcontractors
3. Direct costs for labor (labor hours and average wage)
4. Direct costs for equipment (and rental equipment)
5. Labor burden
6. Overhead costs
7. Per-hour and ratio analysis standards

INTRODUCTION

The budgeting process is actually an exercise in goal-setting. You are really establishing standards or targets to aim for throughout your business.

Budgets reduce the diverse aspects of the business into those symbols known as numbers. The only difference between budgeting goals and goals written in words is that numbers can interact with each other (add, subtract, multiply, and divide), while words cannot.

More than **ANY** other tool, the "numbers" of your business can—if properly formatted and correctly symbolized—help you and your people run your company more effectively. However, to do so, your P&L statement, balance sheet, job costing, and other reports need to be set up properly.

Sadly, most contractors do not understand the budgeting and accounting process well enough to take full advantage of what is available to them.

TYPES OF BUDGETS

Top management needs to be sure that the three types of budgets discussed in the Introduction are in place and used.

They are as follows:

1. The CASH FLOW BUDGET—for projecting cash flow into and out of the company. If a shortfall is projected, management needs to obtain the funding to cover it.

2. The ACCOUNTING (or tax) BUDGET—designed to help you plan for and meet the tax requirements established by law. This budget should be addressed with the direct assistance of your accountant or Certified Public Accountant (CPA).

3. The ESTIMATING BUDGET—designed to establish and to project:

   A. Overhead to recover
   B. Sales goals
   C. Gross and net profit margins (GPM and NPM)
   D. Direct costs (or cost of goods sold—COGS)
   E. Average wage
   F. Labor burden
   G. Field-labor hours
   H. Various ratios and per-hour standards

   Once this budget is established, it should be monitored regularly by means of reports, graphs, and various visual aids.

The estimating budget can and should be one of the key operational tools for establishing standards/goals by which to monitor the daily functions of your business. Unfortunately, in a landscape and irrigation contracting company, this happens all too infrequently.

These three budgets are very similar, but they differ in some important aspects. For example, let's say that next year you are going to produce a color brochure for your company. Because of a quantity discount, you decide to print enough brochures for three years. This will cost you $6,000 for the artwork, printing, etc.

For cash flow purposes, you put the entire $6,000 expense in next year's CASH FLOW BUDGET (because you intend to pay all of the expense in that year). Likewise, if you are on the accrual accounting basis—which is generally best for contractors—the entire $6,000 will be included in your TAX PLANNING BUDGET for this year (when you incur the expense), whether you pay for it or not.

Your ESTIMATING BUDGET, however, will account for only one year's worth of brochures (or $2,000) in the overhead portion, because it is not reasonable to charge next year's clients for brochures to be used two and three years from now. We will also place $2,000 for overhead recovery in the estimating budgets two and three years from now, because that is reasonable. It is not necessarily what is real. What is **real** is the $6,000. However, **what is real and what is reasonable are not necessarily always the same.**

INCOME OR PROFIT AND LOSS (P&L) STATEMENT

Diagram (II-1) is a sample P&L statement. The P&L statement is a very important report that you need to monitor in order to understand what is (or is not) happening in your company. I think of it as a job-cost report for all of your jobs thrown together. If used properly and in conjunction with other data and reports, it can provide considerable insight into many aspects of your business.

MULTIPLE DIVISIONS

If you have multiple divisions, with each doing different types of work, I would provide that division with its own P&L if that division comprises over 20% of the business' gross sales. You may even want to break a division down into separate subdivisions (that may or may not do the same types of work).

For instance, I know of a landscape maintenance company with over twenty-four crews in one division (each crew consisting of four laborers). That one division is broken down into six subdivisions (or departments) with four crews and a foreman in each subdivision. Each subdivision is tracked and compared to the others for profitability and cost.

		MONTH	%	Y-T-D	%
1.	SALES (REVENUES)-------->	83,333	100.0%	1,000,000	100.0%

2. COST OF SALES (DIRECT COSTS):

	MONTH	%	Y-T-D	%
MATERIAL (W/TAX)	27,652	33.2%	320,000	32.0%
LABOR	15,794	19.0%	200,491	20.0%
LABOR BURDEN	5,143	6.2%	65,283	6.5%
SUBCONTRACTORS	5,000	6.0%	50,000	5.0%
EQUIPMENT	5,245	6.3%	66,830	6.7%
EQUIPMENT RENTAL	750	0.9%	3,500	0.4%
MISC. DIRECT JOB COSTS	700	0.8%	6,000	0.6%
TOTAL COST OF SALES----->	60,284	72.3%	712,104	71.2%

3.	GROSS PROFIT MARGIN----->	23,050	27.7%	287,896	28.8%

4. OVERHEAD (GENERAL & ADMINISTRATIVE--G&A OR INDIRECT) COSTS:

	MONTH	%	Y-T-D	%
ADVERTISING	500	0.6%	8,000	0.8%
BAD DEBTS	0	0.0%	5,000	0.5%
COMPUTERS, SOFTWARE	208	0.3%	2,500	0.3%
DONATIONS	0	0.0%	200	0.0%
DOWNTIME LABOR	302	0.4%	2,782	0.3%
DOWNTIME LABOR BURDEN	98	0.1%	906	0.1%
DUES AND SUBSCRIPTIONS	75	0.1%	850	0.1%
INSURANCE (OFFICE)	1,200	1.4%	8,000	0.8%
INTEREST AND BANK CHARGES	275	0.3%	3,300	0.3%
LICENSES & SURETY BONDS	75	0.1%	300	0.0%
OFFICE EQUIPMENT	50	0.1%	1,200	0.1%
OFFICE SUPPLIES	176	0.2%	1,750	0.2%
PROFESSIONAL FEES	300	0.4%	3,500	0.4%
RADIO SYSTEMS	388	0.5%	4,650	0.5%
RENT (OFFICE & YARD)	1,200	1.4%	14,400	1.4%
SALARIES-OFFICE	4,167	5.0%	50,000	5.0%
SALARIES-OFFICER(S)	3,333	4.0%	40,000	4.0%
SALARIES-LABOR BURDEN	959	1.2%	11,502	1.2%
SMALL TOOLS/SUPPLIES	78	0.1%	1,500	0.2%
TAXES (ASSET & MILL TAX)	0	0.0%	0	0.0%
TELEPHONE	324	0.4%	3,300	0.3%
TRAINING & EDUCATION	55	0.1%	2,000	0.2%
TRAVEL & ENTERTAINMENT	250	0.3%	1,200	0.1%
UNIFORMS & SAFETY EQUIP.	55	0.1%	550	0.1%
UTILITIES	28	0.0%	600	0.1%
VEHICLES, OVERHEAD	693	0.8%	8,320	0.8%
YARD EXP./LEASEHOLD IMP.	500	0.6%	1,000	0.1%
MISCELLANEOUS	33	0.0%	250	0.0%
	0	0.0%	0	0.0%
TOTAL OVERHEAD---------->	15,323	18.4%	177,560	17.8%

5.	NET PROFIT MARGIN-(NPM)->	7,727	9.3%	110,336	11.0%
6.	OVER/UNDERBILLING------->	_____	0.0%	_____	0.0%
7.	REVISED (NPM)----------->	7,727	9.3%	110,336	11.0%

DIAGRAM II-1. P&L STATEMENT FORMAT

FORMATTING THE P&L STATEMENT

Formatting the P&L statement is crucial. We encourage you to use a format similar to the one in Diagram (II-1). It is a P&L for one division of a corporation. Sole proprietorships and partnerships should use a similar format except for the officers' salaries (owner's draw) which is equivalent to the net profit margin or bottom line.

1. SALES

The sales for the division are at the top of the report, broken down by month and year-to-date (YTD).

2. DIRECT COSTS

Cost of Sales (or direct costs) follows next and includes everything that can be tied directly to a specific job.

You want to put as much as you can into direct costs, including labor burden (i.e., FICA, FUTA, workers' compensation insurance, general liability insurance, unemployment insurance, field-labor medical insurance, field holiday and vacation pay, etc.). Depreciation for your field equipment should be identified and included in direct costs. However, depreciation for office/overhead, vehicles, and office equipment should be included in overhead.

3. GROSS PROFIT MARGIN (GPM)

The GPM is determined by subtracting all direct costs, to include labor burden and sales tax, from sales.

4. OVERHEAD

Overhead is often referred to as indirect or G&A (General and Administrative) costs. The point is that you cannot directly pin G&A costs to a particular field operation or job; hence, the term indirect or general is used to describe them.

5. NET PROFIT MARGIN (NPM)

Net profit margin, before taxes and interest, follows overhead. Ten percent NPM is a nice, round percentage to shoot for after all costs have been included in the expense categories.

I like to graph the NPM "budget vs. actual" on a monthly or quarterly basis. I prefer that a company be on the accrual basis for their financial statements, because it provides a more consistent and accurate sales and expense picture.

6. PROFIT ADJUSTMENTS

Some companies include an "over/underbilling" adjustment category at the bottom of their P&L statement to reflect profits more accurately—if they are, in fact, overbilled or underbilled on their jobs for a given period. Otherwise, your NPM may show dramatic variations from month to month, because billings do not match expenses for the period in question.

It is important that we **CLEARLY DEFINE** and **CONSISTENTLY TRACK** the items on a P&L statement.

We cannot keep changing the types of information included in categories into which we enter budget (or actual) cost data. Otherwise, we will destroy not only the integrity of the information but its usefulness, as well. If direct costs and/or overhead definitions change from year to year, we will not be able to identify and monitor trends.

HISTORICAL DATA AND ESTIMATING BUDGETS

Estimating budgets start by projecting revenues and expenses for the next twelve months (or the next budget period in question). It is easiest to do so a few months prior to a new fiscal year—which may or may not be the calendar year (January 1st to December 31st).

You should review your accumulated historical data (if it is available) and make adjustments to it that will reflect changes that will occur during the next twelve month period.

If historical data is not available (or if you are in a start-up situation), simply make your best estimate of projected revenues and/or expenses.

CHAPTER 3
SALES AND REVENUE BUDGET PROJECTIONS

PURPOSE

To explain the company and division budget format, to see the reason for establishing divisions, and to establish sales and revenue budget projections

In our sample budget in Diagram (3-1), we have sales projections for both a construction and a maintenance division. Direct costs, overhead (indirect or general administrative costs), and net profit margin are all allocated or identified by division. It is not uncommon for a landscape and irrigation company that is doing over 1-2 million dollars in gross annual sales to have 2-4 or more divisions (e.g., construction, maintenance, nursery, tree trimming, pest control, equipment division, etc.) for accounting, budgeting, and management purposes.

Seldom do companies divisionalize their accounting reports where gross annual sales are under $250,000. It is usually impractical to do so. I would also add that as a general rule, it is not practical to establish a separate division in your accounting and budgeting system for any activity that accounts for less than 15-20% of gross annual sales.

You might be asking, "Why divisionalize in the first place? What difference does it make?". I would encourage you to divisionalize your budgets and your P&L statements, at least for sales and direct costs (Phases I and II on Diagrams (2-1) and (2-2)), for the following reasons:

1. It can help identify where you are making money, thus showing where you may want to allocate more resources and therefore increase sales.
2. It can help identify where you are losing money and possible steps to take in order to stop the "leaks" in your company.
3. Start-up divisions (especially nurseries) can be tracked and monitored, thus identifying trends and future profitability. Unless they are separated from other divisions, start-up divisions will dilute the profit and loss performance of those divisions.
4. Division P&L statements can serve as a division manager's "report card," thus helping top management to decentralize the company and allow division managers to operate more autonomously with less interference from top management.
5. Productivity trends can be identified by division and either corrective or reinforcing actions can be taken. Otherwise, one division may be improving while the other is declining. If they are lumped together, they may

	CONSTRUCTION DIVISION				MAINTENANCE DIVISION				TOTALS	
	MONTH	%	ANNUAL	%	MONTH	%	ANNUAL	%		%
1. SALES (REVENUES)------>	83,333	100.0%	1,000,000	100.0%	41,667	100.0%	500,000	100.0%	1,500,000	100.0%
2. COST OF SALES (DIRECT COSTS):										
MATERIAL (W/TAX)	26,667	32.0%	320,000	32.0%	2,233	5.4%	26,791	5.4%	346,791	23.1%
LABOR	16,708	20.0%	200,491	20.0%	14,884	35.7%	178,608	35.7%	379,099	25.3%
LABOR BURDEN	5,440	6.5%	65,283	6.5%	4,846	11.6%	58,157	11.6%	123,440	8.2%
SUBCONTRACTORS	4,167	5.0%	50,000	5.0%	208	0.5%	2,500	0.5%	52,500	3.5%
EQUIPMENT	5,569	6.7%	66,830	6.7%	4,912	11.8%	58,941	11.8%	125,771	8.4%
EQUIPMENT RENTAL	292	0.4%	3,500	0.4%	125	0.3%	1,500	0.3%	5,000	0.3%
MISC. DIRECT JOB COSTS	500	0.6%	6,000	0.6%	833	2.0%	10,000	2.0%	16,000	1.1%
TOTAL COST OF SALES---->	59,342	71.2%	712,103	71.2%	28,041	67.3%	336,497	67.3%	1,048,601	69.9%
3. GROSS PROFIT MARGIN---->	23,991	28.8%	287,897	28.8%	13,625	32.7%	163,503	32.7%	451,399	30.1%
4. OVERHEAD (GENERAL & ADMINISTRATIVE--G&A OR INDIRECT) COSTS:										
ADVERTISING	667	0.8%	8,000	0.8%	333	0.8%	4,000	0.8%	12,000	0.8%
BAD DEBTS	417	0.5%	5,000	0.5%	0	0.0%	0	0.0%	5,000	0.3%
COMPUTERS, SOFTWARE	208	0.3%	2,500	0.3%	83	0.2%	1,000	0.2%	3,500	0.2%
DONATIONS	17	0.0%	200	0.0%	8	0.0%	100	0.0%	300	0.0%
DOWNTIME LABOR	232	0.3%	2,782	0.3%	71	0.2%	851	0.2%	3,633	0.2%
DOWNTIME LABOR BURDEN	75	0.1%	906	0.1%	22	0.1%	270	0.1%	1,176	0.1%
DUES AND SUBSCRIPTIONS	71	0.1%	850	0.1%	28	0.1%	335	0.1%	1,185	0.1%
INSURANCE (OFFICE)	667	0.8%	8,000	0.8%	333	0.8%	4,000	0.8%	12,000	0.8%
INTEREST AND BANK CHARGES	275	0.3%	3,300	0.3%	63	0.2%	750	0.2%	4,050	0.3%
LICENSES & SURETY BONDS	25	0.0%	300	0.0%	54	0.1%	650	0.1%	950	0.1%
OFFICE EQUIPMENT	100	0.1%	1,200	0.1%	38	0.1%	450	0.1%	1,650	0.1%
OFFICE SUPPLIES	146	0.2%	1,750	0.2%	54	0.1%	650	0.1%	2,400	0.2%
PROFESSIONAL FEES	292	0.4%	3,500	0.4%	125	0.3%	1,500	0.3%	5,000	0.3%
RADIO SYSTEMS	388	0.5%	4,650	0.5%	314	0.8%	3,770	0.8%	8,420	0.6%
RENT (OFFICE & YARD)	1,200	1.4%	14,400	1.4%	500	1.2%	6,000	1.2%	20,400	1.4%
SALARIES-OFFICE	4,167	5.0%	50,000	5.0%	3,167	7.6%	38,000	7.6%	88,000	5.9%
SALARIES-OFFICER	3,333	4.0%	40,000	4.0%	1,667	4.0%	20,000	4.0%	60,000	4.0%
SALARIES-LABOR BURDEN	971	1.2%	11,655	1.2%	626	1.5%	7,511	1.5%	19,166	1.3%
SMALL TOOLS/SUPPLIES	125	0.2%	1,500	0.2%	63	0.2%	750	0.2%	2,250	0.2%
TAXES (ASSET & MILL TAX)	0	0.0%	0	0.0%	0	0.0%	0	0.0%	0	0.0%
TELEPHONE	275	0.3%	3,300	0.3%	125	0.3%	1,500	0.3%	4,800	0.3%
TRAINING & EDUCATION	167	0.2%	2,000	0.2%	100	0.2%	1,200	0.2%	3,200	0.2%
TRAVEL & ENTERTAINMENT	100	0.1%	1,200	0.1%	67	0.2%	800	0.2%	2,000	0.1%
UNIFORMS & SAFETY EQUIPME	46	0.1%	550	0.1%	50	0.1%	600	0.1%	1,150	0.1%
UTILITIES	50	0.1%	600	0.1%	30	0.1%	360	0.1%	960	0.1%
VEHICLES, OVERHEAD	693	0.8%	8,320	0.8%	607	1.5%	7,280	1.5%	15,600	1.0%
YARD EXP/LEASEHOLD IMP.	83	0.1%	1,000	0.1%	63	0.2%	750	0.2%	1,750	0.1%
MISCELLANEOUS	21	0.0%	250	0.0%	13	0.0%	150	0.0%	400	0.0%
TOTAL OVERHEAD---------->	14,809	17.8%	177,713	17.8%	8,602	20.6%	103,227	20.6%	280,940	18.7%
5. NET PROFIT MARGIN-(NPM)->	9,182	11.0%	110,184	11.0%	5,023	12.1%	60,276	12.1%	170,460	11.4%
6. PROFIT ADJUSTMENTS------> (ADD'L SAL'S, BONUSES, ETC.)	_____	0.0%	_____	0.0%	_____	0.0%	_____	0.0%	_____	0.0%
7. REVISED NPM------------->	9,182	11.0%	110,184	11.0%	5,023	12.1%	60,276	12.1%	170,460	11.4%

DIAGRAM 3-1: COMPANY BUDGET

	CONSTRUCTION DIV.		MAINT. DIV.		COMBINED	
8. RATIOS & PER HOUR ANALYSIS:	$	%	$	%	$	%
A. SALES PER HOUR (SPH):	$46.25	100.0%	$23.83	100.0%	$35.21	100.0%
B. DIRECT COST PER HOUR (DCPH):	$32.94	71.2%	$16.04	67.3%	$24.61	69.9%
C. GROSS PROFIT MARGIN PER HOUR (GPMPH):	$13.32	28.8%	$7.79	32.7%	$10.60	30.1%
D. OVERHEAD PER HOUR (OPH):	$8.22	17.8%	$4.92	20.6%	$6.59	18.7%
E. PROFIT PER HOUR (PPH):	$5.10	11.0%	$2.87	12.1%	$4.00	11.4%
F. MATERIAL/LABOR (M/L) RATIO:	1.60	:1	0.15	:1	0.91	:1
G. MATERIAL PER HOUR (MPH):	$14.80	32.0%	$1.28	5.4%	$8.14	23.1%
H. EQUIP/LABOR (EQ/L) RATIO %:	35.1%		33.8%		34.5%	
I. EQUIPMENT PER HOUR (EQPH):	$3.25	7.0%	$2.88	12.1%	$3.07	8.7%
J. AVERAGE WAGE:	$9.27	20.0%	$8.51	35.7%	$8.90	25.4%
K. BILLABLE FIELD-LABOR HOURS:	21,621.0		20,980.0		42,601.0	
9. OVERHEAD RECOVERY (THE TRADITIONAL OR "MORS" METHOD):						
A. MATERIAL:	$32,000	10.0%	$2,679	10.0%	$34,679	10.0%
B. LABOR & BURDEN:	$125,630	47.3%	$85,313	36.0%	$210,943	42.0%
C. EQUIPMENT:	$17,583	25.0%	$15,110	25.0%	$32,693	25.0%
D. SUBCONTRACTORS:	$2,500	5.0%	$125	5.0%	$2,625	5.0%
	$177,713	100.0%	$103,227	100.0%	$280,940	100.0%

DIAGRAM 3-1: COMPANY BUDGET (CONT.)

appear to be unchanged and a damaging trend may be left unnoticed.

6. Bonuses can be given partly based on division performance.

7. It can facilitate the preparation of a budget for the entire company.

8. Feedback can be supplied to division personnel. It should be noted that timely and accurate feedback (TAF) can be **your best motivator.**

9. It can provide vital division "budget vs. actual" performance information to estimators, who can then take corrective actions, if necessary, in the estimating process.

10. It can help you to compare equipment costs to revenues generated by that equipment. Thus it can help you to monitor whether or not equipment is covering its costs (justifying its existence) and/or if you are estimating it accurately.

11. Different types of divisions usually have different GPMs. Therefore, divisions with similar sales but with a higher GPM will be more profitable to a company and contribute more to the bottom line: net profit margin (NPM). This is assuming that the overhead is similar for the two divisions. Such information can be invaluable when planning and budgeting for future growth, cash flow, and profitability of the business.

NOTE: Whenever the term "company" appears in the text, it can refer to either a division or to the company as a whole.

ACTION POINT

Use the Exhibit (1) worksheets at the back of the book, (using one sheet for each division) and fill in the sales projections for each division for the budget period in question. Review last year's sales figures to gain an historical perspective. Increase or decrease sales and revenue projections from the last year as deemed necessary. If past performance figures are not available, or if you are in a start-up situation, use your best judgement to make a sales projection for the budget period.

CHAPTER 4

DIRECT COSTS:
Materials and Supplies, Subcontractors, Rental Equipment, Miscellaneous Direct Costs, and Commissions on Sales

PURPOSE
To explain how to establish direct cost budgets

The direct costs covered in this section are relatively easy to budget. Although actual costs may vary dramatically from your budget projections as you proceed into the year, this does not (necessarily) pose a significant problem in the estimating process.

Let's start with materials and supplies.

MATERIALS AND SUPPLIES

1. USING HISTORICAL DATA

If you have good historical data by division, you can use—for budget purposes—the percent of materials from last year as a percent of the materials that you are going to use for the next budgeting year.

In other words, if materials and supplies were running 30% of your sales last year, next year—if your sales are going to increase substantially—you simply take 30% of your new sales projection and use that number for your materials budget for next year.

2. USING RATIOS

If you do not have good historical data that is broken down by division, the best way to budget materials and supplies is to use a material-to-labor (M/L) ratio to calculate dollar amounts for materials. Simply multiply the projected field-labor payroll budget for the year by the appropriate ratio.

If you are in a start-up situation and have no historical data upon which to base your calculations, you can use these ratios as a good way of projecting your costs for materials and supplies within the next budget period.

A. Plant Material Only (3:1)

Companies that just install plant materials are usually able to install $3.00 of material for every $1.00 of gross field payroll (excluding labor burden) paid.

For a company with an average wage of $7.00 per field-labor hour (FLH), a 3:1 M/L ratio translates into $21.00 of material costs per hour (MPH).

B. Hardscape and Plant Material (2 to 2.5:1)

Companies that install plant materials and hardscape (bender board, boulders, decks, patios, etc.) usually have a 2 to 2.5:1 M/L ratio.

In other words, they install $2.00 to $2.50 in materials for every $1.00 spent on field labor. If the average wage is again $7.00, the MPH would run between $14.00 and $17.50.

C. Irrigation, Hardscape, and Plant Material (1.5 to 1.75:1)

Companies that install plants, hardscape, and irrigation generally have a M/L ratio from 1.5 to 1.75:1. Again, using an average wage of $7.00 per hour, the MPH ranges from $10.50 to $12.25.

D. Irrigation Only (1.5 to 2.0:1 and 3.0 to 4.0:1)

Residential irrigation companies usually have a 1.5 to 2.0:1 M/L ratio. For every $1.00 spent on field labor, $1.50 to $2.00 is spent on material costs. However, irrigation contractors who do large commercial and golf course work often realize M/L ratios of 3 to 4:1.

E. Maintenance (.10 to .15:1)

Lawn maintenance or management companies generally have a M/L ratio ranging from .10 to .15:1. This translates into an MPH ranging from $ .70 to $1.05.

SUBCONTRACTORS

We can project subcontractor costs pretty much the same way that we did materials and supplies. If subcontractors have been running 15% of sales on our P&L statement (and we are going to increase or decrease sales in the next year), then we can multiply projected sales by the historical 15% in order to calculate projected subcontractor costs. If our sales are going to change significantly from the previous budgeting period, this procedure will make the appropriate adjustments.

In many cases, subcontractor costs as a percent of sales will vary dramatically from year to year, and there is no way to project these fluctuations. This is fine because it really is not crucial that we identify this number with absolute precision.

RENTAL EQUIPMENT

Rental equipment is very much like subcontractors, materials, and supplies. Costs usually stay fairly consistent. The best thing to do is to look at what you have done in the past and to make a projection with which you feel comfortable.

If you lack workable historical data, you might put a couple of thousand dollars in the rental equipment category. Track it and make adjustments as you go. It is not that important to project it accurately, but give it your best "guesstimate." That should suffice.

MISCELLANEOUS ITEMS

This is where you account for any cost that can be directly connected to a job but really does not fit into one of the other categories. Include such items as: permits, dump fees, storage containers, dumpsters, bags for grass clippings, blueprints, as-builts, photography expenses, etc.

COMMISSIONS

You may or may not have commissioned salespeople in your company. If you do, you should have a category in "direct costs" where you would account for their commissions.

Usually, residential contractors are the only ones who have commissioned salespeople. Commercial landscaping and irrigation contractors generally do not. This is primarily due to the slim profit margins and to the fact that a job bid this year may not be installed for eight, ten, twelve, or even sixteen months down the road. It simply is not feasible to have someone selling that kind of work on a commission basis.

The key regarding commissions is to think out how you are going to handle the whole process ahead of time and to **identify sales goals** for each individual salesperson.

1. AMOUNT

Commissions paid to salespeople generally run 7% to 9% of the price of a project. For instance, if a project installed is priced at $10,000, the commission to the salesperson would be between $700 to $900. The commission would include car and other related expenses. Therefore, this person would not be reimbursed on top of the $700-$900 for those expenses.

2. SALES GOALS

Determine the SALES GOAL for a salesperson by dividing his desired pay for the year (including all expenses) by the percent of commission (e.g., 8%).

Using an average commission of 8%, a commissioned salesperson wanting to make $30,000 per year would have to sell $375,000 worth of landscaping projects throughout the year in order to earn $30,000. The key is to tie the commission to a sales goal and to an all-inclusive commission percentage. Eight percent plus or minus one percent is a good one to use.

3. RESPONSIBILITIES

The next item to address about a commissioned salesperson is, "What will he do? Does he only sell?". In other words, is it the responsibility of the commissioned salesperson just to sell, or is he to supervise the work and make sure that the client (the customer) is happy throughout the entire process?

A. You have to keep commissioned salespeople involved in the project and interested in keeping the customer happy until the end of that project and collection of the final payment.

The salesperson does not just sell jobs. He must have a vested interest in making sure that each job is carried out the way the customer wants it carried out and that the product delivered to the client is the product that was sold to the client.

B. The commissioned salesperson should also usually be responsible for collecting the final check from the customer.

By insisting that the salesperson collects the final check from the customer, you are establishing essential quality control checks and balances in your company.

C. Multiple Job Responsibilities

(1). Structure

If the salesperson is to have other responsibilities (e.g., designing jobs and performing some administrative duties), keep this entirely separate from the commission.For instance, if he is going to be making 7% to 9% on a project, I assume that he is not going to be designing the project. If he is going to design the project as well, then pay him at an hourly rate for design time in addition to the commission (or according to some other predetermined agreement), but keep design time separate from commissions.

Structure the commissions this way, because if you get a really good salesperson, you probably do not want to tie him down spending a lot of time designing. In all likelihood, you will want him to sell more and design less.This structure also allows you to subcontract design work. The result is that your company becomes more flexible structurally. You can expand and increase sales, or downsize quickly and with minimal difficulty.

(2). Establish Sales Goals for Salespersons with Multiple Job Responsibilities

If the salesperson is going to design (or whatever), subtract the projected salary for designing (or other duties) from the his projected sales goal income. Divide what is left over by your commission percentage (e.g., 8%). The result is the amount of dollars that he needs to sell for the year.

As an example, let's say a salesperson wants to make $30,000 a year. The salesperson is also going to do some design work. Approximately $10,000 of his annual paycheck will come from design work. Take what is left over ($20,000) and divide that by 8%. He would, therefore, have to sell approximately $250,000 in contracts in order to cover his commissions.This number can then easily be monitored in order to see if this person is bringing in enough work to cover his annual paycheck.

4. COMPENSATION OF COMMISSIONED SALESPERSONS

The next question that arises is, "How is the commissioned salesperson going to be paid?". Basically, there are three different methods.

A. Straight Commission on Sales

A straight commission means that he does not get paid until he sells a job and collects the final check when that job is completed.

B. Draw Against Commissions

With a draw, unlike a straight commission, salespeople are paid prior to collecting the final check for the project. This is considered a "draw" against any future commissions to be earned.

Commissions are still a percentage of sales, but they are paid prior to all moneys being collected. It's a little different from the straight commission method; but overall, it is the same concept. The timing of the payment of the commission is the only difference.

C. Salary

The salesperson on salary usually receives a monthly paycheck that is going to be the same from month to month.

Although different than the previous two methods, you still ultimately tie pay to sales. The key thing to remember is that if the salesperson does not generate a certain amount of sales, he does not have a job. The principle that a salesperson has to bring a certain amount of sales into the company in order to earn a certain amount of pay does not change; however, the method of payment does. He gets paid steady, consistent amounts.

We have to keep asking ourselves, "Does the salesperson have a clearly-identified sales goal? Is he bringing that amount into the company?". If he is not, you have to make appropriate adjustments.

D. Comparison and Warnings

The straight commission method is simpler for the company, because if the salesperson does not sell and collect the money, the salesperson does not get paid. The draw against commissions is a little more complex, but the concept is the same.

In reality, the salaried concept is also the same; however, it is a little harder to monitor. Quite often, this is where a company gets into trouble. It puts a person on salary and never ties that salary to a particular amount of sales that this person has to generate. Memories get short; there is no method for comparing the two (salary to revenue generated). Salespeople sometimes go on for literally years until someone wakes up and realizes that an individual is not justifying his or her existence within the organization. I have seen cases where people have been paid for a year or two and (because of the bureaucracy and other complications within the organization) no one ever really compared what the salesperson was being paid to the sales that that person was bringing into the company.

Earlier, I mentioned a 7-9% range for commissions on sales. I have seen cases where when you divided actual commissions paid by actual sales generated, the percentage was 20-25%. It is simply impossible to justify having someone in the organization earn such a high commission. This is so because you must add this 20-25% commission on top of overhead and profit for the company. The market (the ones I have seen) just will not allow you to get away with this.

Once you determine the amount of commissions a person is going to earn throughout the year, based on a sales projection, enter that amount into DIRECT COSTS in the commissions category.

Once again, remember that the commissions category does not include money paid for design fees or for pay for administrative tasks that are going to be included in overhead.

ACTION POINT

Fill in the appropriate direct cost categories on Exhibit (1) for the next budget period in question.

CHAPTER 5
FIELD-LABOR PAYROLL AND HOUR BUDGETS

PURPOSE
To explain and to establish the field-labor budget

The field-labor payroll and hour budget is one of the primary tools that you will use for:

1. **Measuring** your company's and/or division's performance.

2. **Controlling** and **directing** the company as a whole and individual divisions, if you have them so identified.

We will use the field-labor hour (FLH) budget in conjunction with other key information to monitor trends and to track vital company statistics. We could use percentage of sales to do so but the FLH method is more accurate. For instance; company or division sales can be divided by the respective field-labor hours to obtain a company-wide or division Sales per Hour (SPH) figure, which we will use later.

Overhead will be divided by the respective field-labor hours in order to obtain an Overhead per Hour (OPH) figure.

FORMAT

We will use Exhibit (2) in the back of the book to develop your company and/or division field-labor hour budget figures. Fill in the information using historical data, if available. Otherwise, use your best judgement when completing the exhibit. Start with the landscape crew data and fill in the appropriate data. If you do not plan to break down your company into divisions, lump all of your labor data in the landscape crew section of Exhibit (2).

AVERAGE WAGE (AW)

Start with the foreman in "1 A (1)" on the exhibit, and fill in the columns as follows:

(a). Quantity of foremen in the company (and/or division) for a specific projected hourly rate for the budget year. If the foremen's hourly rate varies more than one to two dollars per hour, use two categories or an average. Projected hourly rates should include any anticipated raises for that category.

(b). Enter the average hours per month worked (i.e., 4.3 weeks per month x 9 hours per day x 5 days per week = 194 hours per month).

Adjust appropriately for each category.

(c). Enter the number of months per year that each category will work. Make seasonal adjustments accordingly.

(d). Multiply columns (a) x (b) x (c), and enter the total in column (d).

(e). Enter appropriate hourly rate—or average rate for a category, as appropriate.

(f). Multiply columns (d) x (e), and enter total projected payroll in column (f).

(2-6). Fill in the other labor classifications for the division.

(7). Total the field-labor hours and payroll on row (7).

(8). Compute the company/division average wage (AW) on row (8) by dividing the total payroll by the total field-labor hours.

(9). Complete row (9) on the exhibit after you calculate the OTF for the company/division in section 2 of the worksheet as follows:

OVERTIME FACTOR (OTF)

A. Fill in the average number of hours worked per man per week (by division).

B. Calculate and fill in the average number of overtime hours worked per man per week throughout the entire year.

C. Fill in the average number of overtime hours paid per man per week. This is usually the number in row B multiplied by 1.5 (time-and-a-half). This, however, may differ from state to state.

D. Review the example.

E, F, and G. Fill in your company/division overtime information here.

RISK FACTOR (RF)

The "Risk Factor" is applied to your average wage or your crew average wage (including OTF) on a job-by-job basis and is not used for your field-labor budget. It usually ranges from 10-20%, but it could go higher. If you feel very confident in the amount of labor hours bid into a project, a Risk Factor of 10% is normally applied to your average wage and OTF.

If, on the other hand, you are doing unfamiliar work and/or do not feel confident about the accuracy of the labor hours bid into a job, increase the Risk Factor accordingly and multiply your average wage and OTF by it.

TOTALS

Total the projected hours and payroll for the year:

A. Total your field-labor hours (for all divisions).

B. Total your payroll for the year.

C. Note and compare your results to last year's hours (if available).

D. Note and compare your results to last year's payroll (if available).

ACTION POINT

Fill in Exhibit (2) in the back of the book and calculate your company/divisions field-labor hour and payroll budget, average wage, and overtime factor.

CHAPTER 6
CREW AVERAGE WAGE (CAW)

PURPOSE

To explain how to calculate CAW for individual jobs bid

We have covered how to calculate the company/division average wage in Chapter 5. Such an average wage could be referred to as a "true" average wage. It is the total field-labor payroll for a company/division divided by the total number of field-labor hours in that company/division. In fact, in order to be extremely accurate, we should subtract holiday, vacation, sick time, downtime hours—for each respective labor rate. I actually do this when I prepare a budget for a company, with the aid of a computer spreadsheet program such as Lotus 1-2-3 or Excel. As insightful as this exercise may be, it is not necessary for the purposes of estimating because we will build a more accurate crew average wage (CAW) when bidding a specific job.

CAW is utilized to estimate labor costs on specific jobs. It is very simple to calculate, but before we determine these rates for your jobs, let's discuss *why* they are important. Here is a specific example.

Last week I helped a design/build landscape and irrigation installation contractor bid a residential project. Two other companies were also bidding this project (which my client, by the way, had designed). Because my client had worked with the customer throughout the design process, we felt that we had an edge on the competition. However, we knew that we still had to be competitive with our price.

Our initial estimate was $33,345. There were 585 field-labor hours included in the bid. Initially, my client was going to place a three-man crew on the job: a foreman at $14 per hour; a leadman at $8; and a $6 per hour laborer.

The crew average wage worked out to be $9.33. After putting the bid together, we looked at some options concerning the crew. My client thought that he could do the project with a five-man crew. This would be comprised of: a foreman at $14; a leadman at $8; and three laborers—one at $6 and two at $5 per hour. The crew average wage was then recalculated to be $7.60 per field-labor hour.

Note what this change did to his bid.

Old CAW		$9.33	
New CAW		-7.60	
Savings per FLH		1.73	
Total FLHs in bid		x 585	
	Subtotal	1,012	
Plus 30% labor burden		304	
	Subtotal		1,316
Plus 10% OTF			132
	Subtotal		1,448
Plus 10% RF			145
	Total cost savings		1,593
Plus 10% profit			159
	Total price adjustment		$1,752

6-1. CREW AVERAGE WAGE (CAW) COST SAVINGS

This contractor was able to reduce his price—or increase his profit margin, if he had chosen to do so—by a minimum of $1,752, simply by adjusting his crew composition in the field and his crew average wage in the bid.

There was another savings on the job: general conditions' costs would also decrease. The larger crew would complete the job in approximately 15 days, compared to 24 days for the smaller crew. This would translate into additional cost savings as follows:

Crew pickup truck	$234.00	
Crew drive time	110.00	
Supervisory time	295.00	
Total cost savings	639.00	
Plus 10% profit	64.00	
Total price adjustment		$703.00

6-2. ADDITIONAL CAW COST SAVINGS

Total possible savings on this project could total over $2,400, or a little more than 7% of the initial bid. All of this due to a single adjustment regarding the composition of the crew working on the job and the crew average wage used in the bidding process.

Calculating the CAW

Diagram (6-3) displays a sample crew average wage worksheet that is filled out for a planting crew. It is self-explanatory.

Review the diagram, and then calculate your own crew average wage rates for your most often used crew combinations.

Date: 10 / 22 / 99
Job: Jones Residence
Type of Crew: Planting

Type of Labor	Name	Labor Rate		
Foreman	Ed	$14.00		
Leadman	Bill	8.00		
Laborer	NA	6.50		
Laborer	NA	5.50		
Laborer	NA	0.00		
	TOTAL	34.00		
1. Divide labor total ($34.00) by number of crew members (4)		8.50		
2. Add OTF (10%)		.85		
Subtotal			9.35	
3. Add RF* (10-20% depending on difficulty)			.94	
Subtotal			10.29	
4. Round up to nearest ten cents to obtain CAW			$10.30	

DIAGRAM 6-3. CREW AVERAGE WAGE (CAW) CALCULATION

*For a more complete explanation of the "Risk Factor," refer to Chapter 5.

ACTION POINT

Use Exhibit (3) in the back of the book and calculate your crew average wage rates for your most common crew combinations.

CHAPTER 7
LABOR BURDEN

PURPOSE
To explain how to calculate labor burden

Definition: Labor burden consists of those items that can be directly correlated to both office and field payroll and which vary in direct proportion to it. It is calculated as a percentage that is then applied to field and office payroll.

SET PERCENTAGES

Most labor burden items (FICA, FUTA, SUTA, WCI, general liability insurance-GLI) are assigned to you as a fixed percent that will vary little when calculated into your labor burden. For instance, FICA is currently 7.65%. This does not change, whether you are calculating labor burden for prevailing wage projects or non-rated ones.

ACCUMULATED WORK TIME

Other labor burden items (such as sick pay, paid holidays, or vacation time) are based on accumulated paid non-work time (usually days or hours) and need to be calculated differently when bidding rated or non-rated jobs.

The usual range for labor burden in the United States is between 12–16% for office personnel and 28–37% for field personnel.

The labor burden in Canada generally runs 10-20% (plus or minus 2–3%) for office and field personnel, respectively. This is primarily due to the method by which Canada handles its health insurance, workers' compensation insurance, and equivalent for FICA and FUTA.

REVIEWING LABOR BURDEN "BUDGET-TO-ACTUAL"

Labor burden calculations should be reviewed every four to six months and adjusted whenever your insurance rates or experience modification changes. You should also review and adjust your labor burden if you are bidding either rated or non-rated jobs and have not done so in the recent past.

Some companies actually have accounting software that allows them to produce financial statements that display labor burden "budgeted" for actual field and office payroll compared to "actual" accumulated expenses for labor burden items.

For example, if labor burden is calculated to be 32% for field-labor payroll, the accounting software will make the calculations and then compare the result (in dollars and as a percent) to actual dollars spent for FICA, FUTA, WCI, GLI,

	CONST.	IRRIG.	MAINT.	OFFICE
F.I.C.A. (COMPANY PORTION) RATE	.0765	.0765	.0765	.0765
F.U.T.A. (FEDERAL UMEMP.) RATE	.008	.008	.008	.008
S.U.T.A. (STATE UNEMP.) RATE	.035	.035	.035	.035
WORKERS' COMPENSATION INSURANCE RATE	.1297	.1025	.1297	.01
GENERAL LIABILITY INSURANCE RATE (1)	.030	.030	.030	NA
VACATIONS (2)	.0093	.0093	.0093	NA
HOLIDAYS (3)	.0222	.0222	.0222	NA
SICK DAYS (4)	—	—	—	NA
MEDICAL/HEALTH INSURANCE (5)	.0150	.015	.015	NA
MISC._____	—	—	—	—
	========	========	========	========
TOTALS-------------------------------->	.3257	.2985	.3257	.1295
CALCULATIONS FOR ABOVE: OR	32.6%	29.9%	32.6%	13%

(1). GENERAL LIABILITY INSURANCE:

USE POLICY RATE FOR HUNDRED DOLLARS OF FIELD PAYROLL OR
DIVIDE ANNUAL PREMIUM $ 6,015   BY PROJECTED ANNUAL FIELD
PAYROLL $ 200,491   = . 03   .

(2). FIELD LABOR VACATIONS:

DIVIDE TOTAL FIELD CREW VACATION WEEKS PAID IN THE YEAR   5   BY THE
TOTAL FIELD WORK FORCE WEEKS IN THE YEAR 540   = .0093   .

(3). FIELD LABOR HOLIDAYS:

DIVIDE TOTAL FIELD CREW HOLIDAYS PAID IN THE YEAR   60   BY
THE TOTAL FIELD WORK FORCE MAN DAYS IN THE YEAR 2700   = . 0222   .

(4). FIELD LABOR SICK DAYS:

DIVIDE TOTAL FIELD CREW SICK DAYS EARNED IN THE YEAR   Ø   BY
THE TOTAL FIELD WORK FORCE MAN DAYS IN THE YEAR   Ø   = . Ø   .

(5). FIELD LABOR HEALTH/MEDICAL INSURANCE:

DIVIDE TOTAL ANNUAL HEALTH/MEDICAL INSURANCE PAYMENT (FOR FIELD
PERSONNEL ONLY) $ 3,000   BY PROJECTED ANNUAL FIELD PAYROLL
$ 200,491   = .015   .

DIAGRAM 7-1. LABOR BURDEN CALCULATION

holiday pay, vacations, etc. This is an excellent checks-and-balances method for ensuring that your calculated labor burden percentages are correct.

Another method for accomplishing the same thing is to establish a labor burden checking account into which you deposit and accumulate labor burden funds. The amount is determined simply by multiplying your payroll (e.g., $3,000) by your labor burden percent (e.g., 32%). The result ($960.00) is then deposited into the labor burden checking account. These funds are then accumulated, and all labor burden items (including holidays, vacations, etc.) are paid from it.

If your labor burden percentages are correct, you should be able to pay all labor burden expenses from that account. If you cannot cover labor burden expenses from it, your calculations are probably wrong.

USE OF FUNDS

Do not be too hasty to assume that the excess funds accumulated in this account can be withdrawn and spent for non-labor burden expense purposes. I know of one contractor who thought that he had extra money to spend near the end of the fiscal year, so he then proceeded to spend it (not all of it wisely, I might add) on non-essentials. Shortly thereafter, he received two bills—for a combined total of over $40,000—for workers' compensation insurance and general liability insurance premiums that were due for the previous year.

Although he had charged the correct amount for labor burden on his jobs, he did not collect and accumulate it throughout the year. Once his insurance company did its annual audit, it determined that this contractor's business had almost doubled in the previous twelve months, as did his payroll. Because his insurance premiums were based on a much lower projected payroll amount for the year, he only paid half of what he should have paid; and although he included the correct amount of labor burden in his bids and collected the correct amount of dollars from his customers to cover it, he spent the money on non-labor burden items—thinking that the extra money was profit.

The lesson of this story: **Do not just *charge* the correct amount for labor burden, but *save* and *accumulate* this amount in order to ensure that it is used to *pay labor burden expenses*.**

VIEW LABOR BURDEN AS AN INCENTIVE

Labor burden items (such as vacation pay, holiday pay, medical and/or dental insurance, sick pay, etc.) **must be earned.** I have had contractors ask me if they should provide various types of benefits, and whether these benefits would serve as an incentive to productivity. Let me use some examples as my answer.

One contractor provided virtually everyone in his field crew with a minimum of two weeks vacation, six paid holidays per year, and three sick days per year. Medical and dental benefits were also provided. Labor burden ran at well over 40% for the company. This company was very profitable, yet competitive with its pricing.

The key was that this company's field crew **earned** all of these benefits. The entire crew had been with the company for at least eight years. They were budget- and production-oriented and they produced. An extremely high quality standard had been established—and achieved—throughout the company. Things were done right the first time, and mistakes were not tolerated. This company is going strong to this day, and this company's owner still rewards his people for their results.

Another company provided similar benefits, but the field was never run properly. Project budgets and standards were not clearly identified in the bidding process, nor were they (nor could they be) communicated clearly to the field. The owner thought that he could increase productivity if he provided increased benefits (that is, labor burden). In reality, he actually rewarded his crew for poor results. This company self-destructed and is no longer in existence. Ultimately, it was the fault of the owner of that company, because he did not run his company effectively.

My point is this: vacation pay, holiday pay, health insurance plans, etc., do not significantly increase productivity, in and of themselves. They should be viewed as a reward for a job well-done. You cannot just "throw" money and benefits at people and expect the same to increase productivity.

If you want to **INCREASE PRODUCTIVITY** (any productivity), provide your people with:

- Well-defined, reasonable **GOALS** and **BUDGETS**
- Timely and accurate **FEEDBACK (TAF)**
- Fair market **WAGES**
- Results-oriented **BONUSES**

You will not necessarily increase productivity by increasing your labor burden items.

CALCULATING LABOR BURDEN

Diagram (7-1) displays the labor burden calculations for a company with three divisions: construction, maintenance, and irrigation. Office labor calculations are also included within the diagram.

Perhaps your entire company consists of only construction, or only maintenance, or only irrigation; or perhaps you might choose to lump your entire labor force as one labor division. This is fine, but once a division accounts for more than 20% of your gross annual sales, I would encourage you to separate it out in your financial statements and to calculate its own labor burden.

Notice that office labor burden has no labor burden percentage figure for general liability insurance, vacation pay, holiday pay, sick pay, or medical and health insurance. These items (except for general liability insurance) are included in administrative overhead totals for these people.

In the event that an individual's payroll amount (whether they are on salary or are paid on an hourly basis) is partly in administrative overhead and partly in field payroll, split his burden amounts accordingly. For instance, if an owner spends 50% of his or her time working in the field and the other 50% performing administrative tasks, you would then put half of his vacation time, holiday pay, sick pay, and medical insurance premiums in administrative overhead and the other half in the respective field division labor burden category.

It should be noted that labor burden includes only the portion that is paid by the company for that item. The portion for FICA, FUTA, and SUTA that is taken out of an individual's paycheck is not included in labor burden calculations.

Let's explain labor burden step-by-step. In our calculations, we will display percentages as decimals. For example: 27% is .27 and .8% is .008.

1. F.I.C.A. (FEDERAL INSURANCE CONTRIBUTIONS ACT)

The present FICA rate is .0765 (or 7.65%). Enter .0765 into the respective columns.

2. F.U.T.A. (FEDERAL UNEMPLOYMENT TAX ACT)

The present FUTA rate is .008 (or .8%). Enter .008 into the respective columns.

3. S.U.T.A. (STATE UNEMPLOYMENT TAX ACT)

Enter the appropriate SUTA rate for your respective state and/or country, if applicable.

In the event that there is a ceiling (or cap) on what you have to pay for an individual (e.g., SUTA is paid only on the first $7,000 of an individual's gross annual payroll), use the method below to calculate the correct percentage that will be entered into labor burden.

You have eight people on payroll. The SUTA rate is 3.5%. The SUTA ceiling is $7,000. Therefore, you pay 3.5% on the first $7,000 of payroll for a specific individual. Once payroll for the year exceeds $7,000 for a specific individual, you are exempt from paying SUTA for his payroll (in this example only).

Three employees earn more than $7,000 per year, for a total of $45,000. The remaining five people earn under $7,000, for a total of $25,000. You calculate your labor burden SUTA rate as follows:

A. For the three employees earning more than the cap:

3 (# of people) x $7,000 (the cap)	=	$21,000
Multiply the $21,000 x .035	=	$735

B. For the remaining five employees earning less than the cap:

$25,000 (combined payroll) x .035	=	$875
Total projected SUTA payment: $735 + $875	=	$1,610

C. Divide the total projected SUTA payment by the total projected payroll for the company (or the respective division):

$1,610 + ($45,000 + $25,000) = $1,610 + $70,000 = .023 = 2.3%

You would use 2.3% (not 3.5%) for your SUTA labor burden percentage.

4. WORKERS' COMPENSATION INSURANCE (WCI)

Enter the respective WCI percentage in the appropriate division column.

Factor in your experience modification, if applicable. A WCI rate of 7% would be adjusted to 6.3% if your experience modification rate was .9 or 90%.

5. GENERAL LIABILITY INSURANCE (GLI) AND UMBRELLA POLICIES

These labor burden items are calculated for field division payroll only, and they are indicated as such on your policy.

General liability insurance and umbrella policy rates are sometimes difficult to translate into labor burden percentages. However, with a little effort, or a phone call to your insurance agent, you should be able to obtain the needed information from your policy.

We will combine these two types of insurance into one labor burden percent category:

A. GLI

Your policy will usually display GLI insurance rates in one of two ways:

(1). Rate per thousand. This method indicates your rate as a dollar amount of premium for each $1,000 of field payroll.

For instance, a rate of $29 per thousand will translate into a GLI rate of .029, or 2.9%. That is, you divide $29 by $1,000.

If you have no umbrella premium, you would use 2.9% as your GLI labor burden percent.

(2). Premium per total payroll. This method simply states that your premium is based on a certain amount of payroll.

If your company/division payroll is $50,000 per year and your premium is $1,450 per year, divide the annual premium by the projected payroll amount (i.e., $1,450 + $50,000 = .029, or 2.9%).

If you have no umbrella policy, you would again use 2.9% as your GLI labor burden percent.

B. Umbrella Policy

If your policy includes an umbrella policy, simply divide the premium amount by the projected payroll amount on the policy.

For an annual premium of $600, it would be $600 + $50,000, or .012 (which is 1.2%). Add the .012 to your GLI portion amount (which we determined to be .029) in your GLI labor burden percent. In this case, it would equal .041, or 4.1%.

Contact your insurance agent if your policy does not provide a projected annual payroll amount or if you cannot find the information that you need to calculate your general liability insurance labor burden percent.

C. Average National Rates

Rates for the GLI portion of the labor burden factor usually range from 1.5 to 6% nationally. If your GLI labor burden percent is 5% or higher, it is time to shop around for insurance coverage. Contact four or five insurance agents three to four months before your present policy expires. Be sure to provide the exact, same information to each agent for quote purposes. Otherwise, when you compare the quotes, you will not be comparing apples to apples.

Inform your current agent of what you are doing (if you desire, tell them that I told you to do so). You may be surprised at how creative and how competitive your current carrier can become, once they know that you are shopping around.

6. VACATIONS

To obtain the labor burden vacation percent, determine the total weeks of paid vacation for the field-labor force, for the company as a whole, or for a particular division. Divide that number by the total number of actual weeks worked by the company (or respective division). Do not include downtime (see Chapter 8 for a detailed explanation), equipment repair time, paid holidays, paid vacations, paid sick days, etc., in the actual weeks worked.

Allow me to illustrate (in this illustration we will first make our calculations using hours and then convert them to weeks). ABC Landscape and Irrigation Company has three field laborers who each work forty weeks a year. The average work week per man is forty-five hours (5 days a week x 9 hours per day). Everyone receives four paid holidays. One individual receives two weeks paid vacation during the season, another receives one week.

From reviewing past payroll records (time cards and memory), the owner determines that he will budget forty hours per season per field-crew member for downtime. Equipment repairs will total twenty hours per field-crew member per season.

Actual work weeks are computed as follows:

A. Dividing weeks by weeks

Total paid vacation weeks	3

Approximate total hours paid per year:

45 (hrs/wk) x 40 (wks/yr) x 3 (men)	5,400
Minus paid holiday hrs: 3 (men) x 4 (days) x 8 (hrs/day)	- 96
Minus paid vacation hrs: 3 (wks) x 40 (hrs/wk)	- 120
Minus projected downtime: 3 (men) x 40 (hrs/yr)	- 120
Minus equipment repair time: 3 (men) x 20 (hrs/yr)	- 60
Actual field hours worked	5,004

Convert to actual weeks worked:

Divide 5,004 hours by 45 average hours/week	111

Convert to vacation labor-burden decimal:

Divide vacation weeks by actual weeks worked:

$3 \div 111 = .027$ or 2.7% vacation labor burden

B. Dividing hours by hours (a more accurate method)

Total vacation hours: 3 (men) x 40 (hrs/yr)	120
Total actual hours worked	5,004

$120 \div 5,004 = .024$ or 2.4% vacation labor burden

COMPARISON

Notice that if we had not deducted downtime hours, holidays, etc., from our actual weeks worked, our vacation labor-burden decimal would have been lower in both cases:

Method "A"

   5,400 hours + 45 hours/week = 120 actual weeks worked.

   3 (paid vacation weeks) + 120 = .025 or 2.5% (versus 2.7%).

Method "B"

   Paid vacation hours = 120 + 5,400 (approximate total hours paid) = .0222 or 2.2% (versus 2.4%).

Not a lot of difference, but the lower decimals (.024 and .027) are more accurate calculations.

7. HOLIDAYS

Because we have already done most of the work determining the vacation labor burden decimal, calculating holiday labor burden will be much easier. Let's continue using the above example:

A. Dividing days by days

   Total paid holidays: 3 (men) x 4 (paid holidays)   =   12

   Total man-days worked per year:

   $$\frac{\text{Actual man-hours worked per year}}{\text{man-hours per day per man}} \quad = \quad 5,004 + 9 \quad = \quad 556 \text{ man-days per year}$$

   Holiday labor burden decimal: =   12 + 556   =   .0216 or 2.16%

B. Dividing hours by hours

   Total paid holiday man-hours: 12 days x 8 man-hours/day   =   96

   Total hours worked per year   =   5,004

   Holiday labor burden decimal: 96 + 5,004   =   .0192 or 1.92%

Of the two methods, "B" is the more accurate, although a little more complex; however, either is accurate enough for our purposes.

8. SICK DAYS (sometimes referred to as "well" days)

If you provide paid time off to your field employees when they are sick, compute it into the appropriate company/division labor burden decimal, just as you did for the holiday labor burden calculations.

Utilize this method for any other similar type of benefit provided by the company.

9. MEDICAL/HEALTH INSURANCE

Divide total annual medical health insurance premiums paid by the company, which are not reimbursed by the employees, by the company/division projected total annual field payroll.

Total premiums paid by company + Total projected field payroll = $3,300 + $75,000 = .044 or 4.4%

When I compile a budget for a company, I actually deduct holiday, vacation, and sick pay as well as pay for downtime and equipment repairs—from the total projected field payroll. However, I use a personal computer and a somewhat sophisticated spreadsheet program in order to do so, but the above method will suffice for our purposes.

REMEMBER: Consider only the field labor holidays, vacation time, etc., when calculating the GLI, vacation, holiday, sick days, and medical/health insurance portions of labor burden. Put a portion (e.g., 50%) of someone's paid vacations, premiums, etc., in the labor burden totals if that person is actually in the field working a portion of his total work time.

In the United States, your labor burden percent for field divisions will normally fall between 28% to 37%. In Canada (and outside the USA), it normally runs from the high teens to the mid-twenties.

Those companies outside of the United States can utilize the same methods that we have employed in our examples, making the appropriate category adjustments.

ACTION POINT

Using Exhibit (4) in the back of the book, determine your company/division labor burden percentages.

CHAPTER 8

OVERHEAD OR GENERAL AND ADMINISTRATIVE (G&A) COSTS

PURPOSE

To define the categories contained in the overhead budget

INTRODUCTION

We will turn our attention now to the estimating overhead budget.

Remember, this budget is for the next twelve months. This may or may not be the same as your fiscal year, or for the period designated for tax purposes.

For instance, if your tax year is the same as your calendar year and it is the middle of July, you can still prepare an overhead budget for the next twelve months and use that for estimating purposes. This is perfectly legitimate. In other words, you do not have to be at the very beginning of your fiscal year, or your tax year, to prepare an overhead budget.

Remember, too, that not all categories in your estimating overhead budget will reflect actual cash flow expenses; nor will they be the same as the expenses expressed in your P&L statement. This statement is designed to meet legal tax accounting requirements and/or standards. Subsequently, those categories (as defined by tax law) are not necessarily what we will use for estimating purposes.

We will use historical data as we prepare our overhead estimating budget—but only as a point of reference. Just because there are a certain amount of dollars in last year's P&L statement does not mean that we are going to use that same, exact amount for our projected budget. We will look at each category and ask ourselves: Will that amount **increase** or **decrease** over the next twelve month period? We will then make the appropriate adjustments for purposes of budget projections.

If you are in a start-up situation, you can still prepare an estimating overhead budget—but without the use of historical data. You will simply approximate your expenses and make your projections as accurately as possible.

It is **very important** that we do not include in overhead categories anything that can be directly tied to a job. Job-related expenses on specific bids should be included in direct costs and not in overhead.

Although we will break down overhead into twenty-eight different categories, keep in mind that it is not as important in which category we place an expense item as it is to ensure that it is included in overhead somewhere.

Finally, if necessary, we will divisionalize our overhead. If you have more than one division, and you wish to have separate budgets for each division, we will take each expense category and attempt to place the appropriate amount of overhead expense in the respective division. Sometimes, this can become fairly complicated, and in some cases, you are going to have to make calculated guesses in order to arrive at what you think is a fairly accurate number.

Large corporations, ones generally over 10 million dollars in gross annual sales (or over 7-8 divisions), may desire to have corporate overhead budgets in addition to division overhead ones. The corporate overhead budget would include the same categories as the division ones but would be for costs just for the corporate overhead staff and related expenses. It would then be allocated to divisions and jobs being bid within those divisions as an addition to the division overhead per hour (OPH) cost.

Corporate overhead would be allocated to divisions in one of two ways.

1. Total corporate overhead is allocated to divisions as a lump sum dollar amount which divisions would add to their overhead budget as an additional line item.

2. Total corporate overhead is divided by the sum total of all field-labor hours contained in all divisions. This calculation would produce a corporate OPH (COPH) which would be added to the divisions OPH (DOPH) for each division to produce a total OPH (TOPH). The TOPH (which would be different for each division) would be used by divisions for estimating purposes.

Corporate overhead could also be allocated to divisions a third way as a percent of sales but the above two methods are much more accurate and manageable. Of the two methods described above, I personally prefer the latter.

For this chapter, you will need the following items:

1. Previous year-end P&L statement
2. Your latest P&L statement
3. A calculator
4. Pencils (and an eraser)
5. Scratch pads
6. A copy of the budget format: See Exhibit (1) in back of the book
7. Insurance policy papers for office contents, medical/dental/health, and keyman insurance policies

Let's now explain, in detail, the estimating budget categories.

1. ADVERTISING

Advertising, or promotions, would include the following items:

A. "Yellow Page" advertisements

You would usually find this expense included in your monthly telephone bill. Make sure that you include only that portion pertaining to the advertising section listing.

B. Company brochure (printing costs, artwork, design, etc.)

Be sure to prorate these costs; that is, spread these costs out over a period of time in order to reflect the periods of actual use, not the period in which the entire expenditure is incurred.

For instance, if you purchase a three-year supply of a brochure and the cost is $6,000, you would not place the entire expenditure in this year's estimating overhead budget. That would not be fair to this year's clients. For estimating purposes, you would place only one third of that expense ($2,000) in this year's budget.

C. Newspaper advertisements
D. Help-wanted advertisements

E. Magazine advertisements

F. Billboards

G. Door hangers, flyers, business cards, etc.

H. Bulk and direct mail expenses

I. Garden, trade, or home shows (expenses for materials, labor, booth/space fees, etc.)

J. Company resumé (graphics and printing costs, etc.)

K. All other forms of advertising

I usually limit advertising to no more than 1% of a company/division's gross annual sales.

2. BAD DEBTS

Bad debts are money owed to you by customers; money that you do not anticipate receiving. I usually limit the estimating budget for bad debts to .5-1% of gross annual sales.

If you do a lot of government work, small jobs, or a lot of maintenance work, limit bad debts to one half of one percent of your gross sales. If you do a lot of commercial work for clients that you do not know well or large jobs, raise this to one percent.

Again, base this on historical data. However, if you have a rather large, substantial bad debt ($30,000 to $50,000 from the past year or past few years), you cannot put the entire amount in your overhead budget. You have to "cap" the figure. Otherwise, your bids will not be competitive. You cannot charge your future clients for the "sins of the past," so to speak, in overhead. Some of it will have to be taken out of profits.

3. COMPUTERS AND DIGITIZERS

Included in this category are the purchase of computers, software, and digitizers, and training for the use of same (if not included in item 22, "training and education").

I generally depreciate these amounts over a three- to five-year period. If I have purchased a rather expensive accounting system—which, for software alone, can easily cost you $5,000-25,000—and if that system can be upgraded and continue to grow with the business throughout the years, I would spread that expense over a five- to ten-year period depending upon the anticipated useful life of the software.

Spread the cost of computer hardware over a three-year period. We want to recoup expenses for hardware within three years, and because of technological advances plan on replacing that computer at the end of the three-year period, as well. Included, too, in this category would be training costs, consultants for in-house training of your staff, and any maintenance contracts and repairs on subject equipment.

Digitizers come in three types: table models, electronic arms, or sonic measuring devices. These machines usually last a long time. Depending on what you think is reasonable, spread their expense over a five- to seven-year period.

If the costs for computer-aided design (CAD) hardware and software exceed $10,000, spread this cost over a seven- to ten-year period. Otherwise, depreciate them over a three- to five-year period.

4. DONATIONS

Donations are comprised of cash donations, labor, and/or equipment costs incurred for charitable work that your company performs in your community.

For instance, if you donate cash to the YMCA, civic project, or charity, include that in this category. If you donate a crew to work half a day at a local church or community park, include the labor and equipment costs, as well, in this category.

Another item might be the sponsorship of a little league ball team. If your name is on the uniforms, you could underwrite that cost in this category. Although I think it best to put the cost in "advertising," the important thing is to include it in overhead somewhere.

I would not include in this category donations that are driven by tax accounting purposes. For instance, if at the end of the tax year, your accountant says that you have accumulated excess profits (I can never understand that term, "excess profits") and recommends that you make some donations, this is not an item to include in overhead and attempt to charge to next year's clients. Consider this a donation from profits.

For the purposes of estimating overhead, I like to "cap" the total amount of donations at between $200 to $500 (preferably, around $200). Anything more than that should be deducted out of profits, rather than again charging next year's clients for these donations. If you attempt to put too much into this overhead category, it is really your clients who are making the donations. *You are merely the one taking the tax deduction for it.*

5. DOWNTIME LABOR

These are the hours that field crews are being paid, but when they are not working in the field (or producing billable hours). Tabulate it in hours, then multiply hours by your average wage.

This can include paid labor hours for:

A. Bad weather

During bad weather, field crews may work in the office, nursery, yard, etc. It would also include lost hours when crews are non-productive in the field due to weather conditions, but are still being paid.

B. Equipment breakdown time (to and from jobs)

For instance, if a crew pickup truck breaks down en route to a job, resulting in a two to three hour wait for repairs or replacement transportation, include this in downtime hours.

C. Busywork (fixing small tools, cleaning or repairing the office or yard area, etc.).

D. Field crew meetings, safety meetings, etc.

Total this time and include in downtime. As an example: If you consistently had ten people in the field, and if you held a safety meeting once a week that lasted for fifteen minutes, this would total 2.5 hours per week of downtime (10 people x .25 hours = 2.5 hours /week).

E. Nursery watering time

If your nursery is not a separate division on your financial statement, watering time for nursery stock by your crew would be included in downtime. However, if your nursery is a separate division, the labor expended on it would be included in direct labor costs for that division.

Two important notes:

The time spent on equipment repair is not included in downtime. This is incorporated as a portion of equipment costs.

Drive time (of crew, or the mobilization of equipment to and from jobs sites, material acquisition and hauling, etc.) is excluded from this category, as these expenses are included under "general conditions" in a specific bidding situation.

To calculate the annual cost: total downtime hours for the year and multiply by your company/division average wage.

6. DOWNTIME LABOR BURDEN

Multiply your company/division field-labor burden by the dollar amount in item 5 (downtime labor). Enter this amount in your budget for downtime labor burden.

7. DUES AND SUBSCRIPTIONS

This includes dues, fees, subscriptions, and membership charges for:

A. Various state and national associations or organizations

B. Chamber of Commerce

C. Better Business Bureau

D. Professional magazines

E. Dodge Reports

    F. Plan rooms

    G. Green sheets

    H. Publications that are business-related

    I. Discount outlets and warehouse store dues/memberships

8. INSURANCE

    A. Office contents

    B. Dental and medical insurance for overhead people (owners, secretaries, accountants, bookkeepers, estimators, field supervisors, and anyone else whose hours you cannot directly account for within specific bid calculations).

    C. Included here, too, would be life (or keyman) insurance for corporate officers, as long as the company or corporation is the beneficiary.

Notice that we are not including within this section any policy premiums for vehicle and equipment coverage, general liability or umbrella insurance, or insurance on buildings.

9. INTEREST AND BANK CHARGES

    A. Interest on credit cards

    B. Interest on your line of credit

    C. Start-up fees for establishing your line of credit

    D. Interest on outstanding debts to your suppliers

    E. Bank fees (service charges, penalties, bounced check fees, etc.)

In the event that you have no line of credit, you are then using the company's retained earnings to finance your business and to provide working capital. This amount is normally equal to four or five weeks of payroll.

Since this amount is usually tied-up and being used by the company throughout the year, you should multiply this amount by the current savings interest rate (as though it were deposited in a savings or money market account and collecting that interest) and include that interest portion here as a cost-of-money expense.

This is a legitimate expense item to include within overhead. If you did have a line of credit, you would be paying interest on it, and we would put that interest payment in overhead. Because the company is providing its own line of credit, you cannot place these funds into a savings account; therefore, you are losing the interest that you would normally collect from placing that money in the bank.

Note that this category does not include interest on past-due receivables (moneys owed to you), interest on equipment loans, or interest on your mortgage for your office building (or space).

10. LICENSES AND SURETY BONDS

This includes the fees for your state contractor's license and city business license (if applicable) and the cost of the surety bond (not a performance bond) on your license. However, city licenses for specific jobs should be included in general conditions, not overhead.

11. OFFICE EQUIPMENT

This includes: new purchases, repairs, and service contracts for office equipment. It does not include: any item of (or related to) field equipment, overhead vehicles, computers, telephones, radios, beepers, pagers, or digitizers.

For new equipment which you expect to purchase during the budget period, divide the cost by its useful life expectancy and include the cost for one year in this category.

For office equipment presently on-hand, list the items, their replacement cost at new fair market value (FMV) prices. Then divide the FMV replacement cost by the useful life expectancy. Enter that amount in this category.

    A. File cabinets and durable furniture (desks, tables, sofas, and chairs) usually last ten years. Although some items may last longer, you want to recoup their costs within ten years.

    B. The cost of a copy machine can be spread over three to five years, depending on estimated useful life (supplies and toner are included in "office supplies" costs).

C. Blueprint machines can be depreciated over a ten-year period.

D. Calculators: one to two years

E. Plan racks and drafting tables: ten years

F. Typewriters (manual): ten years

G. Typewriters (electric): five years

H. Refrigerators, microwave ovens, etc.: five to seven years

I. Plants, pictures, artwork, paintings, etc.

Many high-end residential construction companies bring clients to their offices. Considerable expense is often incurred in order to make a proper impression. This cost should be included in overhead as a legitimate cost of doing business.

Depreciate artwork, pictures, etc., over a ten-year period. Plant costs should be spread over just one year.

12. OFFICE SUPPLIES

Office supplies are consumable items that are used on a regular basis. They include:

A. Pens, pencils, paper, paper clips, etc.

B. Stationery

C. Paper for copiers, computers, FAX machines; toner, etc.

D. Printing costs (letterhead, forms, artwork, etc.)

E. Postage

13. PROFESSIONAL FEES

Include the following:

A. CPA or bookkeeper (one not on your regular payroll)

B. End-of-year tax preparation

C. Legal fees

A reasonable legal fee amount is between $500 to $700 per year. This would cover having an attorney scrutinize your policy manuals, write a few routine letters, and costs for your annual corporate meeting, etc.

This fee would not cover costs for litigation or lawsuits. It is unreasonable to put this expense into overhead and subsequently must come out of profits. The premise is that it is not reasonable to charge next year's clients for (what we may call) "sins of the past."

D. Outside payroll services

E. Consultants (other than those for your computer which are in the computer overhead category)

F. Incorporation costs (which can be spread over a three- to five-year period)

14. RADIO AND COMMUNICATIONS SYSTEMS

A. Two-way radio systems

Depreciate purchase price over a five to ten year useful lifetime period. Include monthly repeater and/or line charges.

B. Beepers and pagers.

Include monthly service charges.

C. Car/mobile telephone

Depreciate purchase price over a three-year period. Also include monthly toll and usage charges.

D. Maintenance, maintenance contracts, and repairs for radios, pagers and car/mobile telephones.

15. RENT (OFFICE AND YARD)

Use fair market value, especially if you own the facilities. For instance, if you had to rent a similar facility on the open market, what would you have to pay? Quite often, your monthly mortgage payment may be much lower than fair market value. Even if your facility is paid for, it is reasonable to include a FMV rent in your overhead.

Remember, do not confuse your estimating overhead budget with your tax accounting budget (or financial statement) where you would account for your actual or "real" payment. You want to place a reasonable FMV amount of expense for rent within your overhead budget.

Within this category, too, you want to account for your property (and/or asset) taxes on the facilities and building fire insurance.

If you rent your office and yard space, enter this amount in this category.

16. SALARIES FOR OFFICE PERSONNEL (WHO ARE NOT CORPORATE OFFICERS)

Included are:

A. Bookkeepers

B. Office manager

C. Secretaries

D. Receptionists

E. Estimators

F. Designers—administrative portion of salary only if applicable

G. Field supervisor

Attempt to allocate as much of a supervisor's time as is possible to specific jobs in the field. If a supervisor generally spends half of the day in the field supervising and working on jobs and the other half doing administrative work, then put only 50% of his/her salary in overhead. The other 50% will be bid into jobs. Prorate a supervisor's salary 50/50, 60/40, 25/75, etc. as closely as possible to coincide with where he/she spends time.

H. Anyone in the office, other than salespeople.

Sales commissions are usually included in "direct costs." See Chapter 4 for an extensive discussion on commissions/sales.

17. SALARIES (CORPORATE OFFICERS)

Calculate officers' salaries according to fair market value. That is, if a corporate officer was performing his/her function for another company, what would he/she expect to get paid on the open market for an annual salary? Or, if you were to hire another person to run your company in place of the owner, what should that person expect as adequate, reasonable compensation?

It is important to realize that this amount may not equal actual, "real" pay; however, for estimating purposes, we must place in overhead what is REASONABLE—that is, the fair market value amount for that officer.

For example, some of my clients who are owners prefer to draw less than a fair market value salary (only $200-$300 per week) and keep the extra money working within the company. Other clients choose to take far more than a fair market value salary, but it is not reasonable to put all of this amount into overhead for estimating purposes: it would make bids too high.

Even though officers' "real" pay may be less or more, we still estimate using a fair market value amount in overhead.

NOTE: The total for combined office and officer salaries should equal between 8-12% of gross annual sales.

If yours is a large, commercial operation, you should be running around 8% for these two categories (extremely large companies may be even lower). High-end residential companies will run around 12%.

You should not exceed these percentages in overhead. If actual pay for office staff and officer(s) combined exceed our target percentages, do not put the excess in overhead. Rather, make a profit adjustment at the bottom of your budget format after net profit margin for additional excess salaries or bonuses.

18. SALARY LABOR BURDEN

This amount is determined by multiplying the office labor burden percentage by the total of salaries for office personnel and corporate officers. It covers FICA, FUTA, SUTA, WCI, etc. This percentage is discussed in detail in Chapter 7, entitled "Labor Burden."

19. SMALL TOOLS AND SUPPLIES

This includes any non-motorized tool or supply that you purchase throughout the year for general use and not for particular jobs. For example:

A. Wheelbarrows

B. Hand tools

C. Rakes, shovels, hoes, tarps

D. Paint, nails, wire

E. Miscellaneous hardware

This does not include purchase or repair costs for: lawnmowers, chainsaws, weedeaters, blowers, etc. These are a "direct cost" expense and are included in direct equipment costs.

20. TAXES (EQUIPMENT/ASSET TAX OR "MILL TAX" ON BUSINESS)

Some cities or counties will charge an asset tax, charging a percentage of the total value of your equipment and assets. Asset taxes on field equipment and facilities (office space and/or yards) are not included in this category. Only asset tax on office equipment (furniture, computers, etc.) is included in this category.

Asset taxes on facilities is accounted for within "rent." Asset taxes on field equipment should be included in hourly rates for the same.

Mill taxes are rare, though they do exist. A mill tax is a general tax based on sales. It usually runs no more than $200-$300 per year.

21. TELEPHONE

A. Monthly line charges

B. Long distance charges

C. Telephone rental, if applicable

D. Answering or voice mail services

E. Telephone systems purchase and/or installation price depreciated over a five- to ten-year period

F. Repairs and/or maintenance contracts for telephone equipment

G. Answering machines, and/or FAX machines

22. TRAINING AND EDUCATION

A. Books, audio tapes, videos, etc.

B. Workshops and seminar entrance fees

C. Community college/continuing education classes

D. Convention entrance and registration fees (garden and home shows, etc.)

There is no set and patent answer as to how much should be spent on training and education. The key is to determine whether your people are being trained and educated to run the systems that are needed in your company.

If you are spending your money wisely and getting results, you can invest almost any reasonable amount in training and education.

23. TRAVEL AND ENTERTAINMENT

A. Travel costs to seminars, workshops, or conventions

B. Hotel bills (directly associated with the above)

C. Lunches and dinners (including those for clients)

D. Gifts for clients and employees, the company Christmas party, company season tickets to sporting events, etc.

24. UNIFORMS AND SAFETY EQUIPMENT

A. T-shirts and ball caps for crews

B. Uniform rental or cleaning services

C. Safety equipment or items not included in small tools and supplies: goggles, gloves, ear protectors, hard hats, etc.

25. UTILITIES

A. Monthly water, electric, and sewer bill for office/yard

B. Heating gas (or oil) for the office/yard

C. Trash service or a dumpster that is not job-related

D. Office janitorial service

NOTE: Be careful not to include expenses for nurseries which are a separate division and that incur large wintertime expenses for heating, in another division's overhead expense.

26. VEHICLES, OVERHEAD (See calculation explanations in Chapter 10)

A. Automobiles and pickup trucks used by overhead personnel.

This category does not include crew pickup trucks.

B. Mileage reimbursement for private vehicles used by overhead personnel.

Note: For personnel whose salaries are accounted for in both field labor and office categories, prorate vehicle expenses between the categories as you do salaries for individuals using the vehicles.

27. YARD EXPENSE (LEASEHOLD IMPROVEMENTS)

A. Costs associated with maintaining a yard or nursery area, installing mulch beds or drip irrigation, spreading gravel over a driveway, putting up a fence, etc.

B. Leasehold improvements to your office (depreciated over the life of the improvements) such as:

- carpeting
- painting costs
- partitions
- various tenant improvements, etc.

For instance, if you install carpeting and padding for $1,500, and its projected useful life is five years, divide the $1,500 by five. This would amount to $300 per year to be included in leasehold improvements.

Building an entire building on your property is not a leasehold improvement to be included here. This cost would be accounted for by a FMV increase in monthly rent.

Costs for storage yard or nursery area improvements might involve:

- fences
- gravel spread over driveways
- nursery mulch beds
- drip irrigation for nursery areas

Again, these costs would be depreciated over the useful life expectancy of the improvement involved.

Generally, labor expended for yard expenses and leasehold improvements is included in downtime labor.

28. MISCELLANEOUS

Listings under this category should be kept to a minimum, preferably under $200 ($250, at the most). All, or most, of your expenses should be in the categories listed above.

I once had dealings with a company that had a miscellaneous category of over $57,000. Needless to say, I told that company that they needed a little better idea of where they were spending their money.

Miscellaneous is not a catchall category. You cannot just lump everything together. You need to define overhead expenses in order to have effective control of them.

CONCLUSION

Some of these categories tend to overlap (for instance, advertising and donations). Frankly, this does not matter. The key is to be consistent with your definitions and continue to put the same item in the same category year after year.

NOTE: The format that we are using is a P&L statement for a corporation. Even though you may be a sole proprietorship, for estimating budget purposes you should still use this format and enter a salary into your overhead budget for bidding purposes.

REMEMBER: The items and amounts on your CPA/accountant's P&L statement are going to look a little different from the ones described in this chapter. The format used here is for estimating purposes and is somewhat different than a tax accounting P&L statement.

ACTION POINT

Use Exhibit (1) in the back of the book and fill in the different categories for your overhead budget. Use Exhibit (5), which is a recap of the different categories, to help you in this process.

CHAPTER 9
EQUIPMENT COSTS

PURPOSE

To differentiate between direct vehicle and overhead vehicle costs

**To calculate both direct equipment/vehicle and overhead vehicle costs
for budgeting purposes**

**To calculate equipment cost per hour (CPH) amounts for the various types of
equipment/vehicles in your company/division**

BUDGETING FOR EQUIPMENT COSTS

The best method for budgeting equipment (or any item) is to review historical data as a point of reference and then make the appropriate adjustments for the budget period that is being developed. However, this is not always possible. If you are in a start-up situation or if your historical data is unavailable, inaccurate, or not properly formatted, you must make educated calculations (guesstimates).

It is important that you understand the reasoning behind differentiating direct equipment/vehicle costs, which are bid into jobs based upon its costs per hour (CPH) and the number of hours to be used on a specific job and indirect (or overhead) vehicle costs. The latter is bid into individual jobs as a part of overhead markups.

Some estimating systems place all equipment costs (tractors, pickups, dump trucks, overhead vehicles, etc.) into overhead and allocate to specific jobs in their overhead markup—usually as an overall percent markup or as a specific dollar amount multiplied by estimated field-labor hours on a particular job. This method is inaccurate because whether you are in a construction or a maintenance bidding situation, it invariably under- or overstates the actual equipment costs for the job being bid.

OUR GOALS: (1) to identify correctly specific equipment costs on a per hour (CPH) basis, and (2) to multiply that specific hourly cost by the estimated number of hours that particular piece of equipment is to be used on a particular job. We do this in order to determine the total amount of equipment costs that we will bid into each individual job.

If the job requires only pickup trucks and wheelbarrows, then we only want to include costs for the same in the bid. If, on the other hand, the job requires dozers, large dump trucks, tractors, etc., we want to include those costs in the bid. Including all equipment costs in overhead—and indiscriminately spreading them evenly like peanut butter over all of your jobs—precludes accurate and competitive bidding. Refer to Chapters 11 and 12 for a more detailed discussion of bidding equipment on your projects.

1. If you are in a start-up situation or do not have accurate historical data for equipment costs, use either the percent of sales or the equipment-to-labor ratio method outlined in Chapter 10 to calculate your projected annual equipment cost budget.

2. If you have accurate company/division historical data for both direct equipment and overhead vehicle cost, make appropriate adjustments to your projected equipment budget by using one (or a combination) of the following methods:

   A. Increase or decrease direct equipment costs (including repairs and maintenance costs at market rates, in-house mechanics, fuel, and depreciation) by the same percentage of projected increase or decrease in sales.

   For example, if gross annual sales are projected to increase by 15%, increase direct equipment costs by the same amount.

   B. Increase or decrease budget projections for direct equipment costs by multiplying the historical percent (the percent direct equipment was of previous gross annual sales) by the projected amount of gross annual sales.

   For example, if direct equipment costs have historically been 8% of gross annual sales, multiply (for budget purposes) next year's projected gross annual sales by 8% to obtain projected direct equipment costs for next year's budget.

   C. This third method is possible only if you have established a formal equipment division with its own P&L statement and have accumulated accurate maintenance and repair costs for each piece of equipment. If adding, replacing, or deleting field equipment, increase or decrease your projected budget costs as follows:

   (1). Cost out the piece of equipment that is being added or deleted, using the costing section of this chapter in order to determine its CPH and the cost data accumulated in your accounting software. Divide real costs accumulated by actual meter hours on the piece of equipment.

   (2). Multiply the CPH by the projected increase or decrease in hours used.

   (3). Adjust your projected costs accordingly.

   If you are replacing a piece of equipment with one that has the same CPH, there is no need to adjust your projected costs—unless, of course, projected hours of use will increase or decrease significantly (that is, by 15-20%); in which case, you would adjust the projected cost to reflect the change in hours used.

   If the replacement equipment CPH differs from the original (and the projected use hours are roughly the same), multiply the difference in CPH by the projected hours used, and adjust projected costs accordingly. For example:

   $15.00 (new CPH) - $12.00 (old CPH) = $3.00

   $3.00 (CPH difference) x 1,000 (projected use hours) = $3,000 (to be added to projected costs)

   If hours of use are also projected to change, calculate that change in your adjusted cost.

Once you have calculated total costs for the budget year for each piece of equipment in your company, combine all the totals to determine your total equipment cost budget for the year. Add overhead vehicle costs to the field equipment total. Your new total is now your equipment division's sales or income budget for the year. You will achieve this sales goal as you "rent" equipment at cost to the other divisions throughout the year.

As you accumulate sales, you will also accumulate costs on your equipment division P&L statement. The cost of repair parts, in-house mechanics, mechanics' tools and vehicles, and repair work done outside the company are entered as direct costs: materials, labor, equipment, and subcontractor costs respectively. Overhead for the equipment division is calculated just as it was done in Chapter 8.

The goal of the equipment division is to break even. Sales to other divisions are calculated at the predetermined CPH multiplied by the hours used by that division. This amount is then charged to the respective divisions as an equipment direct cost.

If you have accurately projected your usage hours and your CPH for each piece of equipment, sales in the equipment division should equal the sum of all costs, both direct and overhead. If your equipment division does not break even, review your costs and meter hours for each piece of equipment and determine where you went wrong.

D. Overhead vehicle costs.

Even if you have accurate historical cost data for overhead vehicles and associated costs, these do not necessarily increase or decrease in direct proportion to sales.

You could easily increase sales by 15-25% without adding another overhead vehicle. Subsequently, costs remain the same. You only adjust your overhead vehicle costs if the number of overhead vehicles varies.

To determine overhead vehicle cost:

(1). Identify the specific vehicles to be included in overhead. These will be the company-owned vehicles that are used by persons listed in the "overhead salary" category.

REMEMBER: If a person's salary is only partially included in overhead, then a corresponding percentage of his related vehicle cost will be included in overhead.

(2). Determine the CPH for the overhead vehicle (e.g., $3.00 per hour).

(3). Multiply the CPH for the respective vehicles by the projected hours of use. For example:

$3.00 CPH x 2,080 hours for full-time overhead vehicles

$3.00 CPH x 1,040 hours for 50% part-time overhead vehicle use

(4). Add projected mileage reimbursed to overhead personnel for use of their personal vehicles (and any other overhead vehicle expense) to the totals for company-owned vehicles. Place this amount into your projected overhead budget.

NOTE: For tax purposes, you may have a number of "overhead vehicles" that appear on your P&L statement. Unless these are realistically used in direct support of overhead personnel, you cannot include all of these vehicles in your estimating budget or your bids will not be competitive. Again, what is real is not always what is reasonable to put in your estimating overhead budget.

For instance, if you have only two people in overhead, you cannot list four vehicles (at 2,080 hours apiece) in your estimating overhead budget. It just is not realistic or reasonable to include this in your bids.

If you do not have reliable company/division historical data to use for budget projections, or if you are in a start-up situation for a company or division, use a percent of labor (see Chapter 10) to project direct equipment costs. To calculate overhead vehicle costs, use the procedures to calculate overhead vehicle costs described earlier in this chapter.

CALCULATING EQUIPMENT COST PER HOUR (CPH)

There are two reasons for calculating equipment CPHs:

1. To calculate budget projections for direct equipment and overhead vehicle costs.

2. To calculate the amount of direct equipment costs to include in a job in the bidding process.

OUR GOAL: to recover all of our company/division equipment costs for the year in the jobs that we complete and bill in that year. In order to do so, we must:

1. Accurately determine costs.

2. Allocate the correct amount of these costs to jobs.

3. Complete and bill enough work/jobs in the field to cover all of our equipment costs for the year.

It may help to think of yourself as an equipment rental company, or as a car rental firm (such as Hertz or Avis). They, just like you, need to know vehicle/equipment costs (purchase price, interest, repair and maintenance, etc.) in order to determine appropriate customer rental rates.

These rates are based on an hourly, daily, weekly, etc., cost figure that is marked up to cover corporate overhead and profit. If vehicle/equipment costs are not identified accurately, subsequent rental rates would be either too high (prompting customers to shop elsewhere) or too low (causing the company to lose money).

I know of construction and maintenance companies, both large and small, that lost customers—and hundreds of thousands of dollars—because they did not bid their equipment costs accurately; nor did they track their equipment costs (and revenues generated) in order to cover these costs in jobs completed. They did not ensure that they would, at least, "break-even" for such costs (we will cover this in greater detail in Chapter 24).

We need to make the following distinctions (or assumptions) concerning our method of equipment costing:

1. USEFUL LIFE AND METER TIME

A. For cars and trucks under two tons in size, this will be 8,320 hours.

We want to recover our costs for these "light" vehicles in either four years, at 2,080 hours per year (8 hrs/day x 5 days/wk x 52 wks/yr), or, for seasonal companies, in five years (8 hrs/day x 5 days/wk x 42 wks/yr). Incidentally, in this case, the eight hours per day is not necessarily meter (or engine running) time.

After four or five seasons, these vehicles are usually worn out. If they last longer, great. However, we want to be conservative in recovering our costs. We prefer to recoup our costs in no more than four to five years.

B. The useful life of heavy-duty trucks (two ton and larger), tractors, dozers, trenchers, etc., consists of projected actual meter time/hours prior to that piece of equipment needing a major overhaul.

Heavy-duty trucks usually accumulate 10,000 meter hours before needing such an overhaul. However, 10,000 hours may be spread over ten to twelve years, not just four to five like light-duty trucks.

Useful life meter time/hours for tractors, dozers, trenchers, etc., will generally be much less than those for heavy-duty trucks. They can, however, also last ten to fifteen years prior to needing a major overhaul. This equipment is built to last and to endure the wear and tear. If you do not use it, it does not wear out (for the most part) in four to five years as do the lighter trucks and vehicles.

2. SEASONALITY

Light-duty trucks (under two ton) that are not used in winter have the same useful life in hours (8,320) as ones that are used in winter, but useful life is spread out over five years, rather than four. This is because seasonal use is shortened to (approximately) 1,600 hours per year, sometimes less. However, it generally evens out at a total useful life of 8,320 hours.

3. COSTS FOR NEW VS. USED EQUIPMENT

Even if your plan is to buy used equipment, we will cost it out using the "new" purchase price. The CPH will generally be the same for new as well as for used equipment. Another reason for using the new price is because useful life and repair and maintenance costs are easier to determine for new equipment.

In essence, purchasing new vs. old equipment is not a question of operating savings; it is a question of capital: Can you afford to buy new equipment? Used equipment may be cheaper to purchase, but it is usually more expensive to maintain. Therefore, the CPH will tend to remain the same.

4. RENTING VS. OWNING EQUIPMENT

As a general rule, unless you use (and bill for) a piece of equipment at least 50% of the time (20 hours/week), it is not financially practical to own it.

If, however, a piece of equipment is either not readily available or would be impractical to rent, it usually pays to purchase it. But before purchasing, you should weigh projected revenues and operating savings generated to projected costs incurred to assess the financial feasibility of owning it.

5. LEASING VS. BUYING EQUIPMENT/VEHICLES

Leasing equipment and/or vehicles may or may not provide cost benefits to your company or division. Often, the leasing versus buying option is one driven by tax benefits (concerning which you should check with your CPA or accountant), rather than savings for operational costs.

To determine the costs and benefits of leasing, cost out the leased equipment/vehicles as you would those you might purchase. Be sure that you adjust the "Purchase Price Column" so that it will reflect lease payments and other items that may be appropriate (for example: periodic maintenance included in lease agreements, extended warranties, etc.). Other costs, such as fuel costs and life expectancy, should remain the same.

If the lease-life for the equipment/vehicle is significantly shorter than the normal expected life (should you purchase), adjust the life expectancy to the shorter term. Adjust maintenance and fuel cost to account for the shorter period, as well. For example, a leased vehicle that is to be replaced every two years by a "new" leased vehicle, should have an adjusted life expectancy of 4,160 hours (2 years x 2,080 hours per year) versus 8,320 lifetime hours.

Anticipated maintenance, insurance, and fuel costs should cover only the two-year period. Theoretically, there should be little to no anticipated repair or maintenance costs, especially if periodic routine maintenance (replacing worn belts; lube, oil and filters, etc.) is included in the lease.

Finally, compare the lease CPH to the purchase CPH in order to determine any savings. If there are, adjust your estimating budgets accordingly and weigh the option of using the higher or lower CPH in your bids. While the lower is realistic; the higher would provide a "contingency" (or "fudge") factor that should translate into a slightly increased profit margin (you should be able to detect this increased margin in your job-costing reports) for individual projects.

6. SMALL TOOLS/NON-MOTORIZED ITEMS

Small tools and non-motorized items (such as: shovels, rakes, hoes, wheelbarrows, etc.) are costed in overhead and are not included in this chapter.

7. IN-HOUSE LABOR ON EQUIPMENT/VEHICLES

Unless you have a separate equipment company/division with separate financial statements, calculate your projected maintenance costs by utilizing the prices and rates obtained from outside equipment repair shops. These prices will include their materials, labor, labor burden, overhead, profit, etc.

If you attempt to calculate your cost per hour figures using in-house labor (including labor burden, etc.), you will greatly complicate the costing process. In addition, your rates will probably be more expensive than those on the open market; if they are, you should save the money and have the work done on that open market. If, on the other hand, you can do it cheaper, use the open market rate (which is essentially a FMV rate) in your equipment costing and put the extra money in your pocket.

MILEAGE RATES

Using established mileage rates to reimburse your employees for the use of their own vehicles as overhead transportation is fine on a limited basis. These costs are included in overhead. Reimbursing field employees in a similar manner for using their own vehicles on the job is also fine. However, we do not bid costs into a job that are based on mileage rates for company-owned vehicles/equipment. Rather, we use hourly rates. Let me use an example to explain why.

You have two similar-sized jobs. One is fifty miles from your office; the other is only five miles away. If you want to bid a half-ton pickup truck into the jobs, you would calculate (using sample rates) as follows:

1. Mileage rate method: $\dfrac{\text{lifetime costs}}{\text{lifetime mileage}} = \dfrac{\$24{,}960}{100{,}000} = \$\,.2496 = \$\,.25$

2. CPH method: $\dfrac{\text{lifetime costs}}{\text{lifetime hours}} = \dfrac{\$24{,}960}{8{,}320} = \$3.00$

Job "A" (fifty miles away):

Method #1:	100 (miles/day)	x	$ .25 (per mile)	=	$25.00
Method #2:	$3.00 CPH	x	8 hours/day	=	$24.00

Job "B" (five miles away):

Method #1:	10 (miles/day)	x	$ .25 (per mile)	=	$ 2.50
Method #2:	$3.00 CPH	x	8 (hrs/day)	=	$24.00

Overhead and profit markups will be added to both in order to determine the price to charge the customer.

The mileage method breaks down for the following reasons:

1. Most of the wear and tear (broken mirrors, lights, and windows; flat tires, etc.) on the truck will probably occur traveling around the job site once it arrives there. The highway is probably the safest place for the truck. Costing the vehicle based on miles driven to and from the site does not take this into account.

2. Although both methods are comparable for job "A," you have to drive a vehicle fifty miles (one way) in order to gain equity. If all your jobs are this far away, this method might have some merit. However, this probably would rarely be the case.

3. Hours are easier to job cost than mileage; for example, four hours per day, eight hours per day, forty hours per week, etc., versus so many miles per day actually driven. Too, how do you job cost your costs if the vehicle stays at the job site overnight? Or what if this one particular vehicle is used on three jobs in any one day? It would be extremely difficult to prorate between each job and to job cost the individual projects.

WORKSHEET EXERCISES

Let's turn to the Equipment Costing per Hour Calculation Worksheet in Exhibit (6) and cost out some sample equipment and vehicles.

1. **PICKUP TRUCK (1/2 TON):** Used by owner for administrative purposes and for transportation to and from job sites.

Will last four years and have approximately 100,000 miles on it when traded in.

A. Acquisition CPH

Add purchase price (including tax, registration, racks, tool boxes, special paint, etc.) to total lifetime interest to be paid. If this total interest is not on the contract, use the formula below and subtract the anticipated salvage value when you replace the vehicle.

(Purchase price x number of years of payments x interest rate) ÷ 2

$$(\$13,500 \times 4 \times .12) \div 2 \quad = \quad \$6,480 \div 2 \quad\quad = \quad \$3,240$$

Purchase price	$13,500
Interest	3,240
Subtotal	16,740
Salvage value	- 4,500
Total	$12,240

Divide by lifetime hours (8,320): $12,240 ÷ 8,320   = $1.47

Enter $1.47 in the worksheet column (1).

B. Maintenance CPH (total projected costs)

Insurance:	4 years at $750 per yr. (average)	$3,000.00
License fees:	4 years at $150 per yr. (average)	600.00
Lube, oil and filters every 3,000 miles:	(33 x $25)	825.00
Brake jobs:	2 at $350	700.00
Clutch jobs:	2 at $430	860.00
Tune-ups:	3 at $155	465.00
Smog certification:	3 at $50	150.00
Tires:	2 sets at $300	600.00
Misc. (batteries, mirrors, belts, windshields):		1,000.00
1 blown engine for every 3 vehicles (average):	$2,400 ÷ 3	800.00
TOTAL		$9,000.00

Divide total by lifetime hours: $9,000 ÷ 8,320   = $1.08

Enter $1.08 in the worksheet column (2).

C. Fuel CPH

Determine the fuel CPH for automobiles and trucks under two-ton GVW by using one of the methods below. Then enter the amount in worksheet column (3).

(1). Method using total miles used per useful life or per year

a. Total miles driven per year: 18,000

b. Divide by miles per gallon (15):   18,000   +   15   =   1,200 gallons/year

c. Convert gallons to dollars at $1.23/gallon: 1,200   x $1.23 = $1,476 per year

d. Divide $1,476 by hours per year (2,080): $1,476   +   2,080 = $ .71 per hour

e. Enter $ .71 in the worksheet column (1)

f. Note(s):

[1]. You can use useful life hours (8,320) and respective miles (72,000) for the above example, as well.

[2]. If vehicle is used only part of the season, reduce hours accordingly but increase number of lifetime years for vehicle.

(2). Method using number of fillups per month (or week)

a. Determine number of fillups per month: 5

b. Multiply tank size by approximate number of gallons per fillup:

                                    5 fillups   x   20 gallons   =   100 gallons/month

c. Multiply by average price per gallon (e.g., $1.23):   100   x $1.23   = $123.00

d. Divide total costs by hours per month or week (e.g., 173 hours per month): $123.00   +   173   = $ .71

Enter $ .71 in the worksheet column (1)

D. Total CPH. Add columns (1), (2), and (3). This will give you your total costs per hour (CPH) for a half-ton pickup:

$1.47 + $1.08 + $ .71   = $3.26 (round to $3.25)

Half-ton pickups generally run $3.25 per hour, plus or minus a few cents, depending upon the type of options added to the vehicle purchase price (4-wheel drive, air conditioning, AM/FM radio, deluxe interior package, etc.).

NOTE that the CPH is only your projected cost for the vehicle. It is not the price charged to the customer, because it has not yet been marked up for overhead and profit.

Let's now cost out a piece of field equipment.

2. 21" ROTARY MOWER:

Used approximately 4 hours/day, 5 days/week, 40 weeks/year. Purchase price: $850.00. Life expectancy: 2 seasons (then replaced or rebuilt). Interest: none (paid for in cash). Salvage value: none.

A. Acquisition CPH

Divide purchase price by projected lifetime hours: $850.00 + (4 x 5 x 40 x 2)   =

                                          $850.00   +   1,600 = $ .53

Enter $ .53 in worksheet column (1)

B. Estimate lifetime maintenance costs (blades, plugs, oil, filters, etc.). Check with your equipment dealer as they often have such data available: $608.00

Divide lifetime maintenance costs by lifetime hours: $608.00   +   1,600 = $ .38

Enter $ .38 in worksheet column (2)

C. Determine running time (meter time) fuel consumption by one of the following methods:

(1). Ask: How long does 1 gallon of gas last for this piece of equipment? If all day (for four hours used per day), divide the price of a gallon of gas by four hours. Add oil additive costs for 2-cycle engines:

$$\$1.10/gl \;+\; 4 \;=\; \$.28$$

Enter $.28 in the worksheet column (3).

(2). Ask: If I used this piece of equipment all day (8 hours meter time), how many times would I have to stop and refill the tank? (e.g., 2 times). Multiply number of fillups by size of tank (e.g., 1 gallon):

$$1.0 \text{ gallon } \times \; 2 \text{ fillups } = \; 2 \text{ gallons}$$

Multiply result by price for a gallon of gas and divide by 8 hours:

(2 gls x $1.10) + 8 hours = $2.20 + 8 = .2750 (rounded off to nearest cent) = $.28

Enter $.28 in the worksheet column (3).

D. Total CPH. Add columns (1), (2), and (3). This will give you a CPH for a 21″ rotary mower:

$$\$.53 + \$.38 + \$.28 \;=\; \$1.19$$

NOTE: These figures will vary depending on the type of preventative maintenance (PM) program that you have, if any, and the purchase price of the equipment used.

The best method for determining CPH amounts is to use company historical data. However, most companies do not have such information readily available.

3. SMALL TRACTOR: Purchase price, $23,000. Lifetime hours, 3,000

    A. Acquisition CPH

        Divide purchase price by projected lifetime hours: $23,000 + 3,000 = $7.67
        Enter $7.67 in column (1)

    B. Estimate lifetime maintenance costs (engine costs, gas, tires, oil, etc.). Check with your equipment dealer as they sometimes have such data available.

        Divide lifetime maintenance costs by lifetime hours: $12,000 + 3,000 = $4.00
        Enter $4.00 in column (2)

    C. Determine running time (meter time) fuel consumption. For the sake of brevity, let's say the cost of fuel and oil consumption is $2.46.

        Enter $2.46 in column (3)

    D. Total CPH. Add columns (1), (2), and (3). This will give you a CPH for a small tractor.

$$\$7.67 + \$4.00 + \$2.46 \;=\; \$14.13$$

    Again, these figures will vary depending upon the type of tractor and other variables.

4. TRENCHER, WALK-BEHIND: Purchase price, $1,800. Lifetime hours, 750

    A. Acquisition CPH

        Divide purchase price by projected lifetime hours: $1,800 + 750 = $2.40
        Enter $2.40 in column (1)

    B. Estimate lifetime maintenance costs (gas, chain, spark plugs, oil, filters, etc.).

        Divide lifetime maintenance costs by lifetime hours: $800 + 750 = $1.07
        Enter $1.07 in column (2)

    C. Determine running time (meter time) fuel consumption. In this instance, let's say that the cost is $1.85.

        Enter $1.85 in column (3)

    D. Total CPH. Add columns (1), (2), and (3). This will give you a CPH for a trencher.

$$\$2.40 + \$1.07 + \$1.85 \;=\; \$5.32$$

VERIFICATION

To verify your calculated CPH figures, consider the following:

1. Contact your local equipment dealer and/or your owner's manual to verify maintenance costs, production rates, fuel consumption, lifetime hours, etc.

2. Compare your hourly, daily, weekly, or monthly rates to those of your local equipment rental company. Reduce rental rates by 40-50% (to remove overhead and profit markups) in order to determine the CPH for that piece of equipment. Your CPH rates should be reasonably close to theirs.

3. Contact the offices of your state/local Department of Transportation (DOT). They have manuals that contain CPH data for both maintenance and construction equipment. These offices are usually very helpful, and you can use their CPH figures for the purposes of comparison.

4. Obtain a copy of the *Labor & Equipment Production Times for Landscape Construction* noted in the reference section in the back of this book.

5. Dataquest (a company of the Dun & Bradstreet Corporation) offers a *Contractors Equipment Cost Guide* that can be used for comparison purposes. It is expensive ($435.00) but thorough. Call them at 1-800-669-3282 to order a copy.

You can easily utilize a personal computer spreadsheet program (such as Lotus 1-2-3 or Excel) to cost out your equipment. This can simplify the process and the program will make the necessary adjustments quickly and accurately.

ACTION POINT

1. Compute your projected field equipment and overhead vehicle budget amounts and enter them onto Exhibit (1).

2. Using Exhibit (6) in the back of the book, calculate the CPH amounts for your equipment and vehicles and compare to those in Appendix C.

CHAPTER 10
BUDGET RECAP AND RATIO ANALYSIS

PURPOSE
To provide a quick-reference checklist for recapping vital information and ratios in order to evaluate and to monitor a budget

Review diagrams (3-1), (10-1) and (23-2) as we cover the ratios and the budget recap analysis below.

Percentages can provide you with some significant insight into your operation. The ones below are a percentage of sales except labor burden which is a percentage of direct labor.

It is **IMPORTANT TO NOTE** that all percentages given in this chapter are generalizations and approximations. They should only be used for comparison purposes. Actual figures for your company/division will vary.

BUDGET RECAP ANALYSIS

The purpose of the budget recap analysis is to provide you with general industry standards for the items that follow. The budget recap report is a spreadsheet program that allows you to easily compare budget-to-actual costs for the categories identified on it. Unlike diagram (23-2), diagram (10-1) does not include accounts payable or receivable information.

1. GROSS ANNUAL SALES (GAS) OR "SALES": 100%

2. DIRECT COSTS: GENERALLY RUN 70% (± 5%) OF SALES

A. Materials and Supplies
  (1). Construction company: Generally 25% to 35% of sales
  (2). Maintenance company: Generally 5% to 10% of sales

B. Direct Field Labor
  (1). Construction: 20% (±5%) of sales
  (2). Maintenance: 35% (±10%) of sales

C. Direct Labor Burden (for both construction and maintenance)
  (1). United States: 28% to 37% of direct field-labor cost
  (2). Canada and elsewhere: 20% (±4%) of direct field-labor cost

D. Subcontractors (can vary dramatically)
  (1). Construction: From 0% to 25% of sales (sometimes higher)
  (2). Maintenance: Usually less than 5% of sales

(JAN THRU JUN)

	MONTH 6	JUN BUDGET	JUN ACTUAL	JUN VARIANCE	JUN ACT %	Y-T-D BUDGET	Y-T-D ACTUAL	Y-T-D VARIANCE	Y-T-D ACT %
1. SALES-------------->		115,905	130,688	14,783	100.0%	695,430	723,089	27,659	100.0%
2. COST OF SALES:									
MATERIAL		18,708	32,345	(13,637)	24.7%	112,250	137,752	(25,502)	19.1%
LABOR		13,274	12,898	376	9.9%	79,646	94,227	(14,581)	13.0%
LABOR BURDEN		4,580	2,656	1,924	2.0%	27,480	27,589	(109)	3.8%
* EQUIPMENT		2,967	698	2,269	0.5%	17,800	6,016	11,784	0.8%
EQUIPMENT RENTALS		1,225	888	337	0.7%	7,350	12,705	(5,355)	1.8%
SUBCONTRACTORS		26,400	25,769	631	19.7%	158,400	141,408	16,992	19.6%
MISC.		347	1,222	(875)	0.9%	2,083	2,500	(418)	0.3%
TOTAL------------>		67,501	76,476	(8,975)	58.5%	405,009	422,197	(17,189)	58.4%
3. GROSS PROFIT------->		48,404	54,212	5,808	41.5%	290,422	300,892	10,471	41.6%
4. OVERHEAD TOTAL----->		29,167	29,219	(52)	22.4%	175,000	206,338	(31,338)	28.5%
A. SALARY OFFICE		3,000	3,300	(300)	2.5%	18,000	18,900	(900)	2.6%
B. SALARY OFFICER		5,000	5,000	0	3.8%	30,000	30,000	0	4.1%
5. NET OP. INCOME----->		19,237	24,993	5,756	19.1%	115,421	94,554	(20,867)	13.1%
6. OTHER INC/(EXP)---->		0	0	0	0.0%	0	0	0	0.0%
7. NET INC/(LOSS)----->		19,237	24,993	5,756	19.1%	115,421	94,554	(20,867)	13.1%

8. RATIOS/PER HR INFO:	BUDGET	ACTUAL	VARIANCE		BUDGET	ACTUAL	VARIANCE
A. SALES PER HOUR:	$59.99	$63.72	$3.73		$59.99	$64.97	$4.98
B. DIR COSTS PER HR:	$34.94	$42.04	$7.10		$34.94	$37.93	$3.00
C. GPM PER HOUR:	$25.05	$21.68	($3.38)		$25.05	$27.03	$1.98
D. OVHD PER HOUR:	$15.10	$16.06	$0.97		$15.10	$18.54	$3.44
E. PROFIT PER HOUR:	$9.96	$13.74	$3.78		$9.96	$8.50	($1.46)
F. MAT'L/LAB RATIO:	1.41	2.51	1.10		1.41	1.46	0.05
G. MATERIAL PER HR:	$9.68	$17.78	$8.10		$9.68	$12.38	$2.69
* H. EQUIP/LAB RATIO:	22.3%	5.4%	16.9%		22.3%	6.4%	16.0%
* I. EQUIP PER HOUR:	$1.54	$0.38	$1.15		$1.54	$0.54	$0.99
J. AVERAGE WAGE:	$6.87	$7.09	($0.22)		$6.87	$8.47	($1.60)
K. BILLABLE FLH'S:	1,932	1,819	113		11,593	11,130	463
L. NON-BILLABLE FLHS	20	18	2		120	98	22
M. TOTAL FLH'S:	1,952	1,837	115		11,713	11,228	485

* -ACTUAL FIGURES DO NOT INCLUDE DEPRECIATION.

DIAGRAM 10-1. BUDGET-TO-ACTUAL RECAP FORMAT

E. Direct Equipment and Rental Equipment

(1). Construction: 10% (±5%) of sales

(2). Maintenance: 25% (±10%) of sales

3. GROSS PROFIT MARGINS (GPM): 30% (±5%) OF SALES

4. OVERHEAD G&A COSTS FOR EITHER CONSTRUCTION OR MAINTENANCE COMPANIES

A. Generally: 20% (±8%) of sales

B. Large commercial operations: 15% (±3%) of sales

C. Smaller commercial and residential companies and larger high-end residential companies: 25% (±3%) of sales

D. Overhead salaries: 10% (±2%) of sales—about 50% of overall overhead costs

5. NET PROFIT MARGINS (MINIMUMS TO AIM FOR)

A. Company/division sales are under $2.5 million: 10-14%

B. Company/division sales are $2.5-5 million: 8-12%

C. Company/division sales are $5-10 million: 6-10%

D. Company/division sales are over $10 million: 5%

NOTE: Net profit margin (NPM) is calculated after owner's fair market value salary and before bonuses and corporate taxes are paid.

6. PROFIT ADJUSTMENTS

This category includes the portion of overhead salaries in excess of our 8-12% cap imposed in our estimating overhead budget and anticipated bonuses. There is no specified percent or range for this category.

7. REVISED NET PROFIT MARGIN

This category is simply the result of subtracting item 6 from item 5. Therefore, there is no prescribed target percent or range.

8. RATIO AND PER-HOUR CALCULATIONS

The following per-hour items are calculated by dividing annual amounts by the annual total billable field-labor hours. Ratios are obtained by dividing the cost of the first item by the second. For instance, the material-to-labor ratio is obtained by dividing material costs by labor costs.

A. Sales per Hour (SPH)

(1). Construction company/division: $45.00–$65.00

(2). Maintenance company/division: $15.00–$35.00

B. Direct Costs per Hour (DCPH)

(1). Construction company/division: $30.00–$45.00

(2). Maintenance company/division: $10.00–$25.00

C. Gross Profit Margin per Hour (GPMPH)

(1). Construction company/division: $15.00–$20.00

(2). Maintenance company/division: $5.00–$10.00

D. Overhead per Hour (OPH)

(1). Seasonal construction company/division

a. Up to $1.5 million sales: $10.00 (±$3.00)

b. $1.5 million to $5 million sales: $7.00 (±$3.00)

c. $5 million and over sales: $5.00 (±$2.00)

(2). Non-seasonal construction company/division

a. Up to $1.5 million sales: $7.00 (±$2.00)

> b. $1.5 million to $5 million sales: $5.00 (±$2.00)
>
> c. $5 million and over in sales: $3.00 (±$1.00)

(3). Maintenance-only company/division

> Because the SPH varies dramatically from region to region, there is significant variance in the OPH, as well. Your actual figures will be different. Generally, OPH runs 22% (±6%) of SPH for a specific maintenance company or division. Calculate your OPH and compare it to the range provided here.

E. Profit per Hour (PPH), or Net Profit Margin (as defined above) per Hour

> Your PPH is determined by multiplying your company/division projected net profit margin percent by your projected SPH.
>
> For example: If your projected NPM percent was 10% and your projected SPH was $55.00 (for construction) or $22.00 (for maintenance), you would have a projected PPH of $5.50 and $2.20, respectively.
>
> Actual figures will vary dramatically, depending on type of work (construction or maintenance) and geographical location. All figures here are generalizations, and your actual figures will be different.

F. Material-to-labor (M/L) Ratio (See diagram (10-2).)

> Like equipment, material costs are seldom analyzed by landscape and irrigation contractors. We first need to discuss some material-to-labor ratios, then we can convert them to a material cost per field-labor hour (MPH) average.
>
> Companies that install plants and hardscape (bender board, boulders, decks, patios, etc.) usually have a 2 to 2.5:1 material to labor ratio. In other words, they install $2.00 to $2.50 in materials for every one dollar spent on field labor. If the average wage is again $7.00, the MPH would run between $14.00 and $17.50.
>
> Companies that install plants, hardscape, and irrigation generally fall into an MPH range of from 1.5 to 1.75:1. That translates into an MPH of $10.50 to $12.25, if the average wage is $7.00 per hour.
>
> Irrigation-only companies doing residential and small commercial work usually have a 1 to 1.5:1 M/L ratio. Companies doing larger commercial work will experience a M/L ratio from 1.5 to 3:1. Companies performing golf course work will often experience a M/L ratio from 3 to 4:1. Maintenance companies usually have a material-to-labor ratio ranging from .10 to .15:1.

Material-to-Labor (M/L) Ratios			
Type Work	M/L Ratio	Average Wage	MPH (in $)
Planting (only)	3:1	$7.00	$21.00
Planting & Hardscape	2.00–2.5:1	$7.00	$14.00–17.50
Planting, Hardscape & Irrigation	1.50–1.75:1	$7.00	$10.50–12.25
Irrigation (only)	1–1.50:1	$7.00	$7.00–10.50
Maintenance (only)	.10–.15:1	$7.00	$ .70–1.05

10-2. MATERIAL-TO-LABOR (M/L) RATIOS

Maintenance Equipment Cost per Field-Labor Hour			
Equipment Intensity	Equipment as a % of Labor Costs	Average Wage	Co. Average EQPH (in $)
Light	30–40	$7.00	$2.10–2.80
Moderate	40–50	$7.00	$2.80–3.50
Heavy	50–70	$7.00	$3.50–4.90

DIAGRAM 10-3. MAINTENANCE EQUIPMENT COST PER FIELD-LABOR HOUR (EQPH)

Construction Equipment Cost per Field-Labor Hour			
Equipment Intensity	Equipment as a % of Labor Costs	Average Wage	Co. Average EQPH (in $)
Light	15–25	$7.00	$1.05–1.75
Moderate	25–35	$7.00	$1.75–2.45
Heavy	35–50	$7.00	$2.45–3.50
Extra-Heavy	Up to 100	$7.00	$3.50–7.00

DIAGRAM 10-4. CONSTRUCTION EQUIPMENT COST PER FIELD-LABOR HOUR (EQPH)

Use the material per field-labor hour figure much as you do the EQPH one. It provides a quick method of reviewing your bids to see if they are within reason and to check for obvious mathematical errors.

G. Material per Hour (MPH)

(1). Construction company/division: $17.00 (±$3.00)

(2). Maintenance company/division: $2.25 (±$1.50)

H. Equipment-to-Labor (EQ/L) Ratio

(1). Construction company/division

a. Light (pickup trucks and wheelbarrows): .2 to .3:1 (20-30%)

b. Moderate (light above plus tractors): .3 to .4:1 (30-40%)

c. Heavy (moderate above plus backhoes and moderate-sized dump trucks): .4 to .5:1 (40-50%)

d. Extra heavy (all of the above plus graders, large dump trucks, front-end loaders, and semi-trucks/trailers): .75 to 1.0:1 (75-100%)

(2). Maintenance company/division

a. Light (pickup trucks, push mowers, weedeaters, etc.)—primarily for residential work: .3 to .4:1 (30-40%)

b. Moderate (above plus some walk-behind mowers and small riding mowers), primarily for light commercial work: .4 to .5:1 (40-50%)

c. Heavy (all of the above plus large riding and tractor mowers), primarily for heavy commercial/municipal work: .5 to .7:1 (50-70%)

I. Equipment per Hour (EQPH) (See Diagram (10-3) and (10-4).)

(1). Construction company/division: $4.00 (±$2.00)

(2). Construction company/division doing earth moving and grading: $8.00 (±$2.00)

(3). Maintenance company/division: $3.00 (±$1.50)

J. Company/division Average Wage (AW) or Direct Field-labor per Hour

(1). Construction company/division

a. Seasonal: $10.00 (±$2.00)

b. Non-seasonal: $8.00 (±$2.00)

(2). Maintenance company/division

a. Seasonal: $8.00 (±$2.00)

b. Non-seasonal: $7.00 (±$2.00)

K. Billable Field-labor Hours

Here we compare the billable FLH "budget-to-actual" amounts to determine if they are on target for the year.

NOTE: The above general information is provided only as a point of reference, for purposes of **COMPARISON ONLY.** It should **NOT BE USED** for your budgeting or estimating purposes.

ACTION POINT

Calculate the various ratios and per-hour calculations described in this chapter for your individual company or division, and then monitor those figures throughout the year in order to identify particular trends that apply specifically to you.

Section III
ESTIMATING PROJECTS

PURPOSE
To explain in detail how to arrive at a final price for a job

Now that you have your budgets in place, let's refresh our memories as to how we use them in the estimating process by reviewing our models in Diagrams (2-1) and (2-2).

First, we calculate the material, labor, equipment, and subcontractor costs for the production and general conditions phases of the bid. Once we do that, we can then begin to add markups. After we add sales tax and labor burden, we will have determined the total of direct costs for the job being bid. This is, of course, prior to adding the markups for overhead, profit, and a contingency factor (if used).

We then determine how much overhead cost to add to a specific job. Our estimated break-even point (BEP) is identified once we add the overhead to recover on a job to our previously determined direct costs (which include sales tax and labor burden).

The break-even point (BEP) is the estimated dollar amount equal to all costs incurred on that particular job. Profit and a contingency factor is then added to your BEP to arrive at the final price for the job.It is at this point, determining the BEP, that most estimating systems fail. All systems have their strengths and weaknesses, whether the system is:

- Material-times-two (factoring)
- Total direct costs marked up by a predetermined gross profit margin (GPM) percent. This method is sometimes referred to as the single overhead recovery system (SORS).
- Market-driven unit pricing (MDUP)
- The overhead and profit per hour (OPPH) method
- The traditional multiple overhead recovery system (MORS)

Keep in mind throughout the bidding process that you are attempting to determine two sets of numbers in order to answer two questions:

1. **HOW LOW** can you go and still cover all your costs (your BEP)?

And...

2. What will the market let you get away with? Or, how much net profit margin can you add on top of your BEP?

Basically, you are asking, **"HOW HIGH** can you go and still get the job?".

Remember, also, that a good estimating system not only estimates your BEP consistently and accurately, but it does so in such a way that your profit **floats** on top of all costs (BEP) incurred on a job.

Let's turn our attention to determining our production costs.

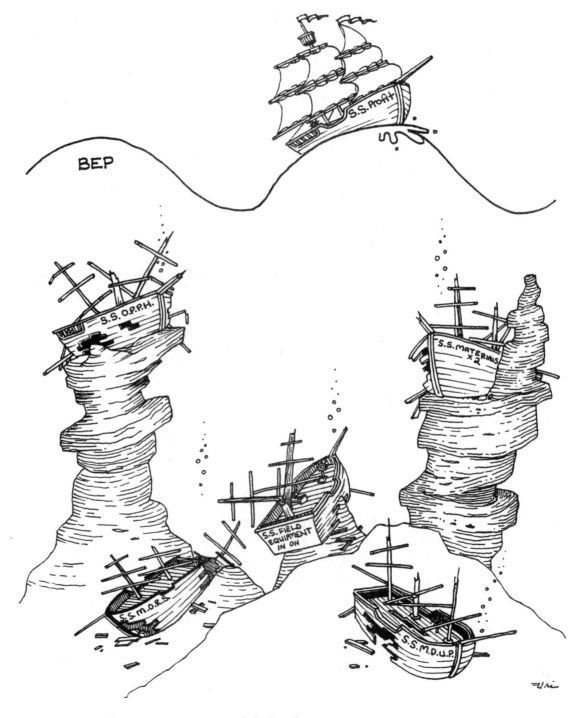

S.S. Profit

CHAPTER 11
ESTIMATING PRODUCTION COSTS

PURPOSE
**To identify and to define production costs and to explain how to
include them in the bidding process**

We begin a lump sum bid by identifying takeoff quantities and estimating the direct costs—material, labor, equipment, and subcontractors (M/L/E/S)—used in the production of the finished product. This is identified as Phase I on Diagrams (2-1) and (2-2).

Finished product/service production costs differ from general conditions (Phase II) costs in that they are the actual M/L/E/S used to produce the finished product or provide the service (i.e., the trees, shrubs, irrigation, patio, mulching, lawn mowing, etc.). Whereas general condition M/L/E/S are required to do the job, but they are not part of the finished product or service. For instance, a portable toilet, a temporary fence, permits, dumpsters, bags for grass clippings, etc., may be required by the job as the related M/L/E/S costs, but these items are not part of the finished product or service. This will become clear when we discuss general conditions in the next chapter.

Make copies of Exhibit (7) for each phase/type of work to be performed (i.e., irrigation, planting, hardscape, maintenance, mulching, etc.). Prepare one marked "Recap" and one marked "General Conditions." Enter the phase/type of work in the "Type of Work" block of the worksheets. Number the pages (1 of 8, 2 of 8, 3 of 8, etc.) in the top-right corner of the worksheets. Staple them together with the recap worksheet first and general condition worksheet second on top of the others.

Build the job in your mind and on paper (your bid worksheets) in the same sequence that you would build it in the field. Keep this general rule in mind as the bid progresses. If questions arise about the job, or if a phase being bid becomes cloudy or overlaps, picture the production steps and the sequences involved. Put these down on paper the way you visualize them. Although not perfect, this method can be very helpful. Even though, some bid items (or phases) will occur simultaneously in the field, you cannot, and need not, show that parallel sequence in the bidding process or on the bid worksheets.

Fill in the recap sheet, as per Diagram (11-1). Start with the first phase of work, proceed to the second, third, and so forth. Skip a couple of lines after the last phase of work (in case we have to add bid items or phases to the project), and then enter general conditions on the recap bid worksheet.

BEFORE you begin, take a moment to collect all of the tools that you will need during the bidding process—bid worksheets, pencils, scales, calculator, scratch pads, etc. Tell your secretary to hold all calls and that you do not want to be interrupted. If you do not have a secretary, turn on your telephone answering machine or notify your answering service to take all your calls. Interruptions that cause you to lose your train of thought during the bidding process can cost you lost time now and lost dollars later.

Enter your first bid item, for the first phase of a project, on the top line in the description column of the bid worksheet pertaining to that phase (see the sample format in Diagram (11-2)). Enter:

- quantity
- unit (EA, SF, LF, CY, LS, etc., as appropriate)
- unit cost or rate
- and items as follows:

MATERIALS

1. WHOLESALE AND REWHOLESALE COSTS

Enter material costs in column (5) at the wholesale price to a contractor.

Contractors who buy materials and resell them to other contractors should enter into their bid the rewholesale cost at which they would resell the materials to other contractors. If feasible, include freight and delivery charges in the price of the materials; otherwise, put them into general conditions at the end of the bidding process.

Companies that grow some of their own nursery stock should also use the rewholesale price in column (5). This is the price that you would resell these items to other landscape contractors or the price you, as a contractor, would have to pay if you bought the materials on the open market.

As a general rule, we will almost always use the rewholesale cost of materials for bidding purposes for the following reasons:

A. The wholesale price paid for materials bought by a nursery for resale to a landscape contractor is usually marked up 30-60% above cost and then sold to the contractor. This is to cover handling, stocking, overhead costs, and profit.

The marked up price, above a nursery's wholesale purchase price, is called the rewholesale price which is charged to the landscape contractor. The rewholesale price is what the contractor should use in column (5)—even if that contractor had a nursery division/operation and sold materials to other landscapers. The original lower wholesale price should not be used in column (5) because it may not include all incurred costs (storage, handling, freight/delivery, etc.).

Unless you have a written quote for materials to be purchased for a specific job, and have incorporated into the bid **all** additional costs to be incurred (handling, re-delivery, etc.) for processing the materials, stick with the rewholesale cost for bidding purposes in column (5).

B. To arrive at a price for retail customers, nursery retail outlets usually mark up their wholesale costs by 100 to 200%. The landscape contractor would buy the same material with only a 30-60% markup added to it.*

Subsequently, retail (residential) customers are accustomed to paying 50% (and more) above the rewholesale price that a landscape contractor would pay for the same material (nursery stock in particular). The bid should reflect this pricing differential in the net profit margin markup, not in the cost of material used on the bid worksheet. You would, therefore, still enter the rewholesale cost on the bid worksheet not the retail price paid by the retail customer; otherwise, you will distort sales tax, overhead, and net profit margin markups, because you are compounding an already inflated cost.

By using a retail versus the more accurate rewholesale cost on the bid worksheet, you are marking up the project an extra 12-13% above your true costs. Profit markup is actually 27.72%, not the 15% indicated in the example.

---

*Since the advent of warehouse chains (e.g., Price Club, Home Depot, etc.) the historical gap between retail, wholesale, and rewholesale is narrowing.

BID WORKSHEET

PROJECT: __JONES RESIDENCE__   PAGE (1) OF (6)

LOCATION: __123 MAIN ST. ANAHEIM__   DATE PREPARED: __1__ / __1__ / __9—__

P.O.C./G.C.: __JERRY JONES__   ESTIMATOR: __KENT ADDETAL__

DUE: __1__ / __15__ / __9—__ PH.: __714-941-7442__ FAX: __N/A__   CREW SIZE: __4__

TYPE WORK: __RECAP SHEET__   HOURS/DAY: __8__   CAW: $ __9.00__

REMARKS:	① DESCRIPTION	② QTY	③ UNIT	④ U/C	⑤ MAT'L	⑥ LABOR	⑦ EQUIP	⑧ SUBS
I. PRODUCE THE PRODUCT								
— SOIL PREP								
— PLANTING								
— SOD								
— MULCH								
II. GEN CONDS								

> Use of Roman and circled Arabic numerals for exhibit purposes; you can use them on your bid worksheet if you desire.

SHI FORM 04-A

DIAGRAM 11-1. BID RECAP WORKSHEET

BID WORKSHEET

PROJECT: __JONES RESIDENCE \<CONTD\>__   PAGE (__3__) OF (__6__)

LOCATION: _____

P.O.C./G.C.: _____   DATE PREPARED: ____ / ____ / ____

DUE: ___ / ___ / ___ PH.: _____ FAX: _____   ESTIMATOR: _____

TYPE WORK: __PLANTING__   CREW SIZE: __4__

HOURS/DAY: __8__   CAW: $__9.00__

REMARKS: ①	②	③	④	⑤	⑥	⑦	⑧
DESCRIPTION	QTY	UNIT	U/C	MAT'L	LABOR	EQUIP	SUBS
– RED MAPLE 24" BOX	4	EA	131 –	524			
– LABOR \<2 MHR/TREE\>	8	HR	9 –		72		
– AMENDMENTS \<3CF/EA\>	.5	CY	20 –	10			
– PLANT TABS \<8/TREE\>	32	EA	.10	3			
– STAKE KITS \<3/TREE\>	4	KIT	10 –	40			
– LABOR \<2/MHR\>	2	HR	9 –		18		
– CRAPE MYRTLE 15 GL	10	EA	29 –	290			
– LABOR \<1/MHR\>	10	HR	9 –		90		
– AMENDMENTS \<1.5CF/EA\>	.5	CY	20 –	10			
– PLANT TABS \<5/TREE\>	50	EA	.10	5			
– STAKE KITS \<2/TREE\>	10	KIT	8 –	80			
– LABOR \<4/MHR\>	2.5	HR	9 –		23		
– SHRUBS 5 GL	30	EA	7 –	210			
– LABOR \<5/MHR\>	6	HR	9 –		54		
– AMENDMENTS \<1CF/EA\>	1	CY	20 –	20			
– PLANT TABS \<2/EA\>	60	EA	.10	6			
– SHRUBS 1 GL	55	EA	3.50	193			
– LABOR \<15/MHR\>	4	HR	9 –		36		
– AMENDMENTS \<100/CY\>	.5	CY	20 –	10			
– PLANT TABS \<1/EA\>	55	EA	.10	6			
				1407	293	Ø	Ø

SHI FORM 04-A

DIAGRAM 11-2. BID EXAMPLE

The example below demonstrates this problem.

	CONTRACTOR'S REWHOLESALE PRICE	RETAIL CUSTOMER'S PRICE	VARIANCE
Base cost/price	$ 1.00	$ 1.50	$ .50
6% Sales tax markup	.06	.09	.03
Subtotal	1.06	1.59	.53
10% Overhead markup	.106	.159	.053
Subtotal	1.166	1.749	.583
15% Profit markup	.1749	.2624	.0875
Subtotal	1.3409	2.0114	.6705
Markup passed on to customer	.3409	.5114	.1705*
Total price with markups	$ 1.3409	$ 2.0114	$ .6705
Markup variance (*) as a % of price	12.72%	8.48%	N/A

DIAGRAM 11-3. REWHOLESALE VS. RETAIL PRICING DIFFERENTIAL

Use true costs on the bid worksheet; otherwise, you are lying to yourself and building inaccuracies into your bids and into the bidding process.

C. I know of large landscape companies that have five to ten divisions (commercial construction, residential design/build, reclamation, maintenance, golf course construction, irrigation service, nursery, equipment, etc.). Many of these divisions sell services and/or materials to one another. To keep intracompany pricing honest, it is best to use fair market value (FMV) prices obtained from the open, competitive market as a transfer price between divisions and to allow (require) estimators to use open market prices in their bidding.

Here's how one of my clients handles this situation. Their estimators are encouraged (not required) to use rewholesale prices from their company's nursery division. However, if they can beat that price by obtaining quotes from outside nurseries, they are to do so. The same rule holds true when buying materials. Purchasing agents are allowed to purchase materials from sources/vendors outside the company, but only if quality standards are maintained and if their own nursery cannot match the price.

2. WARRANTY OF MATERIALS

A. Calculating Warranty Bid Costs

Labor and equipment costs for product warranties will be included in general conditions; however, we will include the material replacement costs in the purchase cost of the material itself. For instance, when bidding a particular size and type of tree, build in your warranty factor as follows:

(1). Ask yourself, "How many of this particular tree can I expect to lose if I plant one hundred of these trees on this particular job site?". Take into account the time of year, location, weather conditions, site conditions, etc. If you think that you would lose two, that is 2%; five would be 5%; six would be 6%, etc.

(2). Increase the unit cost for the tree by the estimated loss percentage.

For instance, a tree that costs you $95.00—with an anticipated loss of five trees per 100 planted (5%)—would be multiplied by 105% or 1.05.

$95.00	Unit cost
x 1.05	1 plus loss percent
$99.75	Total (unit cost with material warranty cost)

Another method would be:

$95.00	Unit cost
x .05	5% (decimal format)
$ 4.75	Material warranty cost
95.00	Add unit cost
$99.75	Total (unit cost with material warranty cost)

Some contractors simply add an across-the-board percent markup to either their entire bid (i.e., 1-2% of the total price), or they increase only the planting portion of the bid by a certain percent (e.g., 5-6%).

Other contractors even have an historical basis for doing so. They track their actual warranty material, labor, and equipment costs, and then arrive at a warranty percent by dividing total actual costs by either the gross sales or the planting costs (costs for material, labor, and equipment).

The problem with this method, even if it is based on accurate historical data, is that it is too vague and general. Although a company-wide aggregate warranty percent may have a certain overall accuracy, jobs usually vary a great deal from one to the other.

We need to be more analytical and job-specific, especially in competitive markets and in times of economic downturns.

There are other methods for guaranteeing plants, products, and/or services. Basically, you are providing a form of insurance for your client and therefore, managing future exposure or risk. I call this "surprise control." You are attempting to ensure predictability, consistency, and certainty for your client by ruling out surprises—both good and bad (and remember, *there are no good surprises in business*).

B. The Lifetime Guarantee

One method of offering your clients "risk control" is to extend a lifetime guarantee (except for "acts of God" and vandalism) on plant materials installed.

This option is offered only on condition that the installer performs the maintenance on the project according to predetermined specifications. If the client breaks the maintenance contract, or does not have the prescribed maintenance performed, the lifetime warranty is voided much as a car warranty would be if servicing requirements were not met.

The reasoning behind the lifetime warranty is three-fold:

(1). The risk to the customer is minimized.

(2). The customer never has to worry about future landscape problems.

(3). The budget for the customer's landscape requirements is predetermined.

In turn, the landscape contractor increases maintenance sales with minimal exposure to risk. If the landscape contractor installed the project and maintains the property properly, there should be a minimal loss of plant materials. It can be a win-win situation for all concerned.

Theoretically, you could actually take out an insurance policy, with your customer as the beneficiary, on plant materials (or other products) in the event that predetermined standards or conditions were not met during a specified period of time.

A client and I briefly considered pursuing this type of "warranty" option in the following situation:

One of his residential customers wanted a 120" box tree installed. The rewholesale cost of the tree to my client (with delivery and crane charges) was $12,000. A four-man crew with a backhoe could install the tree in one eight hour day. My client had $13,000 in costs (M/L/E/S) in the project. The installed price to the customer was a little under $24,000. Net profit margin would be a little over $10,000—not bad for a day's work. There are times when pricing your work by factoring or "material-times-two" is appropriate.

When my client described all of this to me, I said that it sounded like an excellent return for his efforts—as long as he managed and minimized his risk in the event that something happened to the tree. I asked him, "What happens if the tree dies?" and "Are you going to guarantee the tree?". Think about it. Using a traditional one-year warranty method, you can either win big ($10,000), or you can lose big; by spending $13,000 to replace the tree. When you are only dealing with one, very expensive tree, there's not much room for error.

His response was that this was why he was paying me to tell him what to do. He did not know the best thing to do.

My initial thoughts were to recommend a one year warranty except for vandalism and "acts of God." If something happened to the tree, and it appeared that the tree would not survive, you could always pay someone $100 to chainsaw it down on a dark, moonless night and call it "vandalism."

Fortunately, more feasible and ethical options were available. After my client and I discussed it further, we decided to ask the nursery if it would be willing to warranty the tree. My client was well-known by and had an excellent reputation with the nursery. My client's customer was motivated to buy the tree.

The nursery was *very* motivated to sell a $12,000 tree; my client was *extremely* motivated to install said $12,000 tree for $13,000 and be paid almost $24,000 to do so; I was motivated (out of my "socks") to solve the problem for my client, earn my pay, and maintain a client as a future customer. However, *ALL* of us wanted to minimize our "downside" risk, just in case the future did not turn out as we had hoped. That is, what would happen if this very beautiful, *very expensive—VERY PROFITABLE* tree—died?

The nursery had confidence in my client's ability to optimize the chances for the tree's survival. If, however, the tree did die, they could only replace it with two 60″ box trees, as they had no other 120″ box trees available. My client presented this to his customer. The customer said that that was fine with him. Everyone was satisfied, and everyone's risk was minimized.

My point is this: ***warranties are simply a form of risk management.*** It's an insurance policy that should apply as much to you (the contractor) as to your customer. As such, warranties can take many forms. Be creative.

Include warranties as part of the written contract with your client but be sure that your legal bases are covered. Have your attorney review the warranty clause in your contracts.

ACTION POINT 1

Review Diagrams (11-1) and (11-2) and complete portions A and B of Exercise (11-1) in Appendix A.

LABOR

Multiply labor hours for each function by your CAW or specific labor rate for non-crew members and/or non-crew functions (ones performed not as part of a crew).

1. PRODUCTION RATES

Ideally, it is best to use crew production rates based on past performance (historical data); however, this may not always be possible.

There are production rate reference books available. Titles and addresses for ordering these books are listed in the back of this book in the "For Further Reading/Reference Materials" section.

CAUTION: Be aware that the rates provided in these books are *averages* and often tend to be conservative. Be sure to adjust rates in them to reflect specific site conditions and/or crews.

If you have neither past performance data nor a reference manual, ask an experienced foreman or another contractor for input.

Another option would be to do a miniature "time and motion" study. If feasible, have a crew perform part of, or a limited quantity of, a certain function. Otherwise, you will have to use your best "guesstimate." In this case, clearly visualize the steps involved. Be conservative with the amount of time that you allot to a function (owners and managers generally tend to be too optimistic when estimating production rates).

If you bid using unproven production rates, it is best (if practical) that the individual running the crew to perform the task helps to determine the rate used in the bid.

Also, BEFORE starting that particular task (or any task, for that matter) be *sure* that the foreman:
- knows the hours budgeted for it, and that he
- clearly understands the production process/sequence to be used.

Bid worksheet procedure: Review Diagram (11-2) and enter information on the bid worksheet as follows:

A. The word, "LABOR" (or project), in "Description" column.
B. "3/MH" or "3/MHR" (three per man-hour) as the production rate beside description.
C. Quantity of hours for the function in column (2).
D. "HR" in unit column (3) for hour.
E. Labor rate or CAW in column (4).
F. Multiply hours in column (2) by labor rate in column (4) and enter the product (total) in column (6).

2. AVERAGE WAGE

First, determine your crew size and then the CAW by dividing the total sum of the crew labor rates by the number of people in the crew. For example:

1 at $13 + 1 at $8 + 1 at $6 = $27.00 for 3 men. Divide $27.00 by 3 and your CAW is $9.00 per hour.

Use a CAW (including OTF and RF) for bid items when the whole crew is on the site and when the item being bid is performed by a member of that crew.

Use specific rates when an individual is not a member of the crew (e.g., drivers, non-crew member operators, etc.) or for specific individuals in general conditions (e.g., foremen supervisory time, superintendents, etc.).

OVERTIME FACTOR (OTF)

If the crew works more than a forty-hour week and overtime is paid, add the OTF to your labor rates.

If you know that the job being bid will require 50-hour weeks, bid it accordingly. Otherwise, use yearly/season averages (e.g., 45 hours per week throughout the season). Always use the higher of the two numbers.

Divide the number of overtime hours paid by the number of straight-time hours paid. For example, for a 45-hour week it would be:

2.5 + 45 = OT hours paid + straight-time hours paid = .0556 = 5.6% OTF

Increase the CAW by adding the OTF:

CAW	$9.00
OTF ($9 x 5.6%)	.50
Total (CAW + OTF)	$9.50

Do the same for specific labor rates, if they are used, and if such individuals are paid overtime.

Labor Risk Factor (RF): See Chapter 5 for an explanation of the Risk Factor.

Add your RF to the CAW and OTF:

CAW + OTF	$9.50
RF (10%)	.95
CAW + OTF + RF	$10.45
Round up to nearest $ .10.	$10.50

Enter $10.50 in column (4) for your labor rate. I refer to this rate as the "loaded" CAW (i.e., it includes the OTF and RF).

ACTION POINT 2

Complete section C of Exercise (11-1) in Appendix A.

EQUIPMENT

Bid field equipment into the job on an item-by-item basis. This as compared to either lumping it all together in general conditions or accounting for it in overhead.

Use actual meter (or running) time that the piece of equipment will be used for each bid item.

Enter equipment information on the bid worksheet as follows:

1. Enter the description of the piece of equipment in column (1).
2. Beside the description, enter the production rate for this piece of equipment (100′/HR).
3. Enter the number of meter hours in column (2).
4. Enter "HR" in column (3).
5. Enter the CPH in column (4).
6. Multiply the number of meter hours in column (2) by the CPH in column (4). Enter the total in the "Equipment" column (7).
7. Rental Equipment

    If a piece of rental equipment is to be used for only one bid item, enter its entire rental cost (plus fuel expense and delivery, if applicable) in that bid item.

    However, if a piece of rental equipment will be used on more than one bid item—for example, a trencher rented for $130 a day (including fuel and delivery) that is to be used on both main line and lateral line trenching—first determine its CPH.

    To determine your rental cost per hour (rental CPH): divide the rental rate plus fuel costs plus delivery (if applicable) by the anticipated number of actual hours used per day. For example:

    [$85 (rental/day)+$35 (delivery)+$10 (fuel)] ÷ 7 hours/day (meter time) = $130 ÷ 7 = $18.57 CPH

    Round up to nearest $ .25 = $18.75

    The CPH will be the same on each bid item (if it is used on more than one). Be sure that the total in column (7) for the two bid items equals the total cost ($130) for the rental equipment.

8. Crew Pickup Trucks/Company Delivery Truck and Driver Time

    Crew pickup trucks are normally included in general conditions. Company delivery trucks and drivers are also normally included in general conditions. However, you can include these costs in a specific bid item, if you so choose.

For instance, you can include delivery time with materials or equipment mobilization time with delivery time for a trencher in the main or lateral line irrigation bid items. However, it is usually simpler just to put these items in general conditions.

Include equipment operator time on the line directly below the piece of equipment. If the operator's rate is calculated in with the CAW, use the loaded CAW in column (4). If the operator is not part of the crew, enter the operator's rate in column (4).

ACTION POINT 3

Complete section D of Exercise (11-1) in Appendix A.

SUBCONTRACTORS

Subcontractors are added onto the bid sheet at cost.

1. Enter the description in column (1).

2. Enter "1" and "LS" (lump sum) as the quantity and unit for subcontractors for which no other quantity and unit apply.

   (If desired—and if possible—enter actual production units and quantities—e.g., 22,000 SF of hydroseed.)

3. Enter unit cost in column (4).

4. Multiply the quantity in column (2) by the unit cost in column (4) and enter the product (total) in column (8).

Once all bid items and their components are entered on a bid worksheet, put a heavy or double line under the last items in columns (5) through (8).

Total each column and enter the total for each column under the double lines.

Proceed to fill out the bid sheets for the other phases of work to be done.

ACTION POINT 4

Complete Exercise (11-2) in Appendix A.

CHAPTER 12
GENERAL CONDITIONS

PURPOSE
To identify and define general condition items and to explain how to include them in the bidding process

General conditions (GC) constitute Phase II of a bid as outlined in Diagrams (2-1) and (2-2).

BIDDING GENERAL CONDITION COSTS

The costs for GC items are bid the same as Phase I production costs:

- Materials are bid at cost.
- Labor is bid as hours multiplied by a loaded (including the OTF and RF) CAW or labor rate.
- Equipment is bid as hours multiplied by a predetermined CPH.
- Subcontractors are bid at cost.

DETERMINING CATEGORIES

Enter GC items on the general conditions bid worksheet. Costs for items such as dumpsters, dump fees, soil tests, trailers, toilets, etc., can be entered into the "Materials" column. Design fees can be entered as "Subcontractors" or in the case of in-house designers, as actual labor hours multiplied by a labor rate. Lodging and per diem costs should be placed in the "Materials" column although, technically, you could also put it in the "Subcontractors" column. Diagram (12-1) displays the general conditions bid worksheet for the Jones residence project. Diagram (12-2) is a printout of a database for general conditions that we use in our estimating software.

Let's now turn our attention to, and review in detail, the most common GC items and how to bid them.

CALCULATING JOB DURATION

The amount of certain general condition items (supervisory time, crew pickup truck time, foreman administrative time, drive time, clean-up time, etc.) can be directly tied to the number of visits/hits that a maintenance crew makes at a particular site or to the number of job or crew days that a construction crew is on a job site. The number of visits that a maintenance or service crew makes to a given site is usually predetermined by the duration of the season and/or the type of service desired by the client. Therefore determining the number of visits is generally somewhat easier than doing so for a construction project. Construction crews can determine job duration or crew days in two steps using one of two methods.

JOB DAYS = $\frac{52.5}{28.5}$ = 1.85

BID WORKSHEET

PROJECT: JONES RESIDENCE

PAGE (6) OF (6)

LOCATION: _____

DATE PREPARED: ___ / ___ / ___

P.O.C./G.C.: _____

ESTIMATOR: _____

DUE: ___ / ___ / ___ PH.: _____ FAX: _____

CREW SIZE: 4

TYPE WORK: GENERAL CONDITIONS

HOURS/DAY: 8   CAW: $ 9.00

REMARKS:

DESCRIPTION	QTY	UNIT	U/C	MAT'L	LABOR	EQUIP	TOTAL SUBS
GENERAL CONDITIONS	1	LS					
– FOREMAN (.5HR/DAY)	1	HR	12		12		
– CREW PICKUP TRUCK	16	HR	4			64	
– CLEANUP (.25 HR/MAN/DAY)	2	HR	9		18		
– DRIVE TIME (.5HR/MAN/DAY)	4	HR	9		36		
– CALLBACKS (LABOR)	3	HR	9		27		
– CALLBACKS (TRUCK)	3	HR	4			12	
				∅	93	76 =	169

SHI FORM 04-A

DIAGRAM 12-1. GENERAL CONDITIONS BID WORKSHEET (MANUAL)

WORKSHEET

BID ITEM ID	UNIT QTY UNIT	DATABASE TASKCODE	TASK DESCRIPTION	QUANTITY UNIT	MATL. UNIT $COST	LABOR/EQUIP TYPE	QUANT / 1 HR	TASK HOURS	SUB UNIT $$ COST	TASK/BID UNIT $$ COST	BID UNIT $$ PRICE
GEN CONDS	1 LS	GENCOND GC	NUMBER OF DAYS ON JOB-->*	32 EA		*		0.0		0.000	0.00
		GC	SUPERINTENDENT (1HR/DAY)*	32 HR		*S	1	32.0		0.000	31.08
		GC	FOREMAN ADMIN (1HR/DAY) *	32 HR		*F	1	32.0		15.344	21.94
		GC	MOB/DE-MOB EQUIP (L) *	6 HR		*D	1	6.0		10.869	15.54
		GC	MOB/DE-MOB EQUIP (EQ) *	6 HR		*105	1	6.0		12.000	20.77
		GC	CREW PICKUP TRUCK *	256 HR		*100	1	256.0		2.750	4.76
		GC	CREW LOAD/UNLOAD TIME *	16 HR		*C	1	16.0		12.787	18.28
		GC	HAUL MATL-(L) *	8 HR		*D	1	8.0		10.869	15.54
		GC	HAUL MATL-(EQ) *	8 HR		*105	1	8.0		12.000	20.77
		GC	CREW DRIVETIME TO/FM JOB*	32 HR		*C	1	32.0		12.787	18.28
		GC	WARRANTY-(L) *	32 HR		*C	1	32.0		12.787	18.28
		GC	WARRANTY-(EQ) *	16 HR		*100	1	16.0		2.750	4.76
		GC	TRAILER/TOILET/FENCES *	6 WK	25	*		0.0		26.125	40.50
		GC	PERMITS/LICENSES/FEES *	1 LS	200	*		0.0		209.000	323.96
		GC	CLEANUP AT END OF DAY *	32 HR		*C	1	32.0		12.787	18.28
		GC	DUMP FEES *	1 LD	100	*		0.0		104.500	161.98
		GC	DUMPSTERS *	1 EA	150	*		0.0		156.750	242.97
		GC	STORAGE CONTAINERS *	1 MO	175	*		0.0		182.875	283.47
		GC	HAUL DEBRIS (L) *	4 HR		*D	1	4.0		10.869	15.54
		GC	HAUL DEBRIS (EQ) *	4 HR		*105	1	4.0		12.000	20.77
		GC	FREIGHT/DELIVERY FEES *	1 LS	100	*		0.0		104.500	161.98
		GC	SOIL TESTS *	1 LS	55	*		0.0		57.475	89.09
		GC	CREW WATERING TIME *	12 HR		*C	1	12.0		12.787	18.28
		GC	DETAIL JOB *	16 HR		*C	1	16.0		12.787	18.28
		GC	COMMISSIONS *	1 LS	300	*		0.0		313.500	485.94
		GC	DESIGN FEES/AS-BUILTS *	1 LS	250	*		0.0		261.250	404.95
		GC	ESTIMATOR TIME *	3 HR		*E	1	3.0		19.180	27.42
# OVERHEAD #	1.00				###### 1480				0.000	########## ########## 5,740.572	

DIAGRAM 12-2. GENERAL CONDITIONS BID WORKSHEET (COMPUTERIZED)

First, calculate the number of man-hours that the crew is on the job in the Phase I portion of the bid. Do so by simply counting the number of such hours or by dividing Phase I payroll by the CAW for the bid. Second, divide the number of Phase I man-hours by the number of hours worked by the crew per day. Refer to the upper right-hand portion of Diagram (12-1) where we have done such a calculation for the Jones project.

52.5 man-hours divided by 32 (man-hours per day) = 1.6 job or crew days.

or

$473 (Phase I payroll) divided by $9.00 (CAW) = 52.6 man-hours.

52.6 ÷ 32 = 1.6 job or crew days.

A more accurate (but more complicated) method would be to subtract the average daily crew general condition man-hours from the total man-hours worked per day.

32.0 man-hours per day
- 0.5 foreman admin time
- 1.0 cleanup time
- 2.0 drive time
_____
28.5 Phase I man-hours per day

52.6 divided by 28.5 = 1.85 job or crew days

The .25 crew-days (1.85 - 1.6) is not a lot of difference for the Jones residence job. However, for larger projects it would be. Use the second method for jobs lasting more than one week.

MOST COMMON GENERAL CONDITION ITEMS

When bidding GC items, enter the item nomenclature in the description portion of the bid worksheet with the production rate (1 HR/DAY) beside it. The "HR/LABOR," "LS/MATERIALS" on the right indicates the unit and in which M/L/E/S category to total your costs for that particular line item.

NOTE: CAW and other labor rates when used in this chapter and throughout the remainder of the book, refer to a "loaded" rate which includes an overtime and a risk factor, if applicable.

1. SUPERINTENDENT (1 HR/DAY)   HR/LABOR

Superintendent time is the time that an owner or supervisor is actually on the job site supervising it. Put the number of hours (1 hour/day) that an owner or field superintendent will actually be on the job site, supervising the job, beside the description. Usually, a portion of this person's salary is in overhead and a portion in general conditions. Reduce the total amount of their salary that is in overhead accordingly.

For instance, if a superintendent can be bid into jobs 50% of the time (four hours a day), only enter 50% of their salary into overhead. If you put the full salary into overhead and four hours per day are also included in your bids, you are "double-dipping" or charging twice for that particular cost.

*2. FOREMAN ADMINISTRATION TIME (2 HR/DAY)   HR/LABOR

This is the time that a foreman spends on non-production tasks—tasks not included in Phase I on Diagrams (2-1) and (2-2). It would include time to:

• Fill out field daily reports

• Lay out the job

• Order and receive materials

• Coordinate with other subcontractors or the general contractor/owner.

• Coordinate with the home office, etc.

Ask yourself, "On average, how much time does your foreman (or you) spend on items not directly involved with the end-product/service?". The answer usually is about one to two hours each day.

Multiply the number of hours per day by the projected number of days for that particular job and enter the resulting total in the quantity column. Multiply the total number of foreman's administrative hours by the labor rate in the "Unit Cost" column and enter that foreman's hours in the "Quantity" column on the bid worksheet.

3. CREW PICKUP TRUCK (8 HR/DAY)   HR/EQUIPMENT

This is the vehicle that the crew uses on the job site. Multiply the number of days projected for the particular job by eight hours per day—regardless of actual meter/running time hours. Use ten hours per day if you are on the job site ten hours per day and only work four days per week.

If you use a two-ton or larger truck on the job site, use four hours per day or actual/running time per day, whichever is greater.

If you are on a job less than a full day, reduce your hours accordingly.

4. CREW LOAD/UNLOAD TIME (.5 HR/MDAY)   HR/LABOR

This is the time that the crew (or foreman) spends loading/ unloading the trucks at the job site and at the yard in the morning and/or evening.

Multiply the number of crew members by the time each member loads/unloads per day. Multiply this figure by the projected number of days for the job.

For example: On average, each member of a three-man crew spends fifteen minutes in the morning and in the evening on this task.

$$3 \text{ (men)} \times .5 \text{ (hr/mday)} = 1.5 \text{ hrs/mday} \times 10 \text{ (days)} = 15 \text{ hours}$$

For the above example, we would use the CAW for "Unit Cost." If the foreman is the only one doing the loading/unloading, then you would use the foreman's labor rate.

5. DRIVE TIME TO/FROM JOB (1 HR/DAY)   HR/LABOR

If you pay drive time, enter it into GC's in this category. Use your CAW if the entire crew is paid drive time; if your foreman is the only one paid drive time, use your foreman's labor rate.

Whether you should or should not pay drive time is not the issue here. The issue is whether (a) the customer is paying for the drive time, or (b) the company is paying for it out of profit because you are not charging the customer by including it in your bid.

Some companies pay drive time only to individuals who are driving company vehicles to and from the job. Others pay all crew members to and from the job, while still others only pay crews one-way.

My concern is that you bid your work so that it will accurately reflect your company's practice. If you do not, you have a "leak" in your estimating system. That is, you have a cost that is not included in your bid.

You begin by accurately identifying and measuring your drive time. You should then weigh the pros and cons as to what your company policy should be. Make sure that you are aware of the legal ramifications involved. Be sure that you are in compliance with any requirements that may exist in your state.

Depending on the particular job (or competitive situation), it is often best to have a variable policy regarding drive time. In some cases, the amount of drive time can be the difference between winning a bid or losing it. In other cases, landscape and irrigation contractors have realized that it would be cheaper (on certain jobs) to eliminate drive time by accommodating crews in a motel near the job site.

6. MOBILIZATION OF EQUIPMENT (LABOR)   HR/LABOR

Labor time required to transport equipment (tractors, trenchers, backhoes, dozers, etc,) to and from jobs—time that is not included within crew drive time—should be separately identified and bid into jobs under general conditions.

You could include this item in the particular phase of work requiring it (for example, trencher mobilization time could be bid into the irrigation phase of the bid). However, I prefer to identify and include this item in general conditions.

Include labor time to load and unload the equipment onto a trailer (if applicable) and the drive time to and from the job site. Remember, in most cases you will make two separate trips to and from the job for each piece of equipment: one to deliver it to the job and another to remove it.

If there are a number of pieces of equipment to be mobilized on/off jobs, there is the tendency to think that you can reduce mobilization (in half) by removing one piece on the same trip that you take another piece to the job. Good luck! If you genuinely believe that you can do it that way, then bid it so. Frankly, I think that you are probably being overly-optimistic.

Enter the driver's labor rate or an average wage (if more than one person is involved in mobilization) and include an OTF and RF as necessary in the "U/C" (unit cost) column and extend the costs accordingly on the bid worksheet.

7. MOBILIZATION OF EQUIPMENT (EQUIPMENT)   HR/EQUIPMENT

The second component of mobilization is the equipment required to mobilize to and from the job site. Use CPH for the vehicle hauling the equipment in the "U/C" column. Include actual meter/idle time. Do not make the mistake of one contractor who included "meter time" in his bids for a tractor while it was being hauled to the job site on a trailer.

Remember to include two round trips in your bid: one to transport the equipment to the job site and one to remove it later.

8. FREIGHT/DELIVERY CHARGES   LS/MATERIALS

Freight or delivery charges that are not included in the materials cost portion of the bid for planting, irrigation, hardscape, etc., should be identified and included within general conditions.

On the bid worksheet, use a quantity of "1" (one) and "LS" (lump sum) for the unit.

9. HAULING MATERIALS (LABOR)   HR/LABOR

Hauling materials comprises driver and truck time used to obtain materials from nurseries and other suppliers or for employees who deliver materials throughout the day as their primary function.

Enter the driver's labor rate (or an average wage, if more than one person is involved in deliveries) in the "U/C" column. Remember to include both load and unload time, as well as drive time to and from the job site. Beware of duplication if you have already listed these times under (1) crew load and unload time, or (2) drive time to and from the job site.

10. HAULING MATERIALS (EQUIPMENT)   HR/EQUIPMENT

Calculate the hours for vehicles and equipment used to load and haul materials to the job site. Include time for skid steers and other pieces of equipment used in the process.

11. CONSTRUCTION TRAILERS, PORTABLE TOILETS, STORAGE CONTAINERS   MO/WK/MATERIALS

Place these items in the materials column (5); however, you could technically call them equipment and enter them into column (7). Use the appropriate unit (WK, MO, LS) and unit cost in their respective columns.

12. PERMITS, LICENSES, AND FEES   EA/LS/MATERIALS

As in item 11 above, enter these items also in the "Materials" column.

Use "EA" or "LS" as the unit and enter the unit cost in column (5).

13. SOIL TESTS   EA/LS/MATERIALS

Same as item 12.

14. DUMP FEES   LD/LS/MATERIALS

Same as item 12. Use "LD" (load) or "LS" (lump sum) as the unit.

15. DUMPSTERS   EA/LS/MATERIALS

Same as item 12.

16. TRAFFIC CONTROL LS/SUBCONTRACTORS or HR/LABOR

If handled by a subcontractor, enter as such. If company employees are used, enter in the "Labor" column. Enter unit and unit costs and extend/multiply out as appropriate.

17. PLANT WATERING TIME (LABOR) HR/LABOR

Watering-in plants immediately after they are planted usually is included with the planting phase of the bidding process, much as are the staking of trees. You could also place watering-in under general conditions.

In either case, you will probably have additional hand watering after the planting takes place. Enter that time in general conditions.

18. PLANT WATERING TIME (EQUIPMENT) HR/EQUIPMENT

Include any equipment required for watering purposes (trucks, water buffalos, etc.).

19. CLEAN UP AT END OF DAY (.25 HR/MDAY) HR/LABOR

This is the time that the crew uses at the end of the day to put tools away, sweep the site, store materials, lock and secure the site, etc. This usually takes about fifteen minutes for each man per day. Use the CAW unit cost as appropriate.

20. HAUL DEBRIS (LABOR) (2 MHR/TRIP) HR/LABOR

Calculate the amount of time required to load, haul to a dump site or dump, and unload a load of debris (remember to include time for the return trip). Multiply the time required for one trip by the projected number of trips in order to determine total labor hours.

Use the CAW for regular crew labor or a specific rate for non-crew drivers and/or operators.

21. HAUL DEBRIS (EQUIPMENT) HR/EQUIPMENT

Include trucks and equipment (loaders, skid steers, etc.) used to load and unload and haul debris to dumps and/or dump sites and the time used on the return trip.

**22. LODGING ($35/DAY) DAY/WK/MO/MATERIAL

If applicable, include costs for crew quarters. Use the appropriate unit and unit cost (e.g., $35/day; $245/wk; $1,050/mo; etc.) and enter beside "Lodging" in the "Description" column.

**23. PER DIEM ($15/DAY) AND/OR FOOD ALLOWANCE DAY/WK/MO/MATERIAL

Same as Lodging.

24. WARRANTY (LABOR) HR/LABOR

Once the production phases of the bid are completed, ask yourself: "How many trips (or how much time) will be spent on this job to repair the irrigation, replace dead plant material, etc.?". View the job in its entirety. Compare it to other jobs that you have done where the warranty time and conditions were similar.

Think in terms of half or full days and the number of men that will be needed. Bid replacement labor just as you would any other item. For instance, you may anticipate that one job may require two men for two full days during the warranty period. That would be: 2 men x 2 days x 8 hours/day = 32 hours

Another job may only require one person for half a day. That would be: 1 man x 4 hours = 4 hours

If you have limited experience regarding warranties on a particular type of work, seek out some advice from your suppliers, foreman or other contractors. Remember, replacement will generally take longer than the initial installation. You must first remove existing materials, replace them, and then dispose of the debris.

Material replacement costs, you may remember, are included in the costs bid into Phase I (the production phase) of the project. There we simply increased plant materials cost by an anticipated replacement percentage. The attempt, however, to use such a percent (or method) for replacement labor and equipment is too narrow of a procedure. I feel that it is much better (and more realistic) to calculate warranty hours for labor and equipment within general conditions.

Average the labor rates for those who are most likely to perform the warranty work. This rate is usually higher than your CAW. This is because it is usually your foreman (or leadman) that will do replacement work either alone or with the help of a laborer.

25. WARRANTY (EQUIPMENT)   HR/EQUIPMENT

After you have determined labor hours for warranty work, estimate the hours for vehicles and/or equipment.

If you estimate two men for two full days, or thirty-two man-hours, remember that you will need a truck for the same number of days, but for only eight hours a day.

Include all equipment that may be needed for replacement purposes (tractors, mowers, trenchers, etc.). Be sure to factor in disposal time (for both labor and equipment) and disposal fees for debris, if necessary.

**26. DESIGN FEES   HR/LABOR or LS/SUBCONTRACTOR

Design fees can be entered into a bid either within general conditions or into Phase I (production).

In-house landscape architects and/or designers can either be bid as you would other labor, or they can be considered "Subcontractors" and entered into that column as a lump sum expense.

"Legitimate" subcontract landscape architects and/or designers should be bid as such, using a lump sum price.

**27. COMMISSIONS   LS/SUBCONTRACTORS

As a general rule commissions should only be paid on residential projects. Commercial bidding situations do not lend themselves to a commission-type payment structure because of the thin profit margins and the lag time between bidding, signing, and receiving payment for them. In-house estimators should handle such bidding and be put on salaries.

There are a couple of options when including commissions in a bid. The first is to enter commissions in general conditions as a subcontractor cost. However, before you do so, ensure that all payroll taxes, insurance (FICA, FUTA, SUTA, and WCI), and expense reimbursements (auto, mileage, etc.) are taken into consideration and included in the figure entered in general conditions as a commission. This figure will subsequently be marked up with an overhead and a profit markup.

If you do not want to mark up commissions in this manner, use the second option. This method simply adds the commission amount (to include all taxes, insurance, and expense reimbursements) on to the calculated price for the job which already includes overhead and profit.

If commission, including all taxes, insurance, and expense reimbursements, is calculated to be 8% and your normal residential net profit margin is 15%, add the two together (8% + 15% = 23%) and mark up your bid accordingly. If the market will not allow such a high markup (i.e., 23%) you may have to reduce the 23% markup to a more realistic one. Subsequently, some of the 15% net profit margin will have to be used to cover the 8% commission paid to the salesperson.

---

*Convert salaried payroll to an hourly labor rate by dividing the weekly, monthly, or annual salary by the actual amount of hours worked, not the number of hours paid.

For example, a foreman works 45 hours per week and is paid a weekly salary of $500. Determine the hourly rate to be used in bids as follows: $500 + (40 + 5) = $500 + 45 = $11.11. Use $11.11 in your bids. Add a Risk Factor, if necessary. An OTF is unnecessary as the $11.11 already includes the OTF.

**In extremely competitive bidding situations, these items are often included in general conditions, but are not marked up for overhead and profit. I call this "washing through" an item in a bid.

To eliminate these markups (assuming a 10% profit and a 10% overhead markup are being used on the bid), divide such costs by 1 + 10% + 10% (or 1.2).

For example, you are marking up all materials 10% for overhead and 10% for profit on a particular job that you are bidding. Lodging and per diem costs total $1,000, but you do not want to add profit and overhead markup to these two items. How do you include these in your bid without marking them up?

Divide the lodging and per diem costs ($1,000) by 1 plus 10% (the profit markup) plus 10% (the overhead markup): $1,000.00 ÷ (1.0 + .1 + .1) = $1000.00 ÷ 1.2 = $833.33.

Round this up to $834.00. Enter this amount into general conditions for lodging and per diem costs.

Later, when you add the 10% markup for profit and 10% for overhead, the final price in the bid for lodging and per diem will calculate out to $1,000. This is the actual cost "washed through" the bid:

$834.00 x 1.10 (overhead markup)   = $917.00

$917.00 x 1.10 (profit markup)   = $1,009.09

Due to the mathematical methods and the rounding that you have used, you actually add a few dollars above the original $1,000. This is insignificant.

Of course, it goes without saying that if you can add both profit and overhead markups to these items and still get the job, by all means, do so.

CONCLUSION

General conditions constitute the second phase of the bidding process. They usually comprise 6-10% of the price of a job. "Leaks" in a contractor's estimating system can often be traced back to general conditions missed in the bidding process. Study the list of general condition items provided and accurately include in your bids the ones that apply to a particular job.

ACTION POINT

Complete exercises (12-1) and (12-2) in Appendix A.

NOTE: The above list of common items usually included within general conditions is certainly not to be considered complete and/or all-inclusive. Other items such as temporary fences, road signs, traffic barriers, etc., could be added. Include such items to your bids as required. Follow the methods demonstrated in this chapter.

CHAPTER 13

MARKUPS AND THE FIVE MOST COMMON
METHODS OF PRICING JOBS

PURPOSE

To explain the five most commonly used methods of marking up a bid
in order to arrive at a final lump sum price for a job

INTRODUCTION

BIDDING is the science of determining what a project costs. Once we identify those costs, we then add markups to them to arrive at a final price for the project. **ESTIMATING** is the process of adding markups to the identified costs of a project to arrive at a **FINAL PRICE** for it.

There are many methods used to price jobs. We will cover five common ones used today in the market place. The important thing to keep in mind is to first correctly identify all costs (the break-even point) and then add profit and a contingency factor (if used) to it.

As we begin our analysis of the five methods of pricing jobs that are commonly used in the market today, we will refer to the two jobs in Diagram (13-1) that will serve as examples:

	Job "A"	Job "B"
Materials	$100,000	$40,000
Labor with labor burden	15,000	60,000
Equipment	5,000	20,000
Subcontractors	0	0
Total direct costs	**$120,000**	**$120,000**
OPH	$7.50	$7.50
PPH	$5.00	$5.00
Labor hours	1,500	6,000
CAW w/33% labor burden	$10.00	$10.00

DIAGRAM 13-1. DIRECT COSTS

1. THE FACTORING OR "MULTIPLIER" METHOD

Using this method, we simply multiply estimated material costs (or estimated material and labor costs) by a "factor." The factor may be based on past profit and loss statements, or it may be a number "arrived" at as a result of monitoring past competitive bidding situations. The rationale is: If you ended a previous calendar or fiscal year with a sufficient profit, and if material costs were 33% of your gross sales for that year, then all you have to do is multiply material costs for the new year by a factor of 3.0. Supposedly, this will produce prices that will cover all costs and ensure sufficient profit.

Sales taxes, field-labor burden, profit, and overhead are all included in the factor used. A contingency factor may be applied to the job if desired. However, that is not always the case. The flaws in this method are almost too numerous to mention—but it is surprising how many landscape and irrigation contractors bid their work using this "material-times-two" approach.

The only variables that are addressed in this method are:

  • the amount of materials, and…
  • the factor.

Unfortunately, it does not address the multitude of other variables that can, and usually do, apply to the jobs that you bid. Just some of the many items that may change from job to job and that NEED to be dealt with separately in the estimating process are:

  • General conditions
  • Profit markup
  • Site conditions (soil conditions, access, time of year/weather, etc.)
  • Expensive vs. inexpensive materials (with the same labor production rates)
  • Size of crew
  • Types of equipment
  • Subcontractors
  • Labor rates
  • Etc.

The factoring method has its roots in the retail (as well as wholesale) supplier industry. To arrive at the price to charge customers, purchased goods are marked up by a "factor," or "multiplier."

It is commonly used by smaller, unsophisticated nurseries that must purchase the bulk of their nursery stock and irrigation supplies. Because these smaller companies do not experience the wild fluctuations in types of materials, labor, equipment, subcontractors, site conditions, etc.; this uncomplicated method can work in a very non-dynamic situation. Once, however, you leave this extremely controlled environment and enter the landscape and irrigation contracting arena, where chaos tends to be the norm, the factoring method immediately breaks down.

Unfortunately, it is usually the owners of nurseries that fall victim to this mistake when, instead of just selling plants and supplies, they decide to install these plants for their customers. Factoring worked for them in the past, in the nursery business. Why not use it for contracting?

Irrigation suppliers tend to reinforce the application of this pricing method in the landscape and irrigation industry. That is, in some parts of the country, irrigation vendors obtain sets of plans for work being bid in their area. They will then do the takeoff and provide irrigation takeoff quantities and costs to the various contractors bidding on the job. This is bad enough, but pricing advice (in the form of suggested factors) is often offered, as well—saying, for instance, that residential irrigation systems may have a suggested factor of 2.0 to 3.0 while commercial work has one ranging from 1.8 to 2.5, or that the suggested factor for golf course materials may range as low as 1.65 to 2.0. Unfortunately, these factors have little to do with specific jobs. They only address general trends.

CPAs, also, sometimes enter the factor-pricing picture (though, more often than not, their recommendations lean toward the GPM (gross profit margin) markup method).

MY POINT IS THIS: Pricing methods vary in their application. Some methods work well in one environment but not in another. Factoring may be of limited use in a retail or supplier/vendor environment—but not in contracting. **DO NOT** use it to determine your prices.

There is, however, another tool called a "hindsight" markup factor that we use to help evaluate our competition. We call it hindsight because we calculate it AFTER we have priced the job correctly, taking into account all costs and markups.

In your next competitive bid situation:

- Determine your Phases I and II M/L/E/S costs, and...
- add the appropriate markups to determine your price. Then...
- divide the price by the material costs. This will give you your "hindsight" markup factor.

If your competitors are using the factoring method, track their bids and attempt to determine the factors that they are using. A commercial irrigation client tracked a competitor's bids and discovered that the competitor was consistently using a 2.05 multiplier. Once our client had identified the factor, he was able to adjust his own bids to beat that 2.05 factor. He knew that he could reduce his profit margin only slightly and beat his competitor's 2.05 factor and not hurt himself by doing so.

The decision to lower his price was based on sound analysis and on accurate budgeting and estimating systems and methodologies—not on some subjective factor. He knew that he could play his competitor's game, beat him, and survive. He cut his price, not his throat.

Another client, on the East Coast, was bidding a planting job. Plant materials on this job cost $6,000. With a 10% net profit margin (NPM), the estimate would total right at $10,000. It was a good bid, and he was confident of his production rates and his price. However, he knew that his competitors were using factoring for their pricing method (material-times-two). Their price would be around $12,000. Our client doubled his NPM, won the bid, and put an extra $1,000 in his pocket.

See Diagram (13-2) to review the strengths and weaknesses of the factoring method.

Let's turn our attention to Diagram (13-1) at the beginning of this chapter. If we use the mythical method of material times a factor of two, our price for jobs "A" and "B" are $200,000 and $80,000 respectively. As we build upon these two bids and continue our analysis of the other four common methods of pricing work, the flaws of factoring will become quite apparent.

	Job "A"	Job "B"
Material costs	$100,000	$40,000
Factor	x 2.0	x 2.0
Price	$200,000	$80,000

DIAGRAM 13-2. FACTORING MARKUP MODEL

2. THE GROSS PROFIT MARGIN (GPM) OR THE SINGLE OVERHEAD RECOVERY SYSTEM (SORS) METHOD

There are a number of popular derivatives of the GPM/SORS approach to pricing, and there is often some historical basis for using these adaptations. Although it has some merits and applications, virtually all of these positives are only useful when they are incorporated into other, more accurate and flexible, estimating methods. And like factoring the GPM/SORS method is useful for the purposes of "hindsight" analysis.

Some of the popular applications of the GPM/SORS Markup approach are as follows:

A. The 1/3, 1/3, 1/3 Rule

Although the specific fractions may change, their use is the same. The estimator, after reviewing past profit and loss statements (or recollecting from past experience—either his own or that of a trusted "source"), determines that material costs have comprised approximately 33% of gross sales (plus or minus a percent or two). Labor with burden, and possibly equipment costs, accounts for another 33% (more or less). The remaining 33% covers equipment costs (if not combined with labor), overhead, and profit.

Although we **DO NOT** recommend this method, this is how you would implement it:

(1). First, determine material costs and assume that they "should" comprise one-third (or some other predetermined percentage) of the final price for the job.

(2). Then add labor, labor burden, and possibly equipment costs (these "should" be roughly equal to material costs) to material costs. Thus, in this case, you double your material costs, and the result is roughly sixty-six percent of the price of the job.

(3). Equipment costs (if not included with labor), overhead, and profit make up the remaining 34% of the total price for the job.

Using our two jobs as an example, the results would be as follows:

	Job "A"		Job "B"	
Material costs	$100,000	33.3%	$ 40,000	33.3%
Labor burden and equipment	100,000	33.3%	40,000	33.3%
Subtotal	$200,000	66.6%	$ 80,000	66.6%
Overhead and profit	$100,000	33.3%	$ 40,000	33.3%
Total price	$300,000	100.0%	$120,000	100.0%

13-3. GPM MARKUP MODEL (A)

B. GPM Markup Method

Although similar to the previous technique, the GPM Markup method requires you to do more homework. You must first accurately identify specific costs for material, labor, equipment (unless included in labor or in overhead) and subcontractors. Sales taxes are then added to materials and labor burden to field payroll. You then mark up the total according to a predetermined (or desired) gross profit margin.

Using our examples, the process is as follows:

	Job "A"	Job "B"
Direct costs M/L/E/S	$120,000	$120,000
GPM markup (30%)	x 1.3	x 1.3
Total price	$156,000	$156,000

DIAGRAM 13-4. GPM MARKUP MODEL (B)

Another mathematical formula commonly used is:

Jobs "A" and "B"

Direct costs M/L/E/S = $120,000 ÷ (1 - GPM markup desired)
= $120,000 ÷ (1 - .3)
= $120,000 ÷ .7 = $171,429

Or to adjust the formula to arrive at our original price of $156,000:

a. Divide 1 by the sum of 1 plus desired GPM (30%):

$$1 + (1+.3) = \quad 1 + 1.3 = .7692$$

b. Divide total direct costs by .7692:

$$\$120,000 + .7692 = \$156,006$$

The pitfall of the GPM Markup method is what happens to direct costs (M/L/E/S) once you calculate and identify them. Our two examples will help us to understand the inherent error.

Jobs "A" and "B" both have the same direct costs. Assuming that our company field payroll is $15,000 per month (including labor burden), job "A" consists of one (1) month of payroll, while job "B" consists of four (4) months of payroll. If profit is 10% ($12,000) of the 30% GPM markup on both jobs, that leaves only $24,000 for overhead—the remaining 20% of the GPM markup.

Both jobs would then have $24,000 included in the bid to cover overhead costs. Utilizing the company's entire field-labor force, job "A" will last one month. Accordingly, job "B" will last four months—$60,000 job payroll (with burden) divided by the monthly payroll (with burden) of $15,000—but overhead in the bid ($24,000) for a four-month job is the same as a one-month job. Job "B" should have four times the overhead as job "A" because it lasts four times longer. The GPM/SORS Markup method has a serious flaw in this area. A flaw that has proven fatal (or nearly so) to many contractors.

3. THE MARKET-DRIVEN UNIT PRICING (MDUP) METHOD

Do not make the mistake of assuming that there is something inherently wrong with organizing and presenting an estimate in a unit price format. The format is not the issue. The issue is, however, the process (or the lack of a process) used to arrive at your unit price(s).If correctly calculated, unit prices can provide considerable insight into an estimate and plenty of ammunition at the bid table when it's time to negotiate. For this reason, every time I bid a project on the computer, the computer is programmed to simultaneously provide pricing in both a lump sum and a unit price format.

The prices are calculated, however, after all costs for M/L/E/S, general conditions and accurate markups are included in the estimate. These unit prices are then compared to ones normally found on the open market (e.g., so much for a 1, 5, 15 gallon shrub or tree; $1.00 per square foot for sod, hydroseed at 5 cents per square foot; $18.00 per man-hour for maintenance work; walls or fences at so much per linear foot; irrigation systems at $ .60 or $ .90 per square foot, etc.).

However, contractors who rely solely upon the Market-driven Unit Pricing method seriously shortcut the estimating and planning process. In turn, they short-circuit their business systems. Key information and data that is needed to direct and to control individual jobs (as well as the entire company or division) is just not available. As a result, meaningful job costing is not possible, and the company lurches forward in a fog.

It is hard to imagine an "estimating" method that is of less use in helping to run a company than factoring, but the Market-driven Unit Pricing method is. Factoring, at least, requires that you build upon the foundation of material costs. The MDUP system operates totally independent of any relevant data, budgets, costs, or strategic planning whatsoever. Taxes, labor burden, overhead, a contingency factor and profit—all are "supposed" to be included in the Market-driven Unit Price. Unfortunately, that is rarely the case.

To illustrate the problem, let us suppose that you have decided to enter the auto industry (GM, Ford, Chrysler, Toyota, etc., move over). Your first model will be a half-ton pickup truck—of course! In order to cut overhead, you decide to work out of your home and build the first production model in your garage. You have no idea (a) what your costs will be to build a truck, (b) how to determine your costs, or (c) how much you should charge for it. Your solution? Since you have to prove yourself first, you decide to sell it for a little less than comparable models already on the market.

No budgets. No planning. No cost data. No tracking system. No scoreboard. Just how long do you think you'll last?

The question is not whether you can produce the truck (let's assume that you can). It is not whether the other manufacturers can survive by selling trucks at the market-driven unit price for your type of production model. The question is: Can you organize and run your company in such a way that you can survive (and make money) by competing with the established market price?

Survival in the auto industry, any industry, requires much more than a price (especially more than one arrived at by someone else). It requires:

A. An accurate PRICE

B. A well thought-out PLAN

C. A self-correcting PROCESS.

Do not be tricked or lulled into thinking that a landscape and irrigation contractor can survive on anything less.

4. THE MULTIPLE OVERHEAD RECOVERY SYSTEM (MORS) OR THE "TRADITIONAL" METHOD

This method of pricing projects has gained popularity in recent years and is being taught in estimating workshops throughout North America. In fact, I taught it in workshops for over two years when I was employed at Charles Vander Kooi & Associates, Inc.

This method can have distinct advantages over the previous systems, but it does have definite disadvantages. It is overly complex, and it is difficult to make adjustments for varying market conditions. This becomes a particular liability in periods of rapid market change.

The MORS method can (and should) be firmly grounded upon accurate historic data, current financial statements, well thought-out estimating overhead budgets, projected sales, and direct costs for the upcoming budget year.

Let's discuss how it works. Diagram (2-1) provides an overview of the MORS method. M/L/E/S direct costs are clearly identified in Phases I and II on the diagram for both production items and general conditions.

It must be understood, however, that if you do an inaccurate take-off, miscalculate labor or equipment production rates, miss other important site conditions or other bidding variables, then the most perfect of estimating "systems" will be of little help. You MUST DO YOUR HOMEWORK and do it accurately!

Once you have calculated the costs for Phases I and II, you then add the markups:

- SALES TAX is added to materials.

   NOTE: Some states now add sales tax not only to materials, but to the entire job, just as though you were to add another 5-6% profit to the whole job.

- LABOR BURDEN is then added to direct labor costs.

- OVERHEAD is calculated (as described below).

- PROFIT is added to the job based on a straight percent markup on the total of all aforementioned costs.

- Finally, A CONTINGENCY FACTOR is added, if desired.

Up to now, the MORS method is fairly straight-forward. It is with overhead recovery that it begins to break down.

Overhead is recovered (added to a bid) by marking up M/L/E/S direct cost totals for Phases I and II by predetermined percentages:

- Material costs are usually marked up 10%.

- Field equipment costs are usually marked up 25%.

- Subcontractor costs are usually marked up 5%.

These percentages may vary in some applications, but normally remain as stated above.

The cornerstone of the MORS method is the percent that labor (combined with labor burden) is marked up for overhead recovery:

- Large commercial companies (over $1.5M in sales) usually range from 25% to 45%.

- Mid-sized companies ($750K to $1.5M in sales) doing commercial and residential work usually range from 45% to 65%.

- Smaller (under $500K) residential and commercial companies, or larger ($1M +) high-end residential companies, usually range from 65% to 100%.

Keep in mind that these percentage markups on labor will greatly vary from company to company, depending on the type of market served and the structure of the company. In no way should they be utilized without taking these other factors into consideration.

To determine the labor and labor burden overhead markup percent, see the construction division portion of Section 9B at the bottom of Diagram (3-1).

- First, project a complete budget (sales, direct costs, and overhead).

- Second, multiply direct costs (except for labor and labor burden) by their predetermined percentages (5%, 10%, and 25%).

A. Calculate and total projected overhead recovered from materials, equipment, and subcontractors. For example:

	Budget Projection		MORS %		Overhead to Recover
Materials	$320,000	x	10%	=	$ 32,000
Equipment (with equipment rentals)	70,330	x	25%	=	17,583
Subcontractors	50,000	x	5%	=	2,500
Subtotal					$52,083

B. Subtract the subtotal of A above from total company or division overhead to recover:

Total CO/DIV overhead to recover for year	$177,713
Projected overhead to recover from M/E/S	- 52,083
Remaining overhead to recover from labor and burden	$125,630

C. Determine the labor and labor burden markup percent to be used in estimating by dividing remaining overhead to recover by the projected company/division field labor payroll plus labor burden:

$$\frac{\text{Remaining overhead to recover} \quad \$125,630}{\text{Projected annual labor and burden} \quad \$265,774} = .4726 \text{ or } 47.3\%$$

The resulting percentage (in our example, 47%) provides the bidding overhead markup percent for labor and labor burden.

Applying these percentages and the MORS method to our two examples in Diagram (13-1) will produce prices for the jobs as shown in Diagram (13-5).

As you can see, using a 47% versus a 30% overhead markup on labor and labor burden would make a considerable difference on the final price.

The 17% spread would be caused primarily by the economics of scale that a larger company enjoys over a smaller one. The competitive advantage of the larger company is significant.

Job "A"

	DIRECT COSTS		OVHD %		MARKUP
Materials (with tax)	$100,000	X	10%	=	$10,000
Labor with burden	15,000	X	47%	=	7,050
Equipment with rentals	5,000	X	25%	=	1,250
Subcontractors	0	X	5%	=	0
Total	$120,000				18,300

Total direct costs	$120,000
Total overhead to recover on job	18,300
Break-even point (BEP)	138,300
Profit (10%)	15,367
Price for job	$153,667

Job "B"

	DIRECT COSTS		OVHD %		MARKUP
Materials (with tax)	$40,000	X	10%	=	$4,000
Labor with burden	60,000	X	47%	=	28,200
Equipment with rentals	20,000	X	25%	=	5,000
Subcontractors	0	X	5%	=	0
Total	$120,000				$37,200

Total direct costs	$120,000
Total overhead to recover on job	37,200
Break-even point (BEP)	$157,200
Profit (10%)	17,467
Price for job	$174,667

13-5. MORS MARKUP MODEL

A company using a 47% markup is probably one with combined gross sales between $1.0–1.5 million, while a much smaller company's will be from 60 to 75%.

The advantages of the MORS method are many **IF** it is used with accurate, current financial data, accurate takeoff data and production rates, and a thorough job-site condition analysis. Some of the advantages of the MORS method are:

- It can be based on historical data for your specific company.
- M/L/E/S costs are clearly identified.
- Gross profit margin (GPM) is clearly identified.
- Overhead recovery is clearly identified.
- Profit markup is clearly identified.
- It can be budget-driven. Progress can be measured against a predetermined budget/set of standards.
- It can help you to plan and run the field during actual production. Since you are required to build the job step-by-step in your mind and on paper the way that you will build it in the field, the MORS method provides an excellent process which provides the data that is needed to plan and to run the field during the production of the project.
- Meaningful job costing is made possible. The MORS method provides M/L/E/S cost data, as well as hours for labor and for equipment for budget-to-actual comparisons.

The advantages of the MORS method over factoring and the MDUP method should be obvious. The MORS method, also, has a distinct advantage over the SORS/GPM method, an advantage that can be seen in Diagrams

(13-3/4/5). The MORS method compensates for the M/L/E/S ratios within direct costs, while the SORS/GPM method generally does not.

In Diagram (13-4), the price for both jobs "A" and "B" was $156,000 because the direct costs for both were $120,000. The SORS/GPM method considers the M/L/E/S mix or ratios on a job equally. The MORS method varies the overhead markup percent for each separate component of direct costs. It is more flexible and, therefore, more accurate.

Indeed, the MORS pricing method has its strengths—BUT it also has considerable weaknesses that need to be considered and which make it an undesireable method for estimating. Disadvantages of the MORS method:

- The main disadvantage is that it is very complex. Adjusting or adapting it to the various scenarios that you may encounter (T&M, prevailing wage, rapidly changing markets, recessions, etc.) is **tricky,** to say the least.

- While the MORS method addresses the M/L/E/S ratio problem (by using a multiple versus a single markup percent), which the SORS/GPM method does not, it treats ALL jobs throughout the year as if the mix of materials, labor, equipment, and subcontractors were the same as your overall budget.

  Once you determine your labor and labor burden markup percent (47% in our example), it does not change until you recalculate your estimating overhead budget. Subsequently, you do not know until the end of the year if the amount of materials, labor and labor burden, equipment, and subcontractors used in your estimating budget were accurate. If they were not, your overhead percent markup on labor and burden should have been higher or lower, depending upon the actual mix of direct costs.

- Another serious drawback of the MORS method is the "traditional" markups that are applied to materials, equipment and subcontractors: "traditionally" 10%, 25%, and 5%, respectively. These percentages have no clear analytical basis. In and of themselves, they are (at best) "guesstimates." They are like factors and market-driven unit prices: sometime in the past for some job they may have been accurate, but only for a specific company on a specific job within a particular year.

  How do you know that you should mark up materials by 10% for overhead recovery? Answers to that question range from, "It's the customary market markup," to "It should cover your costs for ordering and processing materials," and from "It's what's taught," to "It's convenient."

  These percentages may be close to being accurate, but "accurate" percentages do not remain the same (10%, 25% and 5%) from application to application and from job to job. Nor do they very often come in such neat, rounded numbers, such as: 10.0%, 25.0%, or 5.0%. These numbers are just too "canned."

- Another disadvantage of the MORS method is that the M/L/E/S overhead markup percentages used to bid jobs have little or no analytical justification when applied to "T&M" (time and material) job situations. For instance, if there are no materials involved in a "T&M" job, should you use the same overhead markup percent as determined on your yearly overhead estimating budget (e.g., in our budget 47%)? Or should you increase it? If so, by how much?

- Prevailing wage projects present another problem. If you perform both prevailing wage and non-prevailing wage work, your labor and labor-burden overhead markup percentages for rated (prevailing wage) versus non-rated jobs should be different. For instance, if your overhead markup percentage on labor and labor burden for non-rated jobs with a CAW of $10.00 is 47%, it would drop to 23.5% for rated jobs with a CAW of $20.00.

  This scenario is compounded if your original projected estimating budget contains a yearly field-labor payroll amount consisting of both prevailing and non-prevailing wage labor. If it does, your labor and labor burden overhead markup percentages calculated on your estimating budget become an average percent. Subsequently, it is too high for prevailing wage jobs and too low for jobs performed at non-prevailing wage rates. You have sacrificed either accuracy, competitiveness, or both.

  In addition, if in your projected estimating budget your ratio of non-prevailing wage and prevailing wage labor differs from the ratio of work you actually do for the year, your overhead markup percentages on labor and labor burden would have been wrong, anyway.

- Another shortcoming of the MORS method is that it is difficult to update and keep current. This is due (in part) to the fact that the overhead recovered on jobs is not easily accumulated, tracked, and compared to actual financial statements (that is, compared to the data that is generated by your accounting system). Additional calculations are required, calculations that are not inherent in the MORS estimating process.

- The MORS method is not easily and quickly adjusted. It is difficult to adapt to rapidly changing economies and/or special market situations, that is, with any degree of certainty. You can make adjustments to your overhead markup percentages, but due to the vast number of variables that we have discussed, it is difficult to apply them in such situations accurately or with confidence.

The MORS method actually lost a client of mine $500,000 worth of work. He was using the MORS method before I introduced him to the OPPH method. After we prepared his budgets and had calculated the appropriate numbers/percentages, we rebid some large jobs that he had lost the previous season. Using his numbers and the OPPH method, we found that he could have won those jobs, covered all of his costs, and made money on them. However, because the MORS method that he was using was too inaccurate and complex, he did not take his "best shot" at the bid table.

THE FALLACY OF RECOVERING OVERHEAD UTILIZING PERCENTAGES

The MORS and the SORS/GPM methods contain a **fatal flaw** which renders them both ineffective. They contain a basic mathematical relational error. Both attempt to directly correlate the amount of overhead dollars to add onto a job to the amount of direct costs in the job by means of one or a combination of percentages (or factor/multipliers). However, there is no such correlation (factor) between these two items. Let me explain by using two items that DO directly correlate to one another.

The amount of a workers' compensation insurance (WCI) premium is directly correlated (in direct proportion) to payroll by means of a rate per hundred dollars of payroll (e.g., $7.00 per hundred). This translates to a decimal of .07 or 7%. If annual payroll doubles from $50,000 to $100,000, the WCI premium doubles from $3,500 to $7,000. If payroll decreases, premiums decrease in DIRECT PROPORTION to the decrease. The percentage factor (or multiplier) remains constant but it accurately links premiums to payroll.

This is a *legitimate mathematical relationship* for calculating the cost of WCI. Premiums (actual real costs) are **always** in direct relation to payroll. Premiums and payroll are always linked by means of the same factor—the percentage.

Sales tax on materials and GLI premiums are also examples of costs that are directly correlated to other costs by means of a multiplier. Unfortunately, there is no such direct link (or multiplier) between the components involved in the MORS and the SORS/GPM overhead recovery method. There is simply no *direct correlation* (or factors) linking costs for overhead and direct costs for a job in the field. Subsequently, attempting to tie the two together by means of a single (SORS) or multiple (MORS) multiplier (which never changes because it is so complex and rigid) is automatically flawed because there is no "multiplier" that accurately links the two. Attempting to link overhead to direct costs by means of a percentage multiplier is simply too inflexible, rigid, and inaccurate for the dynamic business of landscape and irrigation contracting.

The MORS and the SORS/GPM methods are simply another version of "materials-times-two" or factoring dressed up in fancy terminology. The MORS method especially is just too complex and intrinsically inflexible. Fortunately, there is an easier method.

The alternative to the previous four methods discussed is to measure and to allocate overhead to jobs in terms of units of whole dollars—the OPH (and not percentage multipliers or factors) in order to determine a break-even point (BEP). A GPM is then calculated once a net profit margin is added to the BEP. This method is far more accurate than any of the other four, and yet it is far simpler and easier to measure, monitor, manage and control by means of per hour and ratio analysis calculated on the BAR worksheet.

Let's get started!

5. THE FIELD-LABOR HOUR RECOVERY OR THE OVERHEAD AND PROFIT PER HOUR (OPPH) METHOD

Although not foolproof, the OPPH pricing method provides considerable advantages over all of the previous methods discussed.

- It begins by adding tax to materials and then adding labor burden to field payroll.

- It then becomes necessary to have a clearly identified overhead amount on a company/division basis to recover for the year.

- The overhead company/division amount is then divided by the projected number of billable field-labor hours in the company/division to determine the overhead per hour (OPH) dollar amount.

   (The OPH for the example in Diagram (3-1) is $8.22 and $4.92. See Section 8 D for the construction and maintenance divisions, respectively.)

- Overhead is then allocated to projects on the basis of the number of field-labor hours estimated in each bid.

   For instance, a construction project that had 1,000 field-labor hours in Phases I and II of the bid would be marked up $8,220 for overhead (1,000 hours x $8.22 OPH).

   There may be some slight variations to this (which we will cover later), but that is essentially how it is done for overhead recovery.

- Profit is then calculated in much the same manner as you have determined overhead.

   You have, however, a choice. You can either:

   Mark up the total of all of the aforementioned costs by a desired percent (e.g., 8%, 10%, 15%, etc.), **or...**

   You can determine your profit markup by multiplying the number of field-labor hours in the bid by a predetermined profit per hour (PPH) dollar amount. This PPH comes from your estimating budget. You will locate it in Section 8 E of Diagram (3-1).

The PPH method has some significant advantages over the conventional wisdom that exists behind the percent markup. However, I always compare both when I bid a project.

First, you establish a predetermined company/division profit for every billable field-labor hour that is to occur in the field. Just as in overhead, we have tied profit to field-labor hours. In Diagram (3-1), PPH is $5.10 and $2.87 for our construction and maintenance divisions, respectively. Personally, for the sake of simplicity, I would probably round these off at $5.00 and $3.00.

Second, we can approach the bid table with our PPH in the back of our mind—based on a total dollar amount of profit that we want to bring into the company or division over a one year period. In Diagram (3-1), this amount is $110,184 for our construction division, or rounded off: $110,000. Our thinking is, "Who cares how we get the $110,000, as long as we get it?".

Third, when we review our bid and finalize our numbers, we may find that the PPH (calculated at a 10% net profit markup) on material-intense jobs is actually quite different from our budget average. For instance, it can easily be $8-10.00 per field-labor hour or almost double your budget average. If you can win the job with this much net profit on it, by all means, do so.

However, we have to realize that (in our example) we could drop our PPH to as low as $5.00 and still be on target regarding profit for the year. In this particular instance, a $5.00 PPH may only be a 5% net profit margin on the job, but we will not hurt ourselves by using the lower amount.

Conventional wisdom cannot see the fallacies and/or the shortcomings of the percent markup method for either profit or overhead. The percent markups that are used in "Factoring," the SORS, and the MORS methods are just too rigid and inflexible. The OPPH method addresses these problems in a very simple, yet effective manner.

Finally, if desired, a contingency factor may then be added to the job.

First, though, there are two critical requisites attached to the OPPH method:

- Projected company/division field-labor hours must be reasonably accurate (within plus or minus 10%) to actual hours for the year.

- Overhead MUST be correctly defined (as we have done in Chapter 8).

NOTE: If you include items in overhead that should be in direct costs for the job being bid, such as field equipment costs and field-labor burden items (e.g., payroll taxes, general liability and workers' compensation insurance, etc.), you will seriously distort the effectiveness and accuracy of any estimating system, including the OPPH method.

If these two requisites are accomplished, the OPPH method can be quite simple, extremely accurate, and—we cannot stress enough this important advantage—adaptable at the bid table.

Although it is easy to use, the OPPH method must be combined with accurate takeoff procedures, along with field labor and equipment production rates that accurately reflect the specific job and site conditions being bid.

If the job is built (planned) in your mind and on paper consistent with the way that it will be produced in the field, the OPPH method facilitates and enhances production planning in much the same way as does the MORS methods. **ALL** data needed for effective job costing is a by-product of the OPPH Method.

Other strengths of the OPPH method:

- It is budget-driven. The OPH dollar amount (for instance, in Diagram (3-1): $8.22 per field-labor hour) is derived directly from your overhead budget.

- It applies to either prevailing wage or non-rated work. You do not have to adjust your OPH amount when bidding different types of work; nor do you have to worry about budget vs. actual variations for one or the other in comparison to your original budget.

- Overhead that is accumulated through bids won for the year is easy to track—because it is so easy to calculate.

- The OPH is determined by only two numbers (billable field-labor hours and overhead costs). It is, therefore, easy to compare your budgeted OPH to your actual OPH on a monthly and/or a year-to-date (YTD) basis.

- Tying your overhead recovery method directly to billable field- labor hours creates another very important advantage: bureaucrats (owners, estimators, bookkeepers, etc.) begin to think of and measure overhead in terms of field labor. Because the concept is so easy to visualize (another advantage), field people and bureaucrats alike can better relate to it.

For instance, if your OPH is $10.00, adding a full-time (2,080 hours per year) laborer in the field will allow you to increase your overhead budget for the year by $20,800 (2,080 hrs/yr x $10 OPH) and not increase your OPH for estimating purposes.

If you add labor to your field crew, and your yearly overhead budget remains the same, your OPH will decrease accordingly. You can, therefore, drop your prices and become more competitive, or you can increase your net profit margin.

By breaking overhead down and thinking of it in terms of OPH "units," so to speak, it becomes more manageable and more meaningful.

- The OPPH method and its resulting data can easily be mathematically converted, reformatted, and compared to the factoring, SORS/GPM, or MORS methods. With the aid of a personal computer (PC) spreadsheet program (or a lot of manual work), the OPPH method can also be converted for comparison to the Market-driven Unit Pricing method. Consequently, using the OPPH method encompasses the benefits of all of the five estimating pricing systems. By comparing one method against the other, you can adjust your bid to incorporate the strengths of each.

- The OPPH system is easily adjusted in order to compensate for fast-changing markets. If sales are plummeting, so will billable field-labor hours. If you are tracking your OPH as faithfully as you should, you will almost immediately see it rise and go through the roof. This is telling you that you need to address the issue (probably by cutting overhead) and get your OPH back in line.

If the economy or a particular market (meaning a specific division in your company) begins to accelerate and displays vigorous growth, you can quickly and easily adjust projected field-labor hours and your overhead budget, thus changing your OPH accordingly. No messy percentages or M/L/E/S ratios to worry about here.

- Gross and net profit margins are both easily identified while using the OPPH method. Accordingly, your break-even point (BEP) is also readily identified.

With so many advantages, one would think that the OPPH method would have few limitations or drawbacks. There are, however, some necessary **PRECAUTIONS.**

- First, unless it is used in conjunction with and in comparison to the other pricing methods, you can become lazy and develop "tunnel vision" and/or "leave money on the table." If your competition is very unsophisticated and is using factors, the MDUP, or the SORS/GPM system, you should recalculate your bids using those methods in conjunction with the OPPH method. This can dramatically sharpen your skills at the bid table.

By tracking bid results (if possible), you can begin to understand your competition's bidding better than they do. Once you do, you can then begin to beat them at their own game as long as you can do so without hurting yourself.

It is easy, too, while using the OPPH process to leave money on the table when you encounter bidding situations with expensive materials, a high material-to-labor ratio, or a large number of subcontractors. These special situations make it necessary for you to monitor your GPM markup on your bids, as well as the MORS markup on labor and labor burden by backing out material, equipment, and subcontractor markups. To do so:

A. Determine your OPPH method total overhead on the bid.

B. Subtract from "A" above: 10% of the material cost, 25% of the equipment cost, and 5% of the subcontractor cost. This is the respective MORS markups on these items.

C. The remaining amount is your overhead markup dollar amount on labor and labor burden using the MORS method for this particular bid.

D. Divide this remaining amount by the total of field labor plus labor burden costs in the bid.

The result is your MORS overhead markup percent for labor and labor burden.

By first determining your total overhead dollar amount to recover on a job using the OPPH method and then "backing into" your MORS labor and labor burden overhead markup percent as we did above, you will be able to compare your OPPH bid method to the MORS one if you desire to do so.

However, you will see that the MORS overhead markup percent on labor and labor burden will wildly fluctuate due to the mixture of M/L/E/S on a particular job, from zero percent (at times) to the high 80-90% range. By comparing the "backed into" MORS percentages to your overall estimating budget percentages (and/or in conjunction with past bidding experience), you will develop a sense, a feel for where the percentage markup on labor and labor burden *ought to be.* If it is too low (that is, below 20%), perhaps you should add extra profit to the job. If it appears inordinately high, you should investigate and check your gross profit margin on the bid.

- Another drawback of the OPPH method is that it is not easily converted manually (that is, without a computer) to unit prices. However, we will cover this in greater detail in Chapter 14.

Finally, as we said earlier, the OPPH method is only as good as the projections for your company's (or division's) overhead budget and billable field-labor hours. If overhead includes field equipment and/or labor burden, taxes, and insurance items, the OPPH method will be rendered inaccurate.

SUMMARY

All five pricing methods have their strengths and their weaknesses. At best, factoring and the MDUP methods are all but useless for determining accurate pricing for your work; at worst, they can cause serious errors. While the SORS/GPM and the MORS methods have some advantages, they should also be used with caution.

The above outlines why I recommend that you estimate your jobs using the OPPH method and then compare its results to the other four methods using the techniques mentioned earlier.

Direct costs for materials, labor, equipment, subcontractors, taxes, and labor burden are rather straightforward to determine for any given job once you take into consideration all site conditions and production rates. Most estimating problems arise from overhead. There is no "direct" or exact way to determine just how much overhead to put on a particular job. By its very nature, it is "vague" and nebulous. That is why it is referred to as a G&A (general and administrative) cost, or an indirect one, because it cannot be directly costed to a job. If it could be directly tied to a job, it would then become a "direct," not an "indirect," cost.

During the estimating process, you are attempting to identify two numbers:

- The first is the total of your direct costs (materials, sales tax, labor, labor burden, field equipment, and subcontractors) plus the amount of overhead you need to recover on the job. These two combined determine your break-even point (BEP). Once you determine this first number, you can then address the second.

- Profit and a contingency factor, if desired, combine to form the second number. How much profit can you put on the job and still get it? And do you want to add a "cushion" (a contingency factor) in the net profit margin?

"HOW HIGH can you go without losing the project?" and **"HOW LOW** can you go to win the bid without hurting yourself (not covering all costs)?" A good estimating system allows you to operate safely within this range.

Remember, a good estimating system produces more than just a **PRICE.** It also produces a **PLAN** and a **PROCESS** that will help you run your jobs and your company, as well.

	Job "A"	%		Job "B"	%
Total direct costs	$120,000	87		$120,000	62
Overhead ($7.50 x 1500 =)	11,250	8	($7.50 x 6,000 =)	45,000	23
BEP	131,250	95		165,000	85
Profit ($5.00 x 1500 =)	7,500	5	($5.00 x 6,000 =)	30,000	15
	138,750	100		195,000	100
Contingency factor	0			0	
Total price	$138,750	100		$195,000	100

DIAGRAM 13-6. THE OPPH MARKUP MODEL

The key is to know your company and your game plan (or strategy) by means of an estimating budget and a solid estimating methodology.

Then, and only then, can you play the estimating game and win.

	Job "A"		Job "B"	
SPH	$92.50		$32.50	
DCPH	$80.00		$20.00	
OPH	$7.50		$ 7.50	
PPH	$5.00		$ 5.00	
GPMPH	$12.50		$12.50	
Overhead %	8.0%		23.0%	
Profit %	5.0%		15.0%	
GPM %	13.0%		38.0%	
BEP %	95.0%		85.0%	
M/L ratio	8.9:1		.9:1	
MPH	$66.67		$6.67	
EQ/L ratio	.44:1	(44%)	.44:1	(44%)
EQPH	$3.33		$3.33	

DIAGRAM 13-7. RATIO/PER HOUR ANALYSIS OF THE OPPH MODEL

Let's turn our attention back to Diagram (13-1) and complete the estimating process utilizing the OPPH Pricing method. The OPPH to be used is $7.50, and the PPH is $5.00.

Our original examples in Diagram (13-1) and (13-2) depict two extremes. Job "A" is material-intensive while job "B" is labor-intensive. The material-to-labor ratio and the material per hour (MPH) identified in Diagram (13-7) so indicate. As you can see, there are ten times the amount of materials per field-labor hour in job "A" as there is in job "B." Although, I purposefully chose these extreme examples to drive home a number of points, you will occasionally encounter such extremes, although they may not be quite so dramatic.

The GPM in Diagram (13-7) is another key indicator. While 13% GPM on job "A" is low for most markets, the 38% GPM on job "B" is probably excessive for most markets, especially the commercial one. Once you have monitored your GPM on a number of bids in a particular market (e.g., commercial maintenance, commercial construction, high-end residential design/build, state or federal prevailing wage, etc.), you will have gained considerable insight into what that particular market will bear. This information can be especially helpful, and it can bolster your confidence at the bid table.

In most markets, you would probably want to move your profit percent closer to 10% for both jobs. However, although I would probably not hesitate to increase the profit on job "A" 3-5 points, I would be cautious about lowering the NPM on job "B" too quickly unless I was confident that 15% NPM was just too unrealistic.

Let's estimate another job—see Diagram (13-8)—that is more the norm for a construction division. Materials total $5,052 while sales tax on materials is 6%. CAW (including OTF and RF) is $7.50. Labor burden is 32%. The job is estimated to take 268 field-labor hours. Due to the requirement of a lot of site work on the job, equipment use is somewhat heavy and projected costs will be $1,275. Subcontractors in the form of a mason and hydroseeder will cost $2,000. OPH is $9.00. The contractor wants profit at either his PPH of $5.00 or 10%, whichever is greater. There is no contingency factor on the job. The job prices out as indicated in diagrams (13-8) and (13-9).

Direct Costs	Totals	%
Materials	$5,052	33.2
Tax on materials (6%)	303	2.0
Labor:268 hrs x Labor CAW ($7.50) =	2,010	13.2
Labor burden (32%)	643	4.2
Equipment	1,275	8.4
Subcontractors	2,000	13.1
Total direct costs	$11,283	74.1
Markups	Totals	%
Overhead ($9.00 OPH)	2,412	15.9
Profit (%/PPH) 10.00% or $5.00 =	1,522	10.0
Contingency factor	0	0.0
Total price	$15,217	100.0

DIAGRAM 13-8. BID USING THE OPPH METHOD

Diagram (13-9) contains the ratio/per hour analysis for this job.

Item	Ratio	$	%
SPH		$56.78	100.0
Overhead (OPH/%)		$ 9.00	15.9
Profit (PPH/%)		$5.68	10.0
GPM ($/%)		$3,934.00	25.9
BEP ($/%)		$13,695.00	90.0
Material/Labor Ratio:	2.51 to 1.0		
Materials (MPH/%)		$18.85	33.2
Equipment/Labor Ratio:	0.63 or 63.4%		
Equipment (EQPH/%)		$4.76	8.4

DIAGRAM 13-9. RATIO/PER HOUR ANALYSIS FOR SAMPLE BID

The only items that appear out-of-line are the equipment/labor ratio and the EQPH. Both are higher than would be expected for a typical landscape construction project. However, these indicators are driven up by the extra equipment needed for the site work.

Diagram (13-10) displays the same project estimated utilizing three of the four other pricing methods.

OVERHEAD RECOVERY

A 75% markup on labor was utilized in the MORS method. This would be a reasonable percentage to expect for a company using a $9.00 OPH. This percent would have to drop to 56% in order to match the OPPH price calculated at $15,217. Such a drop in overhead recovery, however, would be unrealistic for this company.

Diagram (13-11) provides an encapsulation of the pros and cons of the various bidding methods. You can draw your own conclusions from the comparison. The more often that you conduct such an analysis of your bids, the more skilled you will become in distinguishing the nuances of the various estimating methods. Had we used the PPH of $5.00 to price this project, our price would have been lowered by $182.00 to $15,035. It would have met our yearly budget projections for the company and made us more competitive, had it been necessary.

Diagram (13-12) is a printout of a Bid Analysis/Review (BAR) worksheet for the bid using the OPPH method in Diagram (13-8). Due to rounding by the computer, some calculations vary slightly from Diagram (13-8).

MARKUPS FOR THE JONES RESIDENCE PROJECT

Diagram (13-13) displays the Recap Bid Worksheet for the Jones residence. Totals from the Phase I and II bid worksheets are transferred onto it. Phase III markups are then added at the bottom of the worksheet using a $10.00 OPH amount and a 10% net profit margin. Diagram (13-14) is a BAR worksheet for the Jones residence project.

ACTION POINT

Complete exercises (13-1) and (13-2) in Appendix A.

	TOTAL DIRECT COSTS	OH $	OH %	BEP $	PROFIT $	PROFIT %	GPM %	PRICE $	$ AMOUNT OVER LOW BID	FIELD LABOR HOURS	OPH $	PPH $	SPH $	LABOR OH % MARKUP
MATERIAL X 2.0	N/I (*)	N/I	N/I	N/I	N/I	N/I	N/I	10,104	0	N/I	N/I	N/I	N/I	N/I
GPM/SORS	11,283	N/I	N/I	N/I	N/I	N/I	30.0%	16,119	6,015	268	N/I	N/I	60.15	N/I
MARKET-DRIVEN UNIT PRICES (**)	N/I	N/I	N/I	N/I	N/I	N/I	N/I	N/A	N/A	N/I	N/I	N/I	N/I	N/I
MORS	11,283	2,914	18.5%	14,197	1,577	10.0%	28.5%	15,775	5,671	268	10.87	5.89	58.86	75.0%
OPPH	11,283	2,412	15.9%	13,695	1,522	10.0%	25.9%	15,217	5,113	268	9.00	5.68	56.78	56.1%
AVERAGES	$11,283	$2,663	17.2%	$13,946	$1,550	10.0%	28.1%	$14,304	$5,600	268	$9.94	$5.78	$58.60	65.5%

NOTES:
- All percentages are calculated as a percent of the total price.
* - Not identified.
** - Items would be not identified if this method was used.

DIAGRAM 13-10. ANALYSIS USING OTHER BIDDING METHODS

PROS	FAC	U/P	GPM	MORS	OPPH
1. Is fast	X	X			
2. Irrigation suppliers may do T/O	X				
3. Accurately identifies MLES costs			X	X	X
4. Identifies BEP				X	X
5. Identifies GPM markup			X	X	X
6. Identifies overhead markup				X	X
7. Identifies profit markup				X	X
8. Provides sufficient data to job cost			X•	X	X
9. Builds on CO budget/historical data			X•	X•	X•
10. Profit "floats" on BEP costs					X
11. Easily adjusted to rapid market changes					X
12. Easy to monitor and measure key data					X

CONS	FAC	U/P	GPM	MORS	OPPH
1. Ignores CO budget/historical data	X	X			
2. Does not ID labor/burden/equip costs	X	X			
3. GPM not identified	X	X			
4. BEP not identified	X	X	X		
5. Ignores job-site variables	X	X•			
6. Bid process does not pre-plan job	X	X			
7. Bid process provides no documentation	X	X			
8. Does not facilitate job costing	X	X			
9. Cannot job cost labor/equip. hours	X	X	X•		
10. Profit not identified on bid	X	X	X		
11. Overhead not identified on bid	X	X	X		
12. Factor not analytically based	X				
13. Overhead recovery inaccurately tied to direct costs		X	X		
14. Field equipment included in overhead recovery		X•			
15. Profit cannot "float" on BEP			X	X	
16. Extremely complex/inflexible				X	
17. PPH focus can miss extra profit					X

• Not always the case.

DIAGRAM 13-11. BIDDING METHODS PROS & CONS

```
    OPH------->    $9.00   CAW/AVE WAGE  $6.19   OTF------->    11.11%
    PPH------->    $5.00   PROFIT----->  10.00%  RF-------->    10.00%
                           TAX-------->   6.00%  CAW-LOADED>    $7.50
    NO. UNITS->       1.0  LABOR BURDEN  32.00%  CREW TRUCK>    $5.00
===============================================================================

                  MAT        LABOR       EQUIP        SUBS
                =======     =======     =======      =======
I. PRODUCTION OF FINISHED PRODUCT:
                             230 HRS

                  5,052      1,724       1,225        2,000

II. GENERAL CONDITIONS:
                              38 HRS       10 HRS

                      0        286          50            0
                ==========  ==========  ==========   ==========
    SUBTOTALS:     5,052      2,010       1,275        2,000

III.  MARKUPS:

    A. SALESTAX      303                                            1.99%

    B. LABOR BURDEN------>     643                                  4.23%
                ==========  ==========  ==========   ==========
      SUBTOTAL:    5,355      2,653       1,275        2,000

      TOTAL DIRECT COSTS----------------------------->  11,283     74.15%

    C. OVERHEAD RECOVERY:

        268.1 (NUMBER OF HOURS X OPH)    $9.00 -->      2,413      15.86%
                                                      ==========
        "BEP" SUBTOTAL (DIRECT COSTS + OVERHEAD)---->   13,696     90.00%
                                                      ----------
    D. CONTINGENCY FACTOR (IF DESIRED)------------->         0      0.00%
    E. PROFIT:                                          ----------
        10.00%--------------------------> $1,521 -->    1,521     10.00%
        268.1 HOURS X PPH     $5.00 -->   $1,341 -->         0      0.00%
                                                      ==========
    F. TOTAL PRICE FOR THE JOB------------------->   $15,217    100.00%
-------------------------------------------------------------------------------
IV. ANALYSIS:        $          %                    $/RATIO      %
      A. SPH:      $56.76     100.00%   J. MAT/LAB:     2.51 :1
      B. DCPH:     $42.08      74.15%   K. MPH:       $18.84     33.20%
      C. OPH:       $9.00      15.86%   L. EQ/LAB:                63.42%
      D. PPH:       $5.67      10.00%   M. EQPH:       $4.75      8.38%
      E. BEP:     $13,696      90.00%   N. GC:          $335      2.20%
      F. OVHD:     $2,413      15.86%   O. GCPH:       $1.25      2.20%
      G. PROF:     $1,521      10.00%   P. GCH/TH:               14.21%
      H. GPM:      $3,934      25.85%   Q. FACTOR:      3.01 X   MAT'L
      I. GPMPH:    $14.67      25.85%   R. UNIT PRICE-------> $15,217
V. MORS PERCENTAGE MARKUP ON LABOR & BURDEN--------------->    54.98%
    (ASSUMING A MARKUP OF 10/25/5% ON MAT'L., EQUIP., & SUBS.)
```

DIAGRAM 13-12. BAR WORKSHEET FOR DIAGRAM 13-8

BID WORKSHEET

PROJECT: **JONES RESIDENCE** PAGE (**1**) OF (**6**)

LOCATION: **123 MAIN ST. ANAHEIM** DATE PREPARED: **1** / **1** / **199—**

P.O.C./G.C.: **JERRY JONES** ESTIMATOR: **KENT ADDETAL**

DUE: **1** / **15** / **199—** PH.: **714 941-7442** FAX: **N/A** CREW SIZE: **4**

TYPE WORK: **RECAP SHEET** HOURS/DAY: **8** CAW: $ **9.00**

REMARKS:

DESCRIPTION	QTY	UNIT	U/C	MAT'L	LABOR	EQUIP	SUBS
I. PRODUCE THE PRODUCT							
– SOIL PREP				240	72	45	—
– PLANTING				1407	293	—	—
– SOD				280	45	—	—
– MULCH				120	63	15	—
II. GEN CONDS				—	93	76	—
				2047	566	136	0
III. MARKUPS				—	—	—	—
– SALES TAX	6	%		123	—	—	—
– LABOR BURDEN	32	%			181	—	—
				2170	747	136	0
					136		
					0		
					2170		
					3053		
OVERHEAD ⟨OPH × TOTAL HOURS⟩	63	HR	10 –		630		
					3683		
CONTINGENCY FACTOR					0		
					3683		
PROFIT	10	%			409		
FINAL PRICE					4092		

SHI FORM 04-A

DIAGRAM 13-13. JONES RESIDENCE BID RECAP WORKSHEET

```
    OPH------->      $10.00   CAW/AVE WAGE   $8.18   OTF------->       0.00%
    PPH------->       $5.00   PROFIT----->  10.00%   RF-------->      10.00%
                              TAX-------->   6.00%   CAW-LOADED>       $9.00
    NO. UNITS->         1.0   LABOR BURDEN  32.00%   CREW TRUCK>       $4.00
===============================================================================

                      MAT        LABOR        EQUIP          SUBS
                     =======     =======      =======       =======
I. PRODUCTION OF FINISHED PRODUCT:
                                 53 HRS

                     2,047        473           60             0

II. GENERAL CONDITIONS:
                                 10 HRS       19 HRS

                        0          93           76             0
                     =========   =========    =========     ==========
    SUBTOTALS:       2,047        566          136             0

III.  MARKUPS:

     A. SALESTAX         123

     B. LABOR BURDEN------>       181
                     ===========  =========   =========     ==========
        SUBTOT:       2,170        748          136             0

        TOTAL DIRECT COSTS------------------------------>   3,053    74.62%

     C. OVERHEAD RECOVERY:

        62.9 (NUMBER OF HOURS X OPH)    $10.00 -->            629    15.38%
                                                         ==========
        "BEP" SUBTOTAL (DIRECT COSTS + OVERHEAD)---->       3,683    90.00%
                                                         ----------
     D. CONTINGENCY FACTOR (IF DESIRED)------------->           0     0.00%
     E. PROFIT:                                          ----------
          10.00%----------------------->       $409 -->      409
          62.9 HOURS X PPH    $5.00 -->        $315 -->        0     0.00%
                                                         ==========
     F. TOTAL PRICE FOR THE JOB------------------->       $4,092   100.00%
-------------------------------------------------------------------------------
IV.  ANALYSIS:       $           %                       $/RATIO       %
     A. SPH:       $65.02     100.00%   J. MAT/LAB:        3.61 :1
     B. DCPH:      $48.52      74.62%   K. MPH:          $32.53     50.02%
     C. OPH:       $10.00      15.38%   L. EQ/LAB:                  24.01%
     D. PPH:        $6.50      10.00%   M. EQPH:          $2.16      3.32%
     E. BEP:    $3,682.86      90.00%   N. GC:          $169.42      4.14%
     F. OVHD:     $629.36      15.38%   O. GCPH:          $2.69      4.14%
     G. PROF:     $409.21      10.00%   P. GCH/TH:                  16.49%
     H. GPM:    $1,038.57      25.38%   Q. FACTOR:        2.00 X   MAT'L
     I. GPMPH:     $16.50      25.38%   R. UNIT PRICE-------->     $4,092
V. MORS PERCENTAGE MARKUP ON LABOR & BURDEN------------------>      50.61%
   (ASSUMING A MARKUP OF 10/25/5% ON MAT'L., EQUIP., & SUBS.)
```

DIAGRAM 13-14. BAR WORKSHEET FOR THE JONES RESIDENCE

CHAPTER 14
UNIT PRICE BIDS

PURPOSE
To explain how to calculate accurate unit prices which include all costs and a net profit

CALCULATING UNIT PRICES

Calculating unit prices manually (without the aid of a computer) is one of the most difficult aspects of estimating. It involves more arithmetic and more diverse steps—see Diagrams (14-1) and (14-2)—than just about anything else you will do when estimating.

Once you understand, though, how to reflect accurately your company/division overhead, general conditions, and profit into your unit prices, you will be well-armed to negotiate your work with much more confidence.

Because unit prices are so essential and powerful, **IF calculated accurately** (a rather large "if"), virtually every job that I bid, I do so using a personal computer (PC) estimating software program that simultaneously produces both lump sum and unit prices.

Changes in profit margins, crew average wage, general conditions, etc., can be made and recalculated in seconds. Diverse "what if" scenarios can be run and evaluated, providing more intelligent options as you analyze your company/division estimates. Changes that customers want to incorporate into bids can be accommodated, accomplished right there in front of them, and communicated to them in terms of unit prices.

Even though calculating unit prices manually is more complex and time-consuming than doing so on a PC, it is still a powerful tool for the effective running of your company or division.

Before we calculate unit prices, however, we must first complete a company/division overhead budget for estimating purposes. Once we have that in place, we can then determine our Phase I Material/Labor/Equipment production costs for each bid item on a job for which we desire a unit price.

Calculate Phase I subcontractor unit prices on a separate bid worksheet by adding a GPM markup percent for overhead and profit to the prices quoted by the subcontractor. Add the unit and lump sum prices for subcontractors to columns (11) and (12) of the unit price worksheet under the third double-check—see Diagram (14-2).

General conditions are then determined for the entire project and calculated into a markup percentage that includes overhead and profit. This is then added to Phase I production Material/Labor/Equipment costs for each bid item.

JONES RESIDENCE

UNIT PRICE WORKSHEET   SMITH HUSTON, INC.-16-A

ITEM NO.	DESCRIPTION	QUANTITY (1)	UNIT (2)	MATERIALS + SALES TAX (3)	LABOR + BURDEN (5)	EQUIPMENT (7)	TOTALS (8)	MARKUP % .69 (9)	UNIT PRICE (10)	UNIT PRICE ROUNDED (11)	TOTAL PRICES (12)
I.	PRODUCE PRODUCT/SERVICE										
1	SOIL PREP	2000	SF	255	96	45	396	574	.287	.30	600
2	RED MAPLE 24" BOX	4	EA	412	120	|	732	1061	265.25	265.00	1060
3	CRAPE MYRTLE 15 GL	10	EA	408	150	|	558	809	80.90	81.00	810
4	SHRUBS 5GL	30	EA	250	72	|	322	467	15.56	15.75	473
5	SHRUBS 1 GL	55	EA	222	48	|	270	392	7.13	7.25	399
6	SOD	1000	SF	297	60	|	357	518	.518	.55	550
7	MULCH	1000	SF	127	84	5	226	328	.328	.35	350
	SUBTOTAL <#1>			2171	630	60	2861				4242
II.	GEN. CONDS. SUBTOTAL <#2>			2171	124	76	200				
III.	MARKUPS - OVERHEAD (OPH $10.00) SUBTOTAL #3	63	HR		754	136	3061				
							630				
							3691				

Right-side annotations:
- <DOUBLE-CHECK #1>
- <DOUBLE-CHECK #2>
- <DOUBLE-CHECK #3>
- 141
- DIFFERENCE DUE TO ROUNDING DECIMAL
- ADD SUBCONTRACTORS (WITH ALL MARKUPS) HERE →
- TOTAL HERE
- Ø
- $4242

NET PROFIT + CONT. FACTOR   10%   = 3691 ÷ (1.0 + .1) = 3691 ÷ .9 =
TOTAL / FINAL PRICE   4101   = 3691

CALCULATING MARKUP %
1. COL 8 FINAL PRICE ÷ COL 3 SUBTOTAL <#1>   4101
2. SUM OF 1 ÷ COL 8 FINAL PRICE   1240   1240 ÷ ... = 1.240
3. ROUND UP   .3024   .3024 ÷ 4101 = .31
4. SUBTRACT ROUNDED DECIMAL IN #3 FROM 1.0 TO CALCULATE MARKUP DECIMAL/%, WRITE AT TOP OF COL 9.   1.0   1.0 − .31 = .69 = 69%

DIAGRAM 14-1. UNIT PRICE TWELVE-COLUMN PAD WORKSHEET

We begin a unit price bid by first identifying the individual bid items for which we want to calculate unit prices, using a twelve-column accounting pad as displayed in Diagram (14-1).

PRODUCTION OF FINISHED PRODUCT/SERVICE

- Enter the bid items in the "Description" column.
- Enter the quantity and unit in columns (1) and (2), respectively.
- Skip a few lines after the last bid item and enter heavy or double "Subtotal" lines for the columns as indicated in Diagram (14-1).
- Label columns (1) through (12) accordingly. Note that materials in column (3) include taxes and that labor in column (5) includes labor burden.
- Next, take a number of blank bid worksheets and enter the Material/Labor/Equipment information for two to three bid items per page as shown in Diagram (l4-2).
- Number each bid item on the bid worksheet with the same number that it has in Diagram (14-1).
- Line through the "Subs" column and write "Total" over it.
- Enter, step-by-step, the production line items for each bid item in the bid worksheets (remembering to include taxes with materials and labor burden with labor in their respective columns).
- Place a dash (—) in the M/L/E columns that have no numbers to enter in them.
- Enter a circle (in pencil) where you do not yet have, but will obtain later, the quantity or price. This is done to help ensure that you do not forget to enter these items later on.
- Total each M/L/E column after placing a double or heavy solid line after the last M/L/E production line item.
- Add across the three M/L/E column totals. Enter this total into the "Total" column (formerly the "Subs" column) on the right hand side of the worksheet.
- Enter the M/L/E column totals onto the twelve column pad on the respective bid item line.
- Add columns (3) through (7) across and enter the M/L/E bid item total onto column (8) of the pad.
- Repeat the above steps for the remaining bid items.
- Subtotal down columns (3) through (8) for the Phase I M/L/E items. This produces Subtotal #1.
- Perform your **FIRST DOUBLE-CHECK** of your work by totaling the Subtotal #1 amounts for columns (3) through (7) and compare it to the total of column (8) for all bid items. The two totals should be the same. If they are not, check your arithmetic for errors.

You are now ready to calculate the general condition costs for the project:

GENERAL CONDITIONS

- Fill out a bid worksheet for the general conditions as shown in Diagram (14-3).
- Enter the Phase II M/L/E individual line items.

Include general condition subcontractor costs (if any) in with your material costs.

- Total them as you did for the Phase I production bid items.
- Next, enter your general condition column totals onto the appropriate line on your twelve column pad, directly under the Phase I M/L/E items.
- Place a line under general conditions, and subtotal (Subtotal #2) columns (3) through (8) as appropriate.
- Do a **SECOND DOUBLE-CHECK** of your totals. Compare the total of the subtotal line for columns (3) through (7) to the subtotal of column (8).

These two numbers should be the same. If they are not, check your arithmetic as you have made a mathematical error somewhere.

BID WORKSHEET
‹UNIT PRICES›

PROJECT: ___JONES RESIDENCE___

LOCATION: _____

P.O.C./G.C.: _____

DUE: ___ / ___ / ___   PH.: _____   FAX: _____

TYPE WORK: _____

REMARKS:

PAGE (___) OF (___)

DATE PREPARED: _____ / _____ / _____

ESTIMATOR: _____

CREW SIZE: ___4___   W/LB

HOURS/DAY: ___8___   CAW: $ ___12.0C___

DESCRIPTION	QTY	UNIT	U/C	W/TAX MAT'L	W/LB LABOR	EQUIP	TOTAL SUBS
① SOIL PREP	2000	SF					
– AMENDMENTS	10	CY	21.20	212			
– SOIL CONDITIONER	2	BG	5.30	11			
– FERTILIZER	3	BG	10.60	32			
– LABOR	1	HR	12 –		12		
– TRACTOR	2	HR	15 –			30	
– OPERATOR	2	HR	12 –		24		
– ROTOTILLER	3	HR	5 –			15	
– OPERATOR	3	HR	12 –		36		
– FINE GRADE	2	HR	12 –		24		
				255	96	45 =	396
⑥ SOD	1000	SF	.297	297			
– LABOR TO INSTALL	4	HR	12 –		48		
– LABOR TO ROLL	1	HR	12 –		12		
				297	60	Ø =	357
⑦ MULCH 2" DEEP	1000	SF					
– REDWOOD BARK	6	CY	21.20	127			
– LABOR	6	HR	12 –		72		
– TRACTOR	1	HR	15 –			15	
– OPERATOR	1	HR	12 –		12		
				127	84	15 =	226

SHI FORM 04-A

DIAGRAM 14-2. UNIT PRICE BID WORKSHEET

MARKUPS

Calculate your overhead on the line directly under your Phase I and II subtotal line. Write "Overhead" on the line followed by your OPH in brackets. You then...

- First, determine your total field-labor hours included in the bid: Divide the total labor in Phases I and II in column (5) by your CAW rate plus labor burden or simply total all hours in Phases I and II.

    $754 divided by $12 = 63 hours. Enter this amount in column (1).

- Second, multiply the total field-labor hours in the job by your OPH:

    63 x $10.00 = $630.00

- Enter the total overhead amount in column (8) of your twelve column pad.

- Add the overhead amount to your Phase I and II subtotal.

- Enter that total on the subtotal line directly below "Overhead."

Next, write "Profit and contingency" on the line directly under the "Subtotal #3" line (if desired, enter your PPH) and calculate your profit and contingency markup. You can either...

- Multiply your total field-labor hours for the job by your projected budget PPH, or...

- Use the percentage markup method (i.e., 10, 15 or 20%).

To do so, subtract your desired profit markup from 1.0 (e.g., 1 - .10 = .90) and divide your subtotal line above it (#3) by the results. This will provide a true 10% net profit for the example which follows.

    $3,691 ÷ .90 = $4,101.11; round off to $4,101.00.

$4,101.00 is the price for the job with a profit and overhead added to the Phase I and II costs.

In order to determine unit prices for the job, we must first determine a markup percent for general conditions, overhead, and profit to add to the individual bid item totals in column (8). To do so, follow the four steps at the bottom left portion of Diagram (14-1) as follows:

(1). Subtract the subtotal (#1) of the Phase I bid items in column (8) from the total price for the job.

    $4,101.00 - $2,861.00 = $1,240.00

(2). Divide the result obtained in step (1) by the total price for the job. Round up the decimal to two decimal places.

    $1,240.00 ÷ $4,101.00 = .3024 = .31 = 31%

(3). Subtract the rounded decimal in step (2) from 1.0. We have now calculated the markup percent.

    1.0 - .31 = .69 = 69%

Enter your percent (in our example, 69%) at the top of column (9).

We will now calculate the column (9) "Markup" totals for the individual Phase I bid items:

- Divide the column (8) total for the first bid item by the markup decimal point.

    Soil prep: $396.00 ÷ .69 = $573.91

- Round up to whole dollars ($574.00).

- Enter this amount in column (9).

Calculate the column (10) "Unit Price" by...

- Dividing the column (9) total by the quantity in column (1).

    $574.00 ÷ 2,000 = $ .287

- Enter this amount in the column (10) "Unit Price" column.

Display your unit prices to three decimal points for items with low unit price amounts, usually under $1.00 or $2.00 (i.e., soil prep, sod, grading, etc.)

- Round up your calculated unit prices to the amount that you will sell to your customers (i.e., soil prep: $ .287 to

BID WORKSHEET

PROJECT: _JONES RESIDENCE (CONTD)_ PAGE ( ) OF ( )

LOCATION: _____ DATE PREPARED: _1_ / _1_ / _199_

P.O.C./G.C.: _____ ESTIMATOR: _____

DUE: _1_ / _15_ / _199_ PH.: _____ FAX: _____ CREW SIZE: _4_

TYPE WORK: _GENERAL CONDITIONS_ HOURS/DAY: _8_ CAW: $ _9.00_

REMARKS:

DESCRIPTION	QTY	UNIT	U/C	W/TAX MAT'L	W/LB LABOR	EQUIP	TOTAL SUBS
– GEN CONDS	1	LS					
– FOREMAN ADMIN	1	HR	16 –		16		
– PICKUP TRUCK	16	HR	4 –			64	
– CLEANUP	2	HR	12 –		24		
– DRIVE TIME (.5HR/M/DY)	4	HR	12 –		48		
– CALLBACKS (LABOR)	3	HR	12 –		36		
– CALLBACKS (TRUCK)	3	HR	4 –			12	
				0	124	76	= 200

SHI FORM 04-A

DIAGRAM 14-3. GENERAL CONDITIONS BID WORKSHEET

$ .30), and...

• Enter that amount in column (11).

Determine your final "Total Prices" in column (12) by...

• Multiplying the unit price in column (11) by the quantity in column (1).

     $2,000.00 x .30 = $600.00

• Enter the total in column (12).

• Repeat this process for the remaining Phase I bid items.

• Subtotal the column (12) Phase I bid items—(see Diagram (14-1).

The total of the prices in column (12) should be the same as, or greater than, the total price calculated in column (8).

• Conduct a **THIRD DOUBLE-CHECK** of your work by subtracting the total price in column (8) from the total price in column (12).

     $4,242.00 - 4,101.00 = $141.00

NOTE: Due to rounding your unit prices in column (11), the product of the unit prices multiplied by the quantities in column (1) is greater than the total in column (8).

Diagram (14-4) is a computer spreadsheet for our example outlined in Diagram (14-1). Notice that it calculates unit prices more accurately and that I was able to add a few extra columns in order to calculate and display separately Phase I individual bid items unit costs, as well as the same marked up for general conditions, overhead, a contingency factor, and profit.

Note also, the **FOURTH DOUBLE-CHECK** that I have built into the lower-right portion of the spreadsheet.

When bidding irrigation or maintenance, follow the steps, procedures, and formats that we have just used. The service or product may change, but the process is exactly the same.

In order to make sure that you do not make and include a mathematical error in your calculations, remember to include the three "double-checks" in your work.

Compare your calculated unit prices to ones commonly used on the open market (that is, Market-driven Unit Prices), knowing that yours include all of your costs plus profit, that they accurately reflect your company's overhead structure, projected profit, and production rates.

With a little extra arithmetic, you can easily calculate your GPM, overhead, and profit percentages for your unit priced jobs. This information will allow you to then compare these calculations to similar ones using the GPM/SORS, MORS, or the OPPH estimating methods (if you so desire).

CONCLUSION

Unit pricing your jobs manually is, perhaps, the most difficult aspect of estimating. However, by following the right procedures and by formatting your data properly, you can make the process much easier.

Remember to "double-check" your work. This will ensure that you do not make and include an error in your mathematical calculations.

ACTION POINT

Complete Exercises 14-1 in Appendix A.

SALES TAX:	6.00%
CREW AVERAGE WAGE:	$9.00
OPH:	$10.00
DESIRED MINIMUM PPH:	$5.00
LABOR BURDEN:	32.00%
PROFIT AS A %:	10.00%

PRICE:	$4,241	OVHD:	15.36% ACTUAL
SPH:	$67.32	PROFIT:	10.00% "
HRS:	63.0	PPH:	$6.51 "
CREW:	4	GPM:	25.36%
HRS/DAY:	8	GC:	4.88%
		SET----->	

#	DESCRIPTION	QTY	UNIT	MATL+TAX	LABOR+LB	EQUIP	TOTALS	U/C CALCS	+G.C. 6.53%	+OH (BEP) 17.07%	+PROFIT 10.00%	U/P	ROUNDED	TOTAL BID	VARIANCE
		1	2	3	5	7	8	CALCS	6.53%	17.07%	10.00%	10	11	12	
1	SOIL PREP	2000	SF	255	96	45	396	0.20	424	511	568	0.28	0.30	600.00	32.35
2	RED MAPLE 24" BOX	4	EA	612	120		732	183.00	783	944	1,049	262.32	265.00	1,060.00	10.71
3	CRAPE MYRTLE 15 GL	10	EA	408	150		558	55.80	597	720	800	79.99	81.00	810.00	10.13
4	SHRUBS 5 GL	30	EA	250	72		322	10.73	345	415	462	15.39	15.75	472.50	10.93
5	SHRUBS 1 GL	55	EA	222	48		270	4.91	289	348	387	7.04	7.25	398.75	11.72
6	SOD	1000	SF	297	60		357	0.36	382	461	512	0.51	0.55	550.00	38.26
7	MULCH	1000	SF	127	84	15	226	0.23	242	292	324	0.32	0.35	350.00	26.04
8							0	0.00	0	0	0	0.00	0.00	0.00	0.00
9							0	0.00	0	0	0	0.00	0.00	0.00	0.00
10							0	0.00	0	0	0	0.00	0.00	0.00	0.00
11							0	0.00	0	0	0	0.00	0.00	0.00	0.00
12							0	0.00	0	0	0	0.00	0.00	0.00	0.00
13							0	0.00	0	0	0	0.00	0.00	0.00	0.00
14							0	0.00	0	0	0	0.00	0.00	0.00	0.00
15							0	0.00	0	0	0	0.00	0.00	0.00	0.00

	MATL+TAX	LABOR+LB	EQUIP	TOTALS	U/C	+G.C.	+OH	+PROFIT	U/P	ROUNDED	TOTAL BID	VARIANCE
SUBTOTAL:	2,171	630	60	2,861	69.8%	3,061	3,691	4,101	XXXXXXXX	XXXXXXXX	4,241	140
GEN. CONDS.	0	- 124	76	200	6.53%							
SUBTOTALS:	2,171	754	136	3,061	74.64%							
OVERHEAD:	63.0 HOURS X OPH ----->			630	17.07%							
SUBTOTALS:				3,691	10.00%							
PROFIT & CONT. FACTOR: 10.00%				410								
				4,101	100.00%							

DBL-CK 1: +G.C. column $0.00 0.00%
DBL-CK 2: +OH column $0.00 0.00%
DBL-CK 3: +PROFIT column $0.00 0.00%
DBL-CK 4: TOTAL BID column VARIANCE $140.14 3.42%

DIAGRAM 14-4. UNIT PRICE COMPUTER SPREADSHEET

CHAPTER 15
SPECIAL BIDDING SITUATIONS

PURPOSE

To explain how to adapt your estimating methods to price lawn maintenance packages, irrigation service time-and-materials (T&M) work, and tree trimming crew work

INTRODUCTION

We will slightly adjust our estimating technique to accommodate three special bidding situations:

1. Landscape management or maintenance crew and equipment package jobs.
2. Irrigation repair or service T&M work.
3. Tree trimming crew and equipment work.

LANDSCAPE MANAGEMENT OR MAINTENANCE "PACKAGE" ESTIMATING

This is accomplished in the same manner outlined in the previous chapters:

- Calculate Phase I production costs.
- Calculate Phase II general condition costs.
- Markups are added in Phase III in order to arrive at the final price, either in lump sum or unit price formats.

This process is excellent for larger residential and commercial projects, ones billing over $500 per month (or which include more than 15-20 man-hours per month). However, using a maintenance "package" bidding approach for smaller jobs greatly speeds up the estimating process without sacrificing accuracy.

First, we determine our maintenance package crew and equipment requirements for a "generic" day. For purposes of example, we will use the following two-man crew:

LABOR	$/%
Leadman/driver	$7.50/hr
Laborer	$5.50/hr
Labor burden	30.0%
OTF (40 hrs/week, 8 hrs/day)	0.0%
RF (risk factor)	10.0%
OPH	$4.00

EQUIPMENT	CPH
Van	$3.50
Mower 21" (2)	$1.00
Blower (1)	$2.00
Edger (1)	$3.00

BID WORKSHEET

PROJECT: _MAINTENANCE PACKAGE ESTIMATING_   PAGE ( ) OF ( )

LOCATION: _____

P.O.C./G.C.: _____   DATE PREPARED: _____ / _____ / _____

DUE: ___ / ___ / ___ PH.: _____ FAX: _____   ESTIMATOR: _KENT COUNT_

TYPE WORK: _____   CREW SIZE: _2_

HOURS/DAY: _8_   CAW: $_7.15_

REMARKS:

DESCRIPTION	QTY	UNIT	U/C	MAT'L	LABOR	EQUIP	SUBS
I. PRODUCTION/SERVICE							
– LABOR HOURS	13	HR	7.15	—	93	—	—
– MOWER 21"	6	HR	1 –	—	—	6	—
– BLOWER	2	HR	2 –	—	—	4	—
– EDGER	4	HR	3 –	—	—	12	—
SUBTOTAL				Ø	93	22	Ø
II. GEN CONDS							
– DRIVE TIME	3	HR	7.15	—	22	—	—
– VAN	8	HR	3.50	—	—	28	—
SUBTOTAL				Ø	115	50	Ø
III. MARKUPS							
– SALES TAX	6	%	—	—	—	—	—
– LABOR BURDEN	30	%	—	—	35	—	—
SUBTOTAL				Ø	150	50	Ø
					Ø		
					50		
					Ø		
					200		
– OVERHEAD <OPH>	16	HR	4 –		64		
BEP					264		
– PROFIT + CONTINGENCY	10	%			30		
					294		
CURB-TIME RATE	=	$294/	=	$22.62			
		/13 HRS					

> Bid WorksheeT and compuTer ToTals someTimes differ due To Rounding.

SHI FORM 04-A

DIAGRAM 15-1 (A). MAINTENANCE PACKAGE ESTIMATING EXAMPLE

```
     OPH------->     $4.00   CAW/AVE WAGE   $6.50   OTF------->    0.00%
     PPH------->     $2.50   PROFIT----->  10.00%   RF-------->   10.00%
                             TAX-------->   0.00%   CAW-LOADED>    $7.15
     NO. UNITS->      13.0   LABOR BURDEN  30.00%   CREW TRUCK>    $3.50
 ===========================================================================
                    MAT         LABOR       EQUIP         SUBS
                   =======     =======     =======      =======
I.  PRODUCTION OF FINISHED PRODUCT:
                               13 HRS

                      0          93          22         _____

II. GENERAL CONDITIONS:
                                3 HRS       8 HRS

                   _____     21          28         _____
                   ==========  ==========  ==========   ==========
     SUBTOTALS:        0          114         50            0

III.  MARKUPS:

     A. SALESTAX        0

     B. LABOR BURDEN------>       34
                   ==========  ==========  ==========   ==========
        SUBTOTAL:       0          149         50            0

        TOTAL DIRECT COSTS--------------------------------->   199     68.08%

     C. OVERHEAD RECOVERY:

           16 (NUMBER OF HOURS X OPH)     $4.00 -->          64     21.92%
                                                         ==========
        "BEP" SUBTOTAL (DIRECT COSTS + OVERHEAD)---->       263     90.00%
                                                         ----------
     D. CONTINGENCY FACTOR (IF DESIRED)------------->         0      0.00%
     E. PROFIT:                                          ----------

           10.00%------------------------------------->        29     10.00%
                                                         ==========
     F. TOTAL PRICE FOR THE JOB--------------------->       292    100.00%
 ---------------------------------------------------------------------------
IV. ANALYSIS:       $           %                   $/RATIO        %
     A. SPH:       $18.24     100.00%   J. MAT/LAB:    0.00 :1
     B. DCPH:      $12.42      68.08%   K. MPH:       $0.00        0.00%
     C. OPH:        $4.00      21.92%   L. EQ/LAB:                43.71%
     D. PPH:        $1.82      10.00%   M. EQPH:      $3.13       17.13%
     E. BEP:      $262.72      90.00%   N. GC:       $49.45       16.94%
     F. OVHD:      $64.00      21.92%   O. GCPH:      $3.09       16.94%
     G. PROF:      $29.19      10.00%   P. GCH/TH:                18.75%
     H. GPM:       $93.19      31.92%   Q. FACTOR:     0.00 X  MAT'L
     I. GPMPH:      $5.82      31.92%   R. UNIT PRICE-------->  $22.45
V. MORS PERCENTAGE MARKUP ON LABOR & BURDEN------------------>   34.63%
   (ASSUMING A MARKUP OF 10/25/5% ON MAT'L., EQUIP., & SUBS.)
```

DIAGRAM 15-1 (B). BAR WORKSHEET FOR THE MAINTENANCE PACKAGE

Second, we determine the average amount of drive time per day for this crew if they only worked on the smaller type jobs for which the maintenance package would be used.

As you will see, this is a crucial consideration. If the crew would hit eight to ten job sites per day, with an average drive time of ten minutes between them, the drive time for each day would be as follows:

8 jobs at 10 minutes each	=	80 minutes
plus 10 to return to yard	=	10 minutes
Total at end of day	=	90 minutes

Or 1.5 man-hours per day per man, a total of three man-hours.

We are now ready to cost out our maintenance package. Turn to Diagram (15-1).

We calculate our Phase I production costs. Notice that I round up to the nearest whole dollar, when more than a few cents appear in our M/L/E columns.

To obtain our Phase I production labor hours, we must subtract our drive time hours from the total paid labor hours. In our example, there are sixteen total paid labor hours for the day; after we subtract, the result is 13.0 production labor hours.

We then calculate the "average" number of hours that the mowers, blower, and edger will be used during the thirteen production hours.

Subtotal your M/L/E costs as shown in Diagram (15-1).

Phase II general conditions are calculated next. The crucial drive time and the van for the full eight hour day is included here. In this example, it is assumed that lunch and break times are not being paid and are not, therefore, included in the production or general condition labor hours. If, for example, a fifteen minute break in the morning is part of the paid eight labor hours per day per man, include it also in general condition labor hours.

$$2 \text{ men x } .25 \text{ hrs/day} \quad = \quad .5 \text{ hrs/day}$$

• Subtotal Phase II general condition costs.

Draw a double line below this, and...

• Subtotal Phases I and II M/L/E costs.

You are now ready to add Phase III markups to the bid.

• There are no materials involved on these jobs and therefore no markup.

• The thirty percent labor burden is then calculated on the total labor for the job.

• Subtotal your bid once again.

• Add equipment to your labor and labor burden totals.

• Subtotal.

• Calculate your overhead.

• Add overhead to the bid.

• Subtotal the bid.

• Add profit and a contingency factor as a straight percent or as a profit per hour (PPH) amount.

You have now arrived at your final price for a "generic" day for your maintenance package.

To calculate the "curb time" rate used for estimating smaller jobs:

• Divide the total daily price by the production hours: $294.00 + 13.00 = $22.62 per curb man-hour

• Multiply this rate by the number of people on the crew.

This translates into $22.62 x 2 = $45.24 per curb crew hour.

You can now easily bid smaller jobs by first estimating your crew's curb time at a particular job. For a job site that requires 30 minutes curb time: 30 minutes of curb time x $45.24 = .5 x $45.24 = $22.62 per visit to the job site.

Remember, curb time starts when your crew pulls up to the curb (at the site being bid) and ends when they drive away (going to the next job or back to the yard). Note the impact that estimated drive time makes on your crew rate of $45.24 if it drops to one hour per man per day or increases to two hours per day per man.

A. Drive time drops from 1.5 hours to 1.0 hour per day per man.

Step (1): $\dfrac{\text{Total price for a day}}{\text{Total production time}} = \dfrac{\$294.00}{14.00} = \$21.00/\text{man-hour}$

Step (2): Multiply $21.00 by number of crew members: $21.00 x 2 = $42.00

B. Drive time increases from 1.5 to 2.0 hours per man per day.

Step (1): $\dfrac{\text{Total price for a day}}{\text{Total production time}} = \dfrac{\$294.00}{12.00} = \$24.50/\text{man-hour}$

Step (2): Multiply by crew size: $24.50 x 2 = $49.00

As you can see, drive time has a considerable impact on the price that you should charge in a maintenance package bidding situation. The maintenance package bidding approach can easily accommodate any number of formats.

For instance, one contractor charges clients $195.00 per month for a three-person crew being on the site for one crew hour per week (if his crews spend more than one hour per week on a site, he knows that he is in trouble on that job). The mathematics work out as follows:

Curb time rate per man-hour	$15.00
Crew size	x 3
	$45.00
Average number of weeks per month	x 4.33
Total	$194.85
Round up to	$195.00

When calculating your curb time rates, adjust your crew size, equipment mix, and drive time accordingly. If the rates that you calculate look too good to be true, they probably are. Go back and review your work.

Do not include materials or subcontractors in your curb time rates, as they will greatly confuse the matter.

I would encourage you to mark up materials a minimum of twenty percent (10% for overhead and 10% for profit) and subcontractors a minimum of fifteen percent (5% for overhead and 10% for profit).

IRRIGATION/SPRINKLER REPAIR AND SERVICE—"T&M" BIDDING

This type of bidding is accomplished in almost the same manner as the other two applications in this chapter. The difference here is that we are now pricing our work to accommodate a "time-and-material" (T&M) billing situation, one in which we bill the client for actual labor time and materials used for the job.

Before we start, we need to address how to handle the pricing of the materials that we use in our repairs and drive time. I recommend charging for repair materials independent of labor rates. For commercial customers, they should be marked up a **minimum** of 20% above cost (10% for overhead and 10% for profit). Residential irrigation markets will usually allow you to mark up materials 40 to 60% above costs to reach the normal retail price.

Another method, if the market will allow, is to charge the customer the manufacturer's list price (discounted in some instances) for materials.

Drive time can be handled one of three ways:

1. It can be included in the hourly rate charged to the client, the same way it is covered in the maintenance package bidding. The client is, therefore, charged an hourly rate based on curb time.

2. The client is charged for actual drive time to the job site. Essentially, the clock begins to run once the driver leaves the yard and stops when the job is completed. This method has some inherent problems if the driver starts from a location other than the yard, gets stuck in traffic, or has to make other stops along the way. An average time could be allocated to the job, instead, but this puts you into the third method.

BID WORKSHEET

PROJECT: _IRRIGATION/SPRINKLER REPAIR_   PAGE (   ) OF (   )

LOCATION: _____   DATE PREPARED: ____ / ____ / ____

P.O.C./G.C.: _____   ESTIMATOR: _I. M. DUNN_

DUE: __/__/__ PH.: _____ FAX: _____   CREW SIZE: _1_

TYPE WORK: _____   HOURS/DAY: _8_ CAW: $ _8.00_

REMARKS:

DESCRIPTION	QTY	UNIT	U/C	MAT'L	LABOR	EQUIP	SUBS
I. PRODUCTION/SERVICE							
– LABOR HOURS	6	HR	8–		48		
SUBTOTAL				Ø	48	Ø	Ø
II. GEN CONDS							
– DRIVE TIME	2	HR	8–		16		
– PICKUP TRUCK	8	HR	3–			24	
SUBTOTAL				Ø	64	24	Ø
III. MARKUPS							
– SALES TAX	6	%					
– LABOR BURDEN	30	%			19		
SUBTOTAL				Ø	83	24	Ø
					Ø		
					24		
					Ø		
					107		
– OVERHEAD ⟨OPH⟩	8	HR	11–		88		
BEP ⟶					195		
– PROFIT + CONTINGENCY	10	%			22		
					217		

LABOR HOUR UNIT PRICE = $217/ = $36.17
/6.0 HRS

SHI FORM 04-A

DIAGRAM 15-2 (A). SPRINKLER REPAIR "T&M" ESTIMATING EXAMPLE

```
    OPH-------->      $11.00   CAW/AVE WAGE   $8.00   OTF-------->      0.00%
    PPH-------->       $2.50   PROFIT----->  10.00%   RF--------->      0.00%
                               TAX-------->    0.00%   CAW-LOADED>     $8.00
    NO. UNITS->          6.0   LABOR BURDEN  30.00%   CREW TRUCK>     $3.00
==================================================================================
```

	MAT	LABOR	EQUIP	SUBS
	===	=====	=====	====

I. PRODUCTION OF FINISHED PRODUCT:

	6 HRS		
0	48	0	_____

II. GENERAL CONDITIONS:

	2 HRS	8 HRS	
_____	16	24	_____

SUBTOTALS: | 0 | 64 | 24 | 0 |

III. MARKUPS:

A. SALESTAX   0

B. LABOR BURDEN------>   19

SUBTOTAL: | 0 | 83 | 24 | 0 |

TOTAL DIRECT COSTS----------------------------->   107   49.43%

C. OVERHEAD RECOVERY:

8 (NUMBER OF HOURS X OPH) $11.00 -->   88   40.57%

"BEP" SUBTOTAL (DIRECT COSTS + OVERHEAD)---->   195   90.00%

D. CONTINGENCY FACTOR (IF DESIRED)------------->   0   0.00%

E. PROFIT:

10.00%-------------------------> $21.69 -->   22

8 HOURS X PPH $2.50 --> $20.00 -->   0   0.00%

F. TOTAL PRICE FOR THE JOB---------------------> $217   100.00%

```
IV. ANALYSIS:          $          %                      $/RATIO       %
    A. SPH:         $27.11    100.00%   J. MAT/LAB:    0.00 :1
    B. DCPH:        $13.40     49.43%   K. MPH:        $0.00         0.00%
    C. OPH:         $11.00     40.57%   L. EQ/LAB:                  37.50%
    D. PPH:          $2.71     10.00%   M. EQPH:       $3.00        11.07%
    E. BEP:        $195.20     90.00%   N. GC:        $40.00        18.44%
    F. OVHD:        $88.00     40.57%   O. GCPH:       $5.00        18.44%
    G. PROF:        $21.69     10.00%   P. GCH/TH:                  25.00%
    H. GPM:        $109.69     50.57%   Q. FACTOR:     0.00 X    MAT'L
    I. GPMPH:       $13.71     50.57%   R. UNIT PRICE-------->     $36.15
 V. MORS PERCENTAGE MARKUP ON LABOR & BURDEN------------------->     98.56%
    (ASSUMING A MARKUP OF 10/25/5% ON MAT'L., EQUIP., & SUBS.)
```

DIAGRAM 15-2 (B). BAR WORKSHEET FOR SPRINKLER REPAIR

3. You can charge a show-up fee that includes drive time plus a certain amount of time on the job (i.e., the first fifteen to thirty minutes). Time after that is charged at a set hourly rate.

For instance, you might charge $35.00 to show up, knowing that your average job was thirty minutes from your office/yard. The show-up rate would include thirty minutes of drive time plus the first thirty minutes of time on the job. Additional time on the job would be charged out at $27.50 per hour.

Let's look at Diagram (15-2) to see how we determined these numbers.

- Our sprinkler repair man works alone for eight hours a day, forty hours per week which means the OTF is zero.
- All work is on a "T&M" basis, so the "Risk Factor" is also zero.
- Labor hourly rate is $8.00.
- Labor burden is thirty percent.
- This person drives a mini-pickup truck with a CPH of $3.00.
- An average job is thirty minutes away from the office.
- He plans to perform and bill a minimum of four jobs per day.
- Materials are charged to the customer at 25% above costs.
- Approximately $80.00 of materials (at cost) are to be installed per day.
- The overhead OPH is calculated to be $11.00.
- A minimum combined profit and contingency factor of 10% or a PPH of $2.50 is desired, whichever is greater.

Turning to Diagram (15-2):

- We have put six labor hours in Phase I costs.
- General conditions contain the remaining two hours of estimated daily drive time.
- Total price for an average day of sprinkler repair is $217.00 (which is indicated at the bottom of Phase III calculations).

Put another way, total revenue that must be generated per day to cover all costs (including overhead and providing a 10% profit) is $217.00. In other words, we must bill $217.00 per day, excluding materials, to cover all of our costs and to show a ten percent profit. Let's break this down into more meaningful scenarios.

Scenario #1

We bill four jobs per day and keep our repairman busy (billable) all day. Generated revenues are:

4 (jobs) x $35.00 (show-up charge)	$140.00
4 hours billed at $27.50/hour	110.00
Total	$250.00

We have exceeded our goal of $217.00 by $33.00.

Scenario #2

We bill five jobs per day and keep the repairman billable all day.

5 (jobs) x $35.00 (show-up charge)	$175.00
3 hours billed at $27.50/hour	82.50
Total	$257.50

We have exceeded our goal of $217.00 by $40.50.

Scenario #3

We bill six jobs per day and keep the repairman billable all day.

6 (jobs) x $35.00 (show-up charge)	$210.00
2 hours billed at $27.50/hour	55.00
Total	$265.00

We have exceeded our goal of $217.00 by $48.00.

Each of the three scenarios produces an extra $33.00 to $48.00 of profit, in addition to the $21.70 profit built into the rates. The key is to keep your repairman billable all day, and bill a minimum of $217.00 per day. If that occurs, any money billed above the $217.00 for labor only is extra profit.

You should track your irrigation service work on a daily basis. At a minimum, the items that should be monitored are:

• Sales per total billable dollar amounts for the day.
• Labor hours and job tasks. (e.g., Drive to Jones' residence, 15 minutes; repair two heads, 35 minutes; return to shop, 15 minutes, etc.).
• Materials used and billed.

Once you have historical data from which to work, go back and adjust your hourly rates and show-up charge, if desired.

TREE CREW ESTIMATING

We follow the same format to cost out tree work for smaller jobs as we do for maintenance work. Materials and subcontractors, if used, should be handled separately from labor and equipment rates for these smaller jobs, which generally last less than a day or two.

Jobs that last more than a couple of days, and seasonal contracts, should be bid as outlined in Chapter 13.

Our sample job outlined in Diagram (15-3) includes details as follows:

Size crew	2 people
Hours	9 hours/day, 45 hours/week per man
Labor rates	$11.00 and $7.00 per hour
Labor burden	45%
RF (risk factor)	10%
Boom truck CPH	$20.00
Chipper CPH	$15.00
Chain Saw CPH	$3.00
Overhead OPH	$10.00
Profit (desired)	10%
Production man-hours per day	16
Drive time man-hours per day	2

In this scenario, the BEP for our tree crew is $617.00 per day, or $34.28 per man-hour. Anything billed above $617.00, excluding materials and subcontractors, goes directly to profit.

The curb time rate per man-hour for tree work is $42.88. Double the rate to $81.50 to obtain your curb time crew rate. I would round it up to $85 per hour (or even $90 per hour, or more) if the market would let me and I could stay competitive.

Remember, as long as you are billing $686.00 per day, you are making money and covering all costs. Overtime, equipment, labor burden, etc., are all included. You even have a ten percent risk factor built into your crew rate.

Many companies set up their tree crews on a commission basis. This provides production incentive to crews while protecting the company from labor overruns. To do so using our scenario, conduct the following calculation:

1. Divide total daily crew labor by desired daily billable amount.

   $187.00 ÷ $686.00 = .2726 = 27.3%

2. Round up to an even percent that is easy to work with. In this case, I would add $19 out of the daily profit to the $187 in anticipated labor costs...

   $187.00
   <u> 19.00</u>
   $206.00 = 30% commission
   $206.00 ÷ $686.00 = 30%

BID WORKSHEET

PROJECT: _TREE CREW_   PAGE (   ) OF (   )

LOCATION: _____   DATE PREPARED: ____ / ____ / ____

P.O.C./G.C.: _____   ESTIMATOR: _BEN HADD_

DUE: ___ / ___ / ___ PH.: _____ FAX: _____   CREW SIZE: _2_

TYPE WORK: _____   HOURS/DAY: _9_ CAW: $ _10.40_

REMARKS:

DESCRIPTION	QTY	UNIT	U/C	MAT'L	LABOR	EQUIP	SUBS
I. PRODUCTION/ SERVICE							
- LABOR HOURS	16	HR	10.40		167		
- CHIPPER	4	HR	15 -			60	
- CHAINSAWS	8	HR	3 -			24	
SUBTOTAL				Ø	167	84	Ø
II. GEN CONDS							
- DRIVE TIME	2	HR	10.40		21		
- BOOM TRUCK	4	HR	20 -			80	
SUBTOTAL				Ø	188	164	Ø
III. MARKUPS							
- SALES TAX	6	%					
- LABOR BURDEN	45	%			85		
SUBTOTAL				Ø	273	164	Ø
					164	↵	
					437		
-OVERHEAD <OPH>	18	HR	10 -		180		
BEP ————————➤					617		
-PROFIT + CONTINGENCY	10	%			69		
					686		

CURB-TIME MAN HOUR RATE = $686 / = $42.88
                              /16 HRS

DIAGRAM 15-3 (A). TREE CREW ESTIMATING EXAMPLE

```
        OPH------->    $10.00   CAW/AVE WAGE    $9.00   OTF------->      5.56%
        PPH------->     $3.00   PROFIT----->   10.00%   RF--------->    10.00%
                                TAX-------->    0.00%   CAW-LOADED>    $10.40
        NO. UNITS->      16.0   LABOR BURDEN   45.00%   CREW TRUCK>    $20.00
```
===

	MAT	LABOR	EQUIP	SUBS
	======	======	======	======

I. PRODUCTION OF FINISHED PRODUCT:

```
                                16 HRS

                          0       166        84       _____
```

II. GENERAL CONDITIONS:

```
                                 2 HRS      4 HRS

              _____         21         80       _____
              ==========     ==========  ==========  ==========
    SUBTOTALS:        0        187        164          0
```

III. MARKUPS:

```
    A. SALESTAX          0

    B. LABOR BURDEN------>        84
              ==========     ==========  ==========  ==========
       SUBTOT:        0        271        164          0

       TOTAL DIRECT COSTS------------------------------->   435    63.68%

    C. OVERHEAD RECOVERY:

          18 (NUMBER OF HOURS X OPH)    $10.00 -->    180    26.32%
                                                   ==========
          "BEP" SUBTOTAL (DIRECT COSTS + OVERHEAD)---->    615    90.00%
                                                   ----------
    D. CONTINGENCY FACTOR (IF DESIRED)------------->      0     0.00%
    E. PROFIT:                                       ----------
          10.00%------------------------------>   $68.38 -->     68    10.00%
          18 HOURS X PPH      $3.00 -->   $54.00 -->      0     0.00%
                                                   ==========
    F. TOTAL PRICE FOR THE JOB------------------------->   $684   100.00%
```

```
IV. ANALYSIS:        $          %                  $/RATIO        %
    A. SPH:       $37.99     100.00%   J. MAT/LAB:    0.00 :1
    B. DCPH:      $24.19      63.68%   K. MPH:       $0.00       0.00%
    C. OPH:       $10.00      26.32%   L. EQ/LAB:                87.61%
    D. PPH:        $3.80      10.00%   M. EQPH:      $9.11      23.98%
    E. BEP:      $615.44      90.00%   N. GC:      $100.80      14.74%
    F. OVHD:     $180.00      26.32%   O. GCPH:      $5.60      14.74%
    G. PROF:      $68.38      10.00%   P. GCH/TH:               11.11%
    H. GPM:      $248.38      36.32%   Q. FACTOR:     0.00 X   MAT'L
    I. GPMPH:     $13.80      36.32%   R. UNIT PRICE-------->  $42.74
V. MORS PERCENTAGE MARKUP ON LABOR & BURDEN------------------>   51.21%
    (ASSUMING A MARKUP OF 10/25/5% ON MAT'L., EQUIP., & SUBS.)
```

DIAGRAM 15-3 (B). BAR WORKSHEET FOR TREE CREW EXAMPLE

... and round up the commission to thirty percent of whatever was billed. If crews perform $1,000 of billable work per day, they earn $300.00 for the day. If they only perform $500.00 of billable work, they earn $150.00.

This type of commission situation usually works well for tree crews. The key is to ensure that you consistently reach your daily minimum billable amount.

CONCLUSION

We have covered three specialized estimating situations in this chapter. The basic bidding format used in Chapter 13 stays the same. We simply make a few adjustments to it in order to accommodate the crew package and "T&M" rates.

Calculate various scenarios using the models provided. Compare your results to the going rates on the market. Vary your drive times and notice how the crew rates change dramatically, depending upon how much of it is in general conditions.

Once you establish your crew or "T&M" rates, monitor daily sales, labor hours, and the jobs tasks performed. Adjust your rates accordingly.

ACTION POINT

Complete Exercises 15-1 and 15-2 in Appendix A.

CHAPTER 16
REVIEWING COMPLETED BIDS

PURPOSE

To explain the need for and the process of reviewing a bid after it is completed

INTRODUCTION

I review a lot of bids for clients, either over the telephone or those sent to me by FAX. I use a copy of the Bid Analysis/Review (BAR) Worksheet displayed in Exhibit (8) or a computerized version of it. Quite often, the client and I discover that we need to change the bid in order to make it more accurately reflect actual costs. We often make the bid more competitive in doing so.

Even though I may not have all of the plans, specifications, and bid documents in front of me when I review a bid, there are usually plenty of areas where I can make a difference.

Having a SECOND SET OF EYES review the bid can usually catch those mistakes that may cost you a lot of money. Also, it can help you to identify those areas (and/or methods) where you can decrease costs and improve productivity.

Many owners have their field foreman or superintendent review and "sign off" on bids before finalizing the price. In this way, field personnel become involved in the bidding process (brought into the information loop, so to speak), as they observe, recommend, and are given the opportunity to improve the bidding process.

Some companies have very formal bid review procedures. Others have none at all. Some companies have found checklists to be useful. The KEY is to have a set of PROCEDURES that will ensure that oversights (errors in math, forgotten items, misplaced decimal points, etc.) do not occur, oversights that could have been prevented by some simple forethought and planning.

I know of one (larger) company that insists that a bid be completely finalized and ready for submission at least twenty-four hours before it is due; otherwise, it is not submitted. Evidently, someone in that company learned the hard way the necessity of controlling the bidding process through checks and balances.

REVIEWING A BID

1. THOROUGHLY READ AND REVIEW ALL BID DOCUMENTS, SPECIFICATIONS, PLANS, ETC.

Do not rely on memory or make assumptions. Ensure that irrigation materials are bid correctly (e.g., brass vs. plastic fittings, type of heads, controllers, etc.) and that other materials are bid as specified (e.g., number of cubic yards of soil amendments per thousand square feet, correct thickness of mulch, quantity and types of tree stakes, etc.).

Also, VERIFY that ALL PHASES of the project that are supposed to be bid are included in the bid.

Residential projects often do not involve bid documents and blueprints. In this case, the estimator should not be the salesperson (especially one on commission) unless it is the owner of the company. Otherwise, there is the temptation to underprice jobs and make them unprofitable just to get a commission. Compare the finished bid to the desires of the client by talking to the salesperson. Review the estimate with them—and all items in it. Have the salesperson review his/her notes of discussions with the clients and ensure (as much as is possible) that the clients' desires are addressed in the bid.

2. GOVERNMENT/MUNICIPAL CONTRACTS

Much information about government work is public record. Check these records in order to obtain the dollar amounts for existing maintenance contracts that are out for bid.

Review the "invitation to bid" documents carefully. Be sure that you adhere to all DEADLINES and FORMATS and that you obtain all necessary SIGNATURES and CERTIFICATIONS (and if necessary, the services of a notary public).

A company once flew me in to review a $7,000,000 multi-phase landscape and irrigation project. I was told that all the necessary preliminary work had been done. During the review process, I discovered that the required bid format had not been adhered to. Sure, all the information was there, but it took us two days to unscramble it and to compile it correctly.

3. REVIEW THE BID LINE ITEM BY LINE ITEM.

Check material costs and production rates for labor and equipment. Verify your production rates by comparing them to standards provided in the reference manuals listed in the back of this book.

Re-think your equipment use. Consider how you might make more effective use of your own equipment, or how you might increase productivity by renting additional specialty equipment. Subcontractors' prices should be verified in writing.

One of the benefits of our estimating software program is that it allows you to verify a unit price that includes material, labor, equipment, and subcontractor costs—as well as overhead recovery, general conditions, a contingency factor, and profit.

These unit prices provide a fast and accurate means of confirming your bid prices by comparing unit prices from previous bids and those found in the open marketplace. You have, at your fingertips, ready reference to such items as:

• Prices per square foot for such things as seeding, soil prep, sod, ground cover, concrete and flatwork.

• Unit prices for specific sizes and species of plants, irrigation heads, etc.

• Linear-foot prices for wiring, irrigation main and lateral line, and headerboard.

4. REVIEW GENERAL CONDITIONS

General conditions usually constitute 5-10% of the price of a job. There are, however, three ways that you can possibly reduce them, if you ask the right questions.

A. Foreman Supervision Hours

This can be bid at the average wage for the crew (not the actual foreman rate) for those hours that the foreman is supervising and the entire crew is on the job site. This may lower your labor rate $3-4.00 per labor hour (not including labor burden).

B. Drive Time to/from the job site

It is very common for installation crews to be paid for drive time only when they are on the job. This is especially true if crews are working at a particular job site for lengthy periods of time—more than two weeks. Otherwise, drive time (ride and sleep time for non-drivers) is normally paid only one way, rather than both to and from the job site.

If drive time costs on a particular job are too high, consider putting your crew into a motel overnight or into a house that is rented on a month-to-month basis. Some contractors even pay drive time at minimum wage, not the normal hourly employee rate.

If drive time is significant enough that it may mean the difference between winning the bid or not, total it up (add labor burden, overhead, and profit) and show it to your crew. Ask them if they are willing to give it up for the sake of getting additional work. Quite often, crews will work with you, especially if you are running out of work or are looking at a job that takes place toward the end of the season.

C. If necessary, motels, delivery charges, permits, and contingency factors can be "washed through" the bid without adding overhead or a profit markup to them. However, if the market will bear it, add overhead and profit to these items.

D. Prevailing Wage Rates

Labor hours for off-site fabrication of irrigation assemblies and pumps, drive time, hauling material, mobilization of equipment, etc. do not necessarily have to be paid at the prevailing wage rate. Check your bid documents, labor tables, or labor board to be sure.

E. Crew Size

If, for instance, you can effectively double the size of your crew, you should see a doubling of your production, thereby cutting in half the number of days on a job. This, of course, is not always possible, but it will illustrate my point. As a result, your crew pickup truck hours, supervisor hours, dumpsters, etc. (anything where cost is tied to the length of a project) may decrease, as well. Keep this is mind and take advantage of this strategy if at all possible when reviewing your bids.

5. ANALYZE YOUR BID PRICE

After you put your bid together, analyze your price according to the five most common ways of pricing projects in use today. These five methods are discussed in Chapter 13. They are:

- Factoring
- Market-driven Unit Pricing (MDUP)
- SORS/GPM Markup Pricing
- Multiple Overhead Recovery System (MORS)
- Overhead and Profit per Hour (OPPH)

A. Factoring

To determine the factor for a job, divide the total price for the project by total material wholesale costs.

Installation projects will probably have a ratio of 3-4:1. This means that material will comprise 25-33% of the price of a project. Maintenance projects usually have a ratio of 15-25:1, or 4-6% of the price.

If your factor for an installation project is 2:1 or less (meaning that materials comprise over 50% of the price of the job), you had better look at it very carefully. Either you are installing very expensive materials, or you are using very aggressive production rates, or both. Just be sure that you feel confident of your production rates for labor and for equipment.

B. Market-driven Unit Pricing (MDUP)

Unless your estimating is computerized and your software is programmed correctly, deriving unit prices is one of the most difficult aspects of estimating. Compare unit prices for your bid to those for previous jobs. Compare them, also, to those commonly seen in your market. If your unit prices are too low, adjust them up individually by adding a little more net profit margin. If they seem high, check your production rates and material costs and adjust them downward if warranted by either increasing production rates or decreasing net profit margins.

C. Gross Profit Margin (GPM)

The GPM on a bid can be a reliable "truth teller" in many markets. It can be one of your most important indicators to monitor.

Once you have tracked your GPM on a number of jobs for a particular market (commercial or residential construction or maintenance, irrigation for golf courses, commercial or residential installation, etc.), you will have a better understanding of both your market and how your bids relate to it. Review and compare the GPM on a job being bid to historical ones in that market.

D. Multiple Overhead Recovery System (MORS)

Although the MORS method is cumbersome and overly complex, if you have used it in the past (and probably have a feel for how the market responds to prices produced by it), I would encourage you to review your overhead recovery by "backing" into your overhead recovery percentages on labor and labor burden.

This is accomplished by following the steps below.

(1). Determine the total overhead to be recovered in your bid: $3,000

(2). Multiply Materials (including tax) by 10%: $10,000 x .10 = $1,000

(3). Multiply equipment costs by 25%: $1,000 x .25 = $250

(4). Multiply total subcontractor costs by 5%: (There are none in our example.)

(5). Subtract overhead recovered from material, equipment, and subcontractors from total overhead to be recovered.

$$\begin{array}{r} \$3,000 \\ -1,000 \\ \underline{-\ 250} \\ \$1,750 \end{array}$$

(6). The result is the remaining overhead to be recovered off labor and labor burden.

To determine the markup percentage on labor and labor burden, simply divide the amount to be recovered (e.g., $1,750) by the total cost of labor plus labor burden (e.g., $3,250).

$$\$1,750 \div (\$2,500 + \$750) = \$1,750 \div \$3,250 = 53.9\% = 54\%$$

Compare the percentage markup on labor and labor burden for this job to those that were established by the MORS budget for the year (if you have one and if you have used this method in the past).

E. Overhead and Profit per Hour (OPPH)

The OPPH method is the preferred pricing method which you should have used to bid your job in the first place. Subsequently, you have already analyzed your bid using this method. However, after reviewing your bid using some or all of the other methods of pricing, if your PPH (Profit per Hour) is higher or lower than your budgeted PPH for the year, you might want to increase or decrease your profit markup on the bid to bring it more in line with your budgeted PPH.

A safe, minimum amount of profit to put on your jobs is your budgeted PPH for the year. You should not go much below it, unless you are in a "T&M" situation. Your PPH on material-intense jobs, however, may be significantly higher than your budgeted one. An example may help.

Materials at wholesale cost (including tax) for the landscape and irrigation renovation of Mr. Johnson's home total $10,000. Labor and labor burden total another $3,250. Equipment costs are $1,000. No subcontractors are used on the job. Using the BAR worksheet which I developed using a Lotus 1-2-3 computer spreadsheet, the bid analysis appears in Diagram (16-1).

Overhead, net profit margin (NPM), and GPM percent markup on this project are calculated as follows:

(1). Overhead % $= \dfrac{\text{Overhead}}{\text{Total price}} = \dfrac{\$3,000}{\$18,750} = 16.0\%$

(2). Material (minus tax) as a % of the job $= \$ 9,524 \div \$18,750 = 50.8\%$

(3). NPM % $= \dfrac{\text{Net profit}}{\text{Total price}} = \dfrac{\$1,500}{\$18,750} = 8.0\%$

(4). GPM $= \dfrac{\text{Overhead + Net profit}}{\text{Total price}} = \dfrac{\$3,000 + \$1,500}{\$18,750} = \dfrac{\$4,500}{\$18,750} = 24.0\%$

```
        OPH------->      $10.00   CAW/AVE WAGE    $6.88   OTF------->     11.11%
        PPH------->       $5.00   PROFIT----->   10.00%   RF--------->    10.00%
                                  TAX-------->    5.00%   CAW-LOADED>     $8.33
    NO. UNITS->          1.0      LABOR BURDEN   30.00%   CREW TRUCK>     $5.00
==================================================================================

                    MAT          LABOR        EQUIP          SUBS
                   ======       =======      =======        ======
I. PRODUCTION OF FINISHED PRODUCT:
                                 250 HRS

                   9,524        2,083          660             0

II. GENERAL CONDITIONS:
                                 50 HRS       68 HRS

                      0          417           340             0
                   =========    =========    =========      =========
    SUBTOTALS:      9,524        2,500        1,000             0

III.  MARKUPS:

    A. SALESTAX       476

    B. LABOR BURDEN------>        750
                   =========    =========    =========      =========
       SUBTOT:     10,000        3,250        1,000             0

       TOTAL DIRECT COSTS------------------------------>   14,250      76.00%

    C. OVERHEAD RECOVERY:

          300 (NUMBER OF HOURS X OPH)   $10.00 -->         3,000      16.00%
                                                         ==========
       "BEP" SUBTOTAL (DIRECT COSTS + OVERHEAD)---->      17,250      92.00%
                                                         ----------
    D. CONTINGENCY FACTOR (IF DESIRED)------------->           0       0.00%
    E. PROFIT:                                            ----------

          300 HOURS X PPH    $5.00 -->   $1,500 -->        1,500       8.00%
                                                         ==========
    F. TOTAL PRICE FOR THE JOB-------------------->     $18,750     100.00%
------------------------------------------------------------------------------
IV. ANALYSIS:         $            %                       $/RATIO       %
     A. SPH:       $62.50      100.00%   J. MAT/LAB:        3.81 :1
     B. DCPH:      $47.50       76.00%   K. MPH:          $31.75      50.80%
     C. OPH:       $10.00       16.00%   L. EQ/LAB:                   40.00%
     D. PPH:        $5.00        8.00%   M. EQPH:          $3.33       5.33%
     E. BEP:   $17,249.65       92.00%   N. GC:          $756.62       4.04%
     F. OVHD:  $3,000.00        16.00%   O. GCPH:          $2.52       4.04%
     G. PROF:  $1,500.00         8.00%   P. GCH/TH:                   16.67%
     H. GPM:   $4,500.00        24.00%   Q. FACTOR:        1.97 X   MAT'L
     I. GPMPH:     $15.00       24.00%   R. UNIT PRICE--------> $18,750
V. MORS PERCENTAGE MARKUP ON LABOR & BURDEN------------------>         53.85%
   (ASSUMING A MARKUP OF 10/25/5% ON MAT'L., EQUIP., & SUBS.)
```

DIAGRAM 16-1. BAR WORKSHEET FOR JOHNSON RESIDENCE

A GPM of 24% is low for most residential markets. In all likelihood, you would want to increase your GPM upwards to between 30-35%. This would increase net profit from $1,500 to a figure between $3,100 and $4,675, as the market dictates. Subsequently, the PPH for the revised GPM would fall between $10.00 and $16.00.

6. OTHER CONSIDERATIONS

A. Ratio Analysis

Once a bid is complete, you should compare ratios for your bid to those in your budget. If they vary significantly, analyze your bid in order to ascertain why.

In most cases, there is usually a simple explanation (e.g., expensive plant materials in the bid, large amount of equipment—or no equipment—needed for the job, etc.).

Let's examine our most common ratios and per hour calculations.

(1). Sales per hour (SPH)—The result of dividing annual sales per division by total annual projected field-labor hours.

(2). Overhead cost per hour (OPH)—Arrived at by dividing total overhead by total annual projected field-labor hours.

(3). Net profit margin per hour (NPMPH or PPH)—Obtained by dividing net profit margin plus the contingency factor, if used, by total annual projected field-labor hours.

(4). Gross profit margin per hour (GPMPH)—Obtained by adding OPH to NPMPH.

(5). Material to labor (M/L) ratio—The total projected material costs divided by the total projected field-labor dollar costs.

(6). Material per hour (MPH)—Infrequently used, this figure is obtained by dividing total projected material costs by total projected field-labor hours.

(7). Equipment per hour (EQPH)—Obtained by dividing total projected equipment costs (plus rentals) by total projected labor hours.

(8). Equipment to labor ratio (EQ/L)—The percentage obtained by dividing total projected equipment costs (plus rentals) by total projected field-labor costs.

(9). Overhead Recovery (the Traditional or MORS method)—This calculates the dollar amount and the average markup percent that needs to be added to labor and burden in order to recover overhead on the job using the MORS method.

Material, equipment, and subcontractor markup percentages are set at 10%, 25%, and 5%, respectively. The balance is obtained from labor and burden. Note that the four percentages do not have to add up to 100%.

B. Reviewing Net Profit Margin (NPM)

Although we have already discussed profit in Chapter 13, it is worth an even more in-depth analysis.

(1). When considering the NPM for a job:

- Set a profit percent range appropriate for the respective market in which the job falls (e.g., 10-25% for residential installation).
- Determine a percent within that range for each of the five categories listed below.
- Total the five categories and divide by five.

The result should reflect a somewhat accurate profit percent to put on a project.

The five categories are:

a. Need

Gravitate toward the low end of your range (in our example, 10-13%)...

- if this job is likely to be installed during a slow time of year;
- if this is a period when you have no work, and it is not likely that you will get any more; **or...**
- if you need the work immediately.

Otherwise, use a percent toward the higher end (15-25%).

b. Size

Generally speaking, projects that are smaller in size than your normal job should be bid with more built-in profit than usual (e.g., 15-18%). Conversely, ones larger than the norm call for a smaller amount (10-13%) of NPM.

c. Risk

Projects that are familiar to you and to your crews, those that pose a very small risk factor (that is, they should come in ON-BUDGET) generally warrant less profit than those that seem more uncertain.

The less familiar you are with the work, the greater the risk.

Add an insurance policy, of sorts. As unfamiliarity and risk increase, add more net profit margin.

d. Market (what the market will bear)

Scale down your profit if you are in a situation of "low bidder takes all" (where there is no hope of negotiating the job) and if you think that your pricing is in the general ballpark with the other contractors. Otherwise, why waste time bidding the project?

Keep in mind two numbers (especially the second):

- **HOW LOW** can you go and still cover all costs? What is your **BEP?**
- **HOW HIGH** can you go and still get the job?

Do not be afraid to push the high end of the scale.

e. Negotiating Ability

If, on the other hand, you are able to "negotiate" into the project (you are the contractor of choice for the project), gravitate toward the higher end of the range. It is understood, of course, that when someone else is the contractor of choice, the reverse is true.

Your ability to negotiate should be reflected in the markup percent. You should know from past experience what you can get away with and how well you are able to win the trust and confidence of the client and negotiate into the job.

(2.) Contingency (aggravation) Factor (CF)

Although difficult to analyze, add a lump sum dollar amount of NPM (net profit margin) to a job as an added risk contingency factor. Add it to NPM after you have determined an amount for profit using the five categories above.

Add a CF if you feel that the owner, architect, general contractor, etc., just seems to grate you the wrong way. Call it a "bribe." What amount of a bribe would it take to make this project worth putting up with due to the possible hassles and frustrations that might go with it?

C. Prevailing Wage Projects

DO YOUR HOMEWORK. This is a law that must govern all bidding situations, and it is especially true of those using prevailing wage rates.

(1). VERIFY your labor rates.

Because CAW (crew average wages) on rated jobs can be double (even triple) those of non-rated ones, it is imperative that you use correct CAWs in the bid. Erroneously high CAWs can render your bid non-competitive, and erroneously low ones can cause significant underbidding and losses on the project.

Obtain the labor rates **IN WRITING.**

- They are included in the bid documents and Federal and State labor tables.
- If necessary, call your state or federal labor board or consult a labor attorney who specializes in this area.

NOTE: Quite often, you are allowed to have varying types of labor with diverse rates in your prevailing wage crews.

(2). VERIFY your labor burden

Required benefits are not always easily calculated for rated projects. Company ESOP, pension plans, medical benefits, etc., may legitimately substitute for those identified in the bid documents and/or labor specifications.

When calculating labor burden onto labor, ensure that you are neither "double-dipping" (charging twice) nor excluding required benefits.

(3). REVIEW general conditions.

Look for those labor functions that may not require prevailing wage rates. For example: off-site fabrication, drive time, warranty work, owner's on-site labor, etc., may or may not be excluded from prevailing wage rates. Check the wage determination table in the bid documents/job specifications, or call the state labor board or the U.S. Department of Labor if necessary.

D. "Double-dipping" the Owner's Salary

This is a means of dropping your CAW considerably. The scenario is realistic and can be used in extremely competitive situations:

If all of an owner's salary is in overhead, and if he or she is going to work in the field and be part of production hours on a certain project, you could (theoretically) compute your CAW with one member's wage (the owner's) being zero.

More often than not, however, you would probably underprice the project and leave money on the table. To prevent this, enter a "fair market value" (realistic labor rate) for the owner when calculating your CAW.

E. Exclusions and deletions.

"When in doubt, exclude it out." RISK MANAGEMENT is the name of the game. If there are portions of a job where you feel the risk too great and/or your experience level with that type of work is not up to the challenge, do not be afraid to submit a bid excluding that portion of the work or subcontract that portion of the job to reduce your risk.

F. The Old **"Gut Check"**

After you have done your homework (reviewed, analyzed, and scrutinized a bid) sit back and take note of how you **feel** about it. Your "gut" can tell you a lot.

Once a bid is completed, I would not recommend shaving off more than a couple of points of net profit from it. However, I would never be afraid of pushing it higher—embarrassingly high, in some cases. It's always smart to occasionally explore the high limit of a market with an unjustifiably high net profit margin.

Of course, remember to take into consideration how accurate your "gut" has been in the past and make adjustments as appropriate.

CONCLUSION

Reviewing completed bids can be a very quick and informal event, or it can be a lengthy, formal process. The key is to prevent tunnel vision and oversights by using a system of **checks and balances.**

That "extra set of eyes" can prove invaluable. They can save you lost dollars and lost opportunities.

In the event that you would like to discuss this process, review a specific bid, or obtain a copy of the computerized version of the BAR worksheet, contact our offices.

CHAPTER 17
ESTIMATING STRATEGIES

PURPOSE
To outline some key strategies to use at the bid table

INTRODUCTION

Estimating, or bidding your work, and business strategies must go hand-in-hand. What is good at the bid table usually has a positive effect on the business as a whole, and good business strategy will impact your estimating. If your business strategy is not accurately reflected at the bid table and in your bids, you are just asking for trouble down the line.

Business sales may take off, but if work is priced wrong, the more work you get, the more money you may lose. Forgetting to connect estimating with business strategy is like the stagecoach driver who forgets to hitch the horses to the coach. You can bet he won't be short on excitement (or bumps).

As we discuss strategy, the foundational principle upon which we will build comes from a statement by Phillip Crosby in his book, *Quality Is Free*.

"Good things happen only when planned. Bad things happen on their own."[3]

Good sound planning is a process. It is a skill that requires constant learning and adjustment. The more skilled we are at it, the better the results.

We will discuss strategy by breaking it into four sections:

- Your BIDS.
- Your CUSTOMERS.
- Your BUSINESS.
- Your JOBS.

This should help you to compartmentalize and to relate to the process more effectively. We need not go into great detail, as most of the strategies are actually rather simple. Let's get started.

YOUR BIDS

1. Be ORGANIZED and METHODICAL.

   A. Remember **PATT** = **P**lan the **P**lace, the **A**ctivity, the **T**ime, and the **T**ools.

   B. BLOCK OUT the time on a schedule and ask not to be disturbed. Have your telephone machine or secretary record calls which can be returned later.

   C. BE SURE that you have all the tools you need BEFORE you start (plans, pencils, pads, counter, etc.).

2. BID the job STEP-BY-STEP in your mind and on paper, just as you will build the job in the field.

3. Use OBJECTIVE REFERENCE POINTS.

   A. INSIST upon clearly-detailed specifications, bid documents, plans, and/or designs and addendums.

   B. VERIFY your production rates by checking past job-cost data or outside reference books/manuals. See the reference list in the back of the book for further details.

   C. OBTAIN quotes, in writing, from suppliers and subcontractors.

4. If possible, DELEGATE tedious work (takeoffs, costing materials, etc.).

5. BE SELECTIVE.

   A. KNOW your strengths AND your weaknesses.

   B. STUDY your company (division), and do an honest appraisal of what you do best (residential, design/build, commercial construction and/or maintenance, athletic fields, etc.).

   C. PURSUE these markets (what you do best) AGGRESSIVELY.

6. **Crystallize** your thinking. Do not deal in generalities. Think SPECIFICS and be EXACT. Don't just SWAG ("scientific wild anti-analytical guess") it or use "Jesus" factors (ones which you hope will "save" you). Ask:

   A. What size crew will effectively get the job done?

   B. What is the desired GPM on the job being bid?

   C. How many 1, 5, 15 gallon shrubs can your crew plant per hour?

   D. Ad infinitum.

7. Make money EVERYWHERE.

   A. Do not think "loss leader."

   B. Do not think that the amount of profit you are willing to give up on a job is your advertising budget for the year.

   C. Accurately identify "T&M" rates by including labor burden and an overtime factor in them if applicable.

8. Think RISK MANAGEMENT.

Labor production rates and conditions can vary dramatically from job to job. VARIABLES INCREASE RISK.

   A. When bidding work, think in terms of best-case and worst-case scenario. Ask yourself, "What's the worst set of circumstances that could happen on this job? What is the best?"

   B. Determine the most likely set of circumstances and plan accordingly.

   C. Bid according to the scenario chosen by adjusting your Risk Factor on labor.

   D. Subcontract work to another contractor who is more familiar with that type of work if you consider the risk too great.

9. Have SOMEONE check your bid. Guard against tunnel vision. Think "extra set of eyes," be it a foreman, field superintendent, or whoever.

I personally review a lot of bids for contractors over the telephone. By asking certain questions, I can sometimes dramatically change the pricing.

10. Check your GUT.

   A. Do you "have" to get this job?

   B. Are you being too pessimistic or too optimistic about the risk factors involved?

   C. Is a developer or general contractor trying to manipulate you or back you into a corner?

   D. Are you rushing or "forcing" the bidding process and overlooking key items?

   E. Are you impatient, stressed-out, and/or overwhelmed?

   F. Do you often feel remorse (bidder's remorse) after you get a job?

Your emotions before, during, and after the bidding process can tell you a lot. Be aware of your emotions, and take the necessary steps that will make the bidding process a positive, rather than a negative, event.

11. Let your PRICE do your TALKING.

Do not get caught up in long, undesirable discussions—especially when you do not really want the job (at any price) in the first place.

If the job is too small, inconvenient, not worth the hassle with an owner, architect, etc., add 20-30% (or more) net profit to your bid.

Give the client your price, step back, and watch. Do not attempt to justify your price. Just say that this is what you need to get in order to cover your costs on the job, realizing that 20-30% of the price is the emotional cost to you.

YOUR CUSTOMERS

1. See the job (product or service desired) through the customer's eyes.

   A. TAKE NOTES.

   B. ASK dozens of QUESTIONS.

   C. Realize that most residential clients really do not entirely understand what they are getting into.

   Get on their "wavelength" by asking insightful questions, and by using the next step.

   D. BE OBSERVANT—Stop, look, and listen!

2. UNCOVER and identify "HIDDEN" agendas.

   Most of us have already met "the customer from hell." Some people have hidden agendas. The earlier we identify them, the better our chance of managing the situation, the customer's expectations, and the job (should we decide to take it).

   Clients may be asking for your bid merely as a part of their process of shopping around with leads from the yellow pages and other such directories. They are usually looking for the lowest price. Be aware of this.

   Some people will attempt to grind you to get your prices down. Others will nickel-and-dime you to death.

   One of my clients did work for an attorney. Halfway through the job, my client noticed the attorney's license plate. It read, "SUE EM." This is what I call a "hidden agenda." Identify them early, before you sign a contract, if you can.

3. MANAGE the EXPECTATIONS of your customers.

   Customers (people) want "certainty." In business, "certainty" is achieved when the performance standard achieved is equal to the one that was desired and planned.

   Many expectations (or standards) are NON-negotiable: ones concerning ethical treatment, courtesy, safety, legalities, etc.

   Other expectations ARE negotiable: pricing, schedules, the production process, etc.

   There is usually a lot of room for negotiation and flexibility. However, once expectations are set and communicated to the client, you should MEET or EXCEED those expectations. Learn how to identify, direct, and control (manage) your clients' expectations in order to meet or exceed them.

   Meeting your customers' expectations will build credibility, trust, referrals, and/or repeat business.

   Surprises do just the opposite. Remember: in business, **there is no such thing as a good surprise.**

4. EDUCATE your customer.

   Whenever possible, educate your client about the product and/or service provided. Explain irrigation timing, maintenance and fertilization requirements, plant material, etc.

   Nurture and cultivate your leads and customers and make them more INFORMED CONSUMERS.

5. SOLVE your clients' problems.

   Teach everyone in your business to see themselves as problem-solvers. Train them to look for, identify, and solve customers' problems.

Remember, if customers could (had the time, talent, tools, and/or desire to) solve their own problems, they would not need you.

6. ADD VALUE (in your client's eyes).

    A. Be aware of the studies and statistics that will help your client be confident of the value of his/her investment. If appropriate, share this information with your client.

    For instance, according to a brochure published by the Associated Landscape Contractors of America (ALCA), the American Association of Nurserymen, the Garden Council, and the Professional Plant Growers' Association:

    - *MONEY* (the magazine) reports that landscaping can have a recovery value of 100–200%.

    - The Gallup Organization reports that, overall, new home buyers and buyers of previously-owned homes estimate that landscaping, on the average, adds 14.8% to the value or selling price of their home.[4]

    This brochure, and other similar literature, is available through The Garden Council. See Associations listing in the back of this book.

    Use this type of information to your and your clients' advantage.

    B. Large and small things can add value to a product or service by making a client's life simpler and less encumbered.

    Large items like:
    - Automatic controllers
    - Full-service maintenance
    - Lifetime warranties of plant materials

    Smaller items like:
    - Conveniently placed stepping stones
    - Strategically located annual plants that enhance the grounds and make the site "pop" out at you

    All these (and more) can add value in the eyes of the client. Some items may have direct monetary value, others may not.

7. NEGOTIATE from a **position of strength.**

    Information is power. If you know your "numbers" concerning a bid, this will translate into a sense of confidence, and confidence engenders trust.

    TRUST wins the negotiating battle. As George Schultz (former United States Secretary of State) said, "Trust is the coin of the realm."

8. LISTEN, LISTEN, LISTEN.

    Most problems and disputes boil down to improper communication.

    If you would analyze the disagreements, arguments, and surprises that you have experienced, you would probably find that 99% of them were caused because somebody did not listen, intently and carefully.

    A. Play back (REPEAT) to the other party what you think you heard.

    B. Ask him if "such-and-such" was what he meant.

    C. Stop and ask questions such as: "Does this make sense to you?" or "Do you mean…?".

    D. Listen and TAKE NOTES.

    E. Observe BODY LANGUAGE.

9. Add CREDIBILITY to your information.

Use references to bolster your position:

    A. A newspaper or magazine article can help to build the trust factor between you and your clients.

    B. Use association-produced materials wisely. They can build trust.

C. COMPUTER PRINT-OUTS can, if used properly, add immense credibility to your position.

I am amazed at the number of my clients who have told me that they won jobs, in large part, because prospective customers were so impressed by the spreadsheets and the information generated by our estimating software. Properly formatted, accurate information builds credibility.

YOUR BUSINESS

1. MANAGEMENT

A. Develop and pursue WRITTEN GOALS and BUDGETS. Remember what Tom Peters says in his book, *Thriving on Chaos:* "What gets measured gets done..."[5]

If you cannot measure it, you cannot direct and control it. Management by mysticism just does not get results.

B. Think and communicate in MEANINGFUL MANAGERIAL TERMS (MMTs).

If you want to excite and motivate your people, you must:

- Be SPECIFIC.
- Use SOLID (concrete) TERMINOLOGY.
- Speak in terms of OBJECTIVE, MEASURABLE STANDARDS.

Vague terms, nebulous goals, and an almost mystical sense of direction will never provide the direction, or drive, necessary for a successful business.

Explorers make charts and maps of their discoveries so that others can follow in their tracks. Doctors, researchers, and scientists employ the scientific method (with all of its shared documentation) in order to unravel the unknown.

So too, business people must use MMTs as they strive to communicate and improve the process of providing excellent products and services.

C. Encourage CREATIVITY.

Adapt and develop new markets, products, and services.

D. Link TQM (TOTAL QUALITY MANAGEMENT) to measurable standards, goals, and results.

Implement TQM, and measurable standards, throughout the company, not just in the field.

E. See the BIG PICTURE.

Legitimate business is the process of accumulating wealth while providing worthy products and services that assist others in the pursuit of worthwhile goals.

Business, good business, is really the people-growing business. While sound estimating and business strategy can help you make money, there is more. See yourself as a coach who develops (nurtures) his/her team (the people) and strives for newer and taller heights.

2. MARKETING

A. KNOW your market.

- Know where you fit in.
- Be selective.
- Understand what you do best and be aware of your weaknesses. Be very cautious when you approach work that has been trouble for you in the past.
- Pursue your strong suit with a vengeance.

B. See the LONG-TERM cash flow consequences of REPEAT business and REFERRALS.

Due to poor quality control, a commercial landscape and irrigation maintenance company lost a municipal contract, one that had brought in $500,000 a year for a number of years. Not only did he lose a half million dollars of revenue for that year, but for ____ number of years in the future. You fill in the blank.

Your customers are your best advertisement.

Whereas a satisfied customer may tell 2-3 other potential clients, statistics show that a dissatisfied customer will tell 9-10 potential customers.[6]

It is up to YOU to determine if these "ads" are going to be good or bad ones for your business.

A gas company was putting in a seven-mile long, four-inch gas main through a town where one of my clients lived. The gas company had to trench and install the line across lawns and backyards. The areas had to be resodded and/or hydroseeded.

When asked for suggestions as to who should perform the work, the homeowners insisted that my client receive the contract.

When my client was invited to bid the work, he did not even know that the job was up for bid. He asked the gas company who he would be bidding against. When told, he informed the representative for the gas company that his (my client's) prices were always higher than the other two bidders. My client was told that as long as his price was reasonable, he would get the work.

Thanks to customer referrals, my client got the job and is doing similar work for the gas company all over his part of the country.

Imagine it. This contractor had the job before he even knew that there was a job to be had. He got the referral, and this will probably lead to even more "repeat" business.

Do you see the long term implications of how you conduct your business?

C. Do not fall into the trap of having to be "low bidder."

A commercial landscape and irrigation contractor once told me that if he was not the low bidder, he did not get the job. He rarely, if ever, negotiated work. After seeing him and his operation in action, I knew why: no one in their right mind would want to repeat the experience that they had had with this guy.

If you have to be low bidder to get a job, you are probably doing something wrong. See it as a symptom of a larger problem.

D. Don't wrestle with a pig. He likes it, you get dirty.

Smart generals choose their battles wisely. The battles not fought, the ones you would probably lose, are the most important ones of all.

E. Selectively EXPERIMENT with MARKETING.

There are many, many marketing ideas and methods that you could pursue. I know contractors that have lost thousands of dollars (and hundreds of hours invested)—and even their company—due to poorly thought-out marketing methods.

• Conduct small experiments that cost you only a few dollars.

• Monitor the results, and then throw gasoline where the fire is.

In other words, see success before you commit large amounts of money or time to a new, unproven approach.

F. TARGET your markets.

I encourage landscape and irrigation contractors to identify:

- Landscape architects
- Developers
- Designers
- Purchasing agents
- Base purchasing offices and officers
- City parks and recreation departments

- Ad infinitum

Enter their names, addresses, phone numbers, and FAX numbers into a computer database.

Begin to develop relationships with these potential clients (and sources of projects) by sending them a cover letter and some basic information about your company and its capabilities.

- Keep it simple.
- Ask about opportunities to bid work on any future projects.
- Identify specific projects coming up for bid and refer to them in future telephone conversations.
- Take notes.
- Follow up the letter with a phone call.
- Use a computer database.
- Do not give up!

Consistently groom this list as you develop relationships with the individuals on it.

Use this approach in conjunction with the BID BOARD that is discussed in Chapter 19.

3. YOUR FINANCES

A. Train bureaucrats to support the field.

Insist that bureaucrats support the field operation. Help them understand that their job is doing so. Put them to work out in the field one or two days a month (or quarter).

Set a goal to have your office staff prepare and provide DAILY job-cost reports for the field. This is an excellent goal that will focus their attention and their energies where it should be, supporting people in the trenches with timely, accurate feedback.

B. Train your field personnel to support the bureaucrats.

- Purchase orders should be used throughout the company. Insist that field personnel use them.
- Insist that Field Daily Reports (FDRs), delivery tickets, and related paperwork is delivered to the office each and every day.

YOUR JOBS

1. DO DAILY JOB-COST REPORTS

Timely, accurate feedback (TAF) can be the number-one motivator for the whole company if used properly.

At a minimum, provide the field and key staff with daily job-cost reports for labor hours, comparing labor hours bid to actual performance.

2. "RE-" BID THE JOB

Once you have won the job, go through it again from beginning to end. Look for mistakes and for hidden opportunities to make, or to lose, money.

I once reviewed a $170,000 bid for a client. The contractor had misplaced a decimal point that cost him $6,000 in labor costs that should have been included in the bid.

This contractor had neglected to account for the labor involved in the placement of a granite pathway. His unit cost was calculated to be $2.93 a square foot. It should have been $4.50.

The owner of the project was already leaning toward the installation of a concrete-type walkway. My client could subcontract this out for $2.50 a square foot. He got on the phone, and the owner was persuaded to tilt his lean into a change order. Rebidding the job turned a potential $6,000.00 loss into a nice gain.

Rebid your jobs after they are won. IT PAYS!!

3. PLAN, PLAN, PLAN!

Murphy's law tends to prevail in this industry: If something can go wrong, it will—and at the worst possible time. It is up to YOU to impose corporate willpower upon circumstances that will naturally tend to disintegrate and self-destruct.

Planning is the never-ending process of FOCUSING the CORPORATE WILL on chaos and confusion.

4. MANAGE YOUR RISK BY USING SUBCONTRACTORS

"When in doubt, sub it out." If the situation warrants, do not be afraid to pass the risk along—or, at least, to share it.

CONCLUSION

We have only touched upon some of the key strategies that you might use. Of course, the possible strategies are endless. Study your industry and your business. Become a student of both.

ACTION POINT

Review the list of reference material located in the back of this book. It will help you to locate more sources of sound business strategy.

CHAPTER 18

COMMON MISTAKES,
or "Who's Through in L&I Contracting?"

PURPOSE
To identify common bidding mistakes

INTRODUCTION

Businesses usually become dysfunctional for one of three (or a combination of these three) reasons.

1. There is a serious (perhaps even fatal) flaw in the estimating system regarding:

   - OVERHEAD recovery
   - Estimating and tracking EQUIPMENT use
   - Estimating and handling FIELD LABOR

     (We will only address equipment and field labor as we have already covered overhead in previous chapters.)

2. PERSONAL PRIDE (ego) interferes with the daily implementation of sound business principles and practices.

3. The owners are simply NOT PREPARED to be in business for themselves.

   They are overwhelmed because:

   - They do not have the know-how (at this stage of their careers) and TRAINING, **and/or...**
   - They lack the emotional STAMINA required to meet and to address the challenges of running a business.

SEVEN SCENARIOS

The following seven scenarios will help illustrate the most common mistakes made by contractors. I have labeled them as follows:

- "Iron Man" Mike
- The Sugar Daddy High
- Too Much Caffeine (or Testosterone)
- The Vampire Syndrome
- Egomania
- Looking for "Foxes" in All the Wrong Places...
- In God We Trust (All Others Use Purchase Orders)

1. "IRON MAN" MIKE

(Mike is not his real name.) There was a construction company that intended to employ enough field labor to total approximately 50,000 field-labor hours for the year. Total equipment costs were estimated at about $450,000 for the year. Indirect overhead was estimated at another $300,000.

For bidding purposes, direct costs (material, labor, labor burden, and sub-contractors) were included in the bid at cost. Equipment costs were combined with overhead and added to the bid at $15 per estimated field-labor hour ($450,000 equipment costs plus $300,000 overhead, both divided by the 50,000 projected field-labor hours).

Sounds simple enough, but there was a **fatal flaw** in the process: equipment costs were not bid based on what would be needed for a particular job. All equipment costs were averaged and bid into

"Iron Man" Mike

all jobs the same regardless of how much or how little equipment was required on each particular job. Jobs requiring nothing more than pickup trucks and wheelbarrows were charged the same ($15 per hour) as ones needing bobcats, trenchers, and backhoes.

It is not uncommon for contractors to estimate their jobs in this manner. The consequences are subtle and eventually can be disastrous:

- Labor-intense jobs (requiring only pickup trucks and wheelbarrows) are estimated far too high with inflated equipment costs. Subsequently, you do not get these jobs in a competitive market, because your price for the job is too high.

- Jobs that are extremely equipment-intense are charged too little for equipment costs. Because the bids are underpriced, you get these jobs—and you keep getting these underpriced projects. The result is that you are using all of your equipment, but you are charging your customers for only a fraction of its actual cost.

Two things need to happen in order to correct this scenario:

A. Equipment bid into every job should be bid the same as you would bid labor.

Equipment should be included in bids only as the job requires.
ACTUAL USAGE, in hours, should be multiplied by a predetermined cost per hour (CPH) figure.

B. Equipment usage must be MONITORED through job costing on a job-by-job basis. This will ensure that equipment costs bid into specific jobs is compared to equipment costs actually incurred on the job.

2. THE SUGAR DADDY HIGH

A commercial landscape and irrigation contractor and a large home builder developed a close relationship. The home builder provided the landscape irrigation contractor over one million dollars of work a year.

This contractor was the envy of other landscape and irrigation contractors in his market. The pricing of the work was reasonable; the builder paid the contractor within ten days of being invoiced. Yet, the contractor managed to go broke. Why? For a couple of reasons:

A. There were no INTERNAL CONTROLS.

Jobs were bid accurately and competitively, but they were not job costed. There was no effective planning or quality control in the field. The crews were not directed properly, and the clients' problems were not addressed quickly and effectively. As a result, jobs would drag on as the crews did not develop a sense of urgency to complete the jobs.

The company kept digging itself a larger and larger hole as it scrambled to **"rob Peter"** (bill new work) to **"pay Paul"** (pay off old bills for jobs completed six to twelve months prior).

B. The landscape and irrigation firm became a "CAPTIVE" subcontractor.

It was subtle at first, but in the end, the consequences were inevitable. Eventually, the contractor relinquished control of his company to the builder.

The builder began to dictate schedules and precluded the landscaper from working for other clients. The home builder began to ask the contractor to do a few free "favors" (i.e., landscape his home, his secretary's home, his purchasing agent's home, a local church, etc.). The home builder also did not expect to be charged for legitimate extras.

When the economy in that area began to go into recession (in 1990) and the home builder had no work to give to the landscaper, the landscaper, who had not pursued other work, had no other clients to turn to. His cash flow stopped, but past-due payables and payroll taxes did not. Because of his entrenched bad habits, he was unable to turn his business around. He skipped town owing over $145,000 to suppliers, the IRS, etc.

My point is this: Do not get intoxicated on a "sugar daddy high." It is not a question of **IF** you will lose your sugar daddy but **WHEN.**

3. TOO MUCH CAFFEINE (OR TESTOSTERONE)

A few years back, a hard-charging, concrete contractor in the Southwest was feeling pretty good about having a year that saw $1.5 million in gross sales with a net profit of well over 10%. He decided to "put the pedal to the metal," so to speak, and grow even more.

He had office help, but he decided that he needed a full-time estimator. So he hired one—one with very little experience. Over the next twelve months, his sales increased to $2.5 million. Unfortunately, his bottom line went from a 10% net profit for the year to a 10% loss even with the increased sales.

What went wrong? I boiled it down to three main items.

A. Bids were not REVIEWED.

The new (and unproven) estimator was allowed to bid work without the owner or someone else reviewing his work. This cost the company about $150,000.

B. Field COMMUNICATION and AUDIT TRAILS were not in place.

People in the field made decisions and changed the product without proper approval or documentation from inspectors or the owner.

One retaining wall had to be replaced and this cost the contractor over $50,000. Other similar mistakes cost another $40-50,000.

C. Proven SYSTEMS and office STAFF were not in place prior to growth.

This company could just about handle the pace of $1.5 million in gross annual sales. There were problems (the flow of paperwork would become congested, job costing would be late or not done at all, change orders would not be adequately documented, etc.), but these problems were not insurmountable.

Unfortunately, the owner did not try to resolve these problems in his systems before he decided to increase sales. Management was soon overwhelmed. Lack of documentation, litigation, and disorganization cost another $150,000-200,000. The owner (and the company) never did recover.

Simply put, the owner did not manage either himself or his company. He thought that he could run his company on adrenaline (of which, by the way, he had plenty).

The moral of the story is: "You take care of the systems, and the systems will take care of you." Or, "You take care of the business, and the business will take care of you." If you don't, it won't.

4. THE VAMPIRE SYNDROME

It would be so easy if all you had to do was to put bureaucracy in place (systems and procedures) and then watch things take off.

There is, however, one added ingredient: people.

I'm convinced that some people, given the chance, could destroy a McDonald's fast food restaurant. They would change the menu, eliminate the hamburgers, fire the help. They would find or invent some way to make it fail. Show them a hundred times how to do it right, and they would change it and do it wrong.

Vampires are like that. They love to live in the dark. If you show them the light of day, they immediately run back into the dark and back to their habitual coffins. And a coffin is simply a rut with ends installed.

I'm convinced that some contractors are so ego-driven that they eventually deceive themselves. They live in a "coffin" of sorts. The very thought of implementing good procedures and sound systems (the "light of day") makes them run for cover (the dark). No amount of coaching can draw them out of their cavernous lifestyles, and, unlike bats, they cannot see in the dark.

5. EGOMANIA

A rather large landscape management and construction company self-destructed. At its height, this firm employed upwards of 400-500 people. In a well-publicized merger, the company almost doubled in size overnight. This company showed up in markets everywhere.

In their eyes, top management in this company could do no wrong. Marketing, advertising, and obtaining market

share became paramount. Size went to their heads, and they began to believe all the attention being paid to them (attention that was primarily the result of their own marketing efforts).

Three years later, they disintegrated. Why?

Image and marketing (a three-piece-suit-and-tie mentality) became the focal point of the company, BUT the systems necessary for the DAILY DIRECTION and CONTROL of the company were not in place nor were they developed. Estimating was not well-developed; job-costing reports for specific jobs and crucial financial reports were never available. Some employees went for almost a year without knowing who they worked for or to whom they should report.

In essence, top management did not understand that tried-and-proven systems, combined with bureaucrats and a bureaucracy forced to focus upon supporting field operations, **ARE THE HEART** of any good landscape and irrigation construction or services operation.

Forget "image" and the marketing of that image if you do not have good systems. Without the proper systems, you will only dig a deeper hole in which to bury yourself.

6. LOOKING FOR "FOXES" IN ALL THE WRONG PLACES

A large landscape construction and services company ($4-5,000,000 in gross annual sales) spent almost four years trying to get an accounting and job-costing system into place. They went through a couple of software programs and as many comptrollers and bookkeepers.

They just could not get it together, and the recommendations of competent consultants always fell on deaf ears. The delay cost them hundreds of thousands of dollars—most noticeably in the field. The estimator could produce good bids, so

Fixin' the Right Problem

that was not the problem. Unfortunately, the field was incapable of bringing in a job on-budget or on-schedule.

Everyone knew there was a problem (*multiple* problems), but no one would listen to or implement any viable solutions. Everyone "hopscotched" around the real problem. Activity, not results, became the focus.

Field production turned into a *Three Stooges* scene because there was not a well thought-out game plan. Things never improved because there was no job costing to identify specific problems. Labor-hour budgets were not clearly spelled out to field personnel, nor was there any timely, accurate feedback to management or field crews—no scoreboard.

Because of all the confusion and lack of controls that would indicate if jobs were coming in on budget or not, it appears that the field superintendent was able to take advantage of the situation. He "ghosted" the company (put fictitious people on the field payroll) to the tune of about $50,000.

Things went from bad to worse. Wrong solution followed wrong solution in an attempt to fix the problems. It was like the chicken farmer who had a fox in his chicken coop. The farmer's solution was to throw more chickens into the coop.

And when this didn't work, he decided that what he needed was a larger chicken coop ("let's see... if we move this wall out to here, add another down there...buy some more chickens...").

Like the saying goes, "Some days chicken, some days feathers."

7. IN GOD WE TRUST (ALL OTHERS USE PURCHASE ORDERS)

A landscape contractor was grossing approximately $500-600,000 a year in the residential and commercial installation market.

Purchase orders (the PAPER TRAIL that identified who was ordering what for which particular job) were not used; neither was a system for job costing in order to compare bid-to-actual material costs. Invoices from suppliers were paid without proper documentation.

"Blind faith" replaced the CONTROL that a sound system of checks and balances (purchase orders) would have provided.

Of course, abuse was almost inevitable. Key (previously good) employees ran materials for "side" jobs through the company and embezzled approximately $40,000. I'm convinced that had purchase orders been used properly, this problem would have been prevented. The temptation was just too much for employees to resist.

Remember: "A lock is not meant to keep a thief out. It is meant to keep an honest man honest."

CONCLUSION

It is my hope that these common mistake scenarios will provide you with some added insight into the management quagmires into which some people place themselves. To the adage, "Experience is the best teacher," we should add, "Preferably, someone else's."

Section IV

KEEPING ON TRACK
Running Your Company from the Bid Sheet

PURPOSE
**To explain some of the systems and methodologies necessary to
direct and control a Landscape and Irrigation Company**

Just about anyone can produce a bid (i,e., put a price on a job). However, the challenge is not merely to put the price on the job, but to put the job in the ground and to put money in your pocket while you are doing it. To do so, you need to **constantly monitor two things:**

1. INDIVIDUAL JOBS

You must compare budget-to-actual performance on a weekly (I prefer daily) basis.

2. THE COMPANY/DIVISION AS A WHOLE

You must compare yearly and monthly budgets with their respective actual performance data. In essence, you should view your yearly budget as one big job for the whole company or for that particular division.

KEEP IN MIND: As you build the systems which form the backbone of your company, remember these crucial concepts:

1. Timely, accurate feedback (TAF) is powerful and dynamic.

   TAF is the best motivator for changing and improving performance—any performance.

2. The main purpose of the systems in your company is to collect data, format it in easily communicated reports, and get it in front of the people who need it. That is what TAF is all about.

3. Everyone in the company must think in terms of clear, precise, easily communicated written GOALS.

   Crystallize your thinking and set objective targets and standards throughout your operation. Remember, a goal well-defined is a goal half-achieved. Clearly defined goals are self-fulfilling prophecies. Fifty percent of the battle is won once you clearly define and commit to writing the goals that you desire to achieve.

The following chapters contain some simple methods and tools which should help you to monitor and control your company, division and individual jobs. Adapt them to your operation as needed.

CHAPTER 19
SALES

PURPOSE

To explain various methods of monitoring sales performance

Annual budgets establish your sales goals for the year. It is necessary, however, to MONITOR your "budget-to-actual" sales performance during the course of the year.

SALES CATEGORIES

Sales should be broken down into three categories:

1. Contracts signed
2. Work installed and billed
3. Work installed but not billed

TOOLS FOR MONITORING YOUR SALES

1. THE BID BOARD—see Diagram (19-1)

A. Purpose

The Bid Board displays and tracks much of the process of bringing business into the company. It helps you to gain more CONTROL over and to ACCELERATE the bidding process. In turn, the bid board will motivate your people by providing effective feedback and vital, continuous information.

Without a doubt, it can be your most powerful tool. You will obtain a grasp on what I call the "faucet," or the handle, that controls your company. Once you firmly grasp the handle, you can then turn it on or off in order to control the flow. Unfortunately, most contractors never find the handle.

The Bid Board is designed to FOCUS your attention at crucial pressure points in the process of obtaining new work (jobs). It helps you to quantify and to measure the dynamic involved in this process and it establishes goals relating to the sales process.

Your management attention (or energy) can then be focused where it needs to be, when it needs to be.

B. Construction

Start with a 3′ x 4′ dry-ink erasable, white board (available at an office supply store). Use a black dry-ink marker to fill in most of the board. Use red sparingly. If you clutter it up with other colors, it will lose its effect. Make entries in neat, one-inch high, block-style lettering.

C. Use

   (1). Potential New Business

Enter in column (1) the names and telephone numbers of all possible leads, contacts, from whom you might obtain sets of plans for bidding, etc. List in this column:

- Developers
- Owners
- General Contractors
- Government Agencies (city, state, federal, DOD, DOT, etc.)
- Dodge Reports, green sheets, plans rooms
- New projects coming up
- Geographical areas/markets to tap into, etc.

If it might lead to more work, put it here.

The information is right there and at eye level. You know immediately where to look to get new ideas to find new work, because the information is right in your face.

If an item is really HOT and you want to pursue it aggressively, get out the red marker and put an asterisk (*) beside it. Mark only five or six items with red, or the impact will be greatly diminished.

Rarely will you ever erase an item from column (1).

   (2). Projects in Office to be Bid/Designed

Column (2) is for sets of plans already in the office and ready for bidding and for plans that need to be designed and then bid (usually residential ones). Enter the PROJECT or client's NAME here.

Strive to keep 5–10 projects in this column at all times. If you need more work, increase this number to 12–15 (if possible).

The important thing is that you begin to see the natural flow of how to bring business into your company and see that you can often turn it on or off by taking some well-defined course of action.

Once you have designed and bid the project, erase it from column (2) and put it in column (3).

   (3). Bids/Designs Completed and Awaiting Award

Column (3) lists all projects for which you have submitted bids and have yet to be awarded. Include the project NAME, the dollar AMOUNT, and any REMARKS (dates, names, phone numbers) that you feel necessary.

   (4). Keep a running total of bids pending on the "TOTAL" line.

I like to keep as many projects as possible in this column at all times. You might want to put these projects on a computer spread-sheet and tape it to this section of the Bid Board. I actually tape the BID STATUS REPORT (Diagram (19-2)) here and update it continually.

   (5). GOAL

The GOAL identifies the dollar amount of bids that should be pending at any one time. This figure may be what you "feel" is necessary to keep enough work coming into the business, or it may be a more precise calculation.

For instance, if your sales goal is $500,000 for the year, and if you get roughly one out of ten jobs that you bid (10% or .10), divide $500,000 by .10 and the result is $5,000,000. In other words, you need to bid about $5,000,000 of work that year in order to get your $500,000. My residential clients get 50-80% of everything they bid. They would then need to bid $625,000 to $1,000,000 in order to get their $500,000 for the year.

Determine the TAT (turnaround time). Guesstimate how long it is from the time you turn in a bid until you find out if you have the job. This may be one week (for residential work) to four to five months or longer (for commercial work).

Divide the goal for the year by the number of TAT periods in a year. Using a TAT of one month, there are

THE BID BOARD

POTENTIAL NEW BUSINESS	PROJECTS IN OFFICE TO BE BID/DESIGNED	BIDS/DESIGNS COMPLETED AWAITING AWARD		
		PROJECT	REMARKS	AMOUNT
(1)	(2)		(3)	

TOTAL $ _____

GOAL $ _____

DIFFERENCE $ _____

FY-9 _____ SALES GOAL $ _____

CONTRACTS WON $ _____

ADDITIONAL SALES NEEDED $ _____

WORK COMPLETED & BILLED $ _____

ADDITIONAL WORK TO COMPLETE & BILL $ _____

36"

48"

DIAGRAM 19-1. THE BID BOARD

		BUDGET	ACTUAL	VARIANCE			BUDGET	ACTUAL	VARIANCE	
	SALES BUDGET FOR YEAR ------------>	$1,000,000	$150,000	$850,000	: BIDS PENDING GOAL --->		$512,821	$50,000	$462,821	: PERCENTAGE OF BIDS WON-----> 15%
	FIELD HOUR BUDGET FOR YEAR ---->	20,000	3,000	17,000	: WORK BILLED ------------->$1,000,000			$75,000	$925,000	: AVERAGE BID TURN-AROUND-TIME-> 4.0
	AVERAGE SALES PER FIELD HOUR --->	$50.00	$50.00	$0.00	: ADDITIONAL WORK NEEDED TO COMPLETE/BILL--->$925,000					(IN WEEKS)

Codes:
1-PENDING
2-LOST
3-WON

#	BID#	#	GC/OWNER	PROJECT	DATE	CONTACT	PHONE	BID HRS	AMOUNT BID	PEND-1	LOST-2	WON-3	BILLED	BACKLOG	HRS YTD	SPH	OH-%	PRO-%	AVE.	AVE.	REMARKS
SUB :								5,000	$250,000	$50,000	$50,000	$150,000	$75,000	$75,000	3,000	$50.00				: XXXXXXXXX
%								%	100.0%	20.0%	20.0%	60.0%	14.6%	14.6%	0.3%	XXXXXX	20.2%	10.0%		: XXXXXXXXX
1	001	1	ABC BLDRS.	AVALON GARDENS	8/10	BEN HADD	619 967-1881	1,000	50,000	50,000					0	50.00	17.0%	10.0%			:
2	002	2	XYZ CO.	PACIFIC CREST	8/14	I. M. DUNN	714 954-1244	1,000	50,000		50,000				0	50.00	22.0%	10.0%			:
3	003	3	WALT HOMES	SWEET GLEN	8/17	B. GONE	619 723-0815	1,000	50,000			50,000		50,000	0	50.00	24.0%	12.0%			:
4	004	3	HUNT BROS.	ROLLING HILLS	8/23	HARDLY HEER	619 941-1479	1,000	50,000			50,000	25,000	25,000	1,000	50.00	19.0%	8.0%			:
5	005	3	J&G BLDRS.	LOOKOUT POINT	8/28	DUSTY LANE	619 892-4156	1,000	50,000			50,000	50,000	0	1,000	50.00	19.0%	10.0%			:
6								0	0	0	0	0	0	0	0						:
7								0	0	0	0	0	0	0	0						:
8								0	0	0	0	0	0	0	0						:
9								0	0	0	0	0	0	0	0						:
10								0	0	0	0	0	0	0	0						:
11								0	0	0	0	0	0	0	0						:
12								0	0	0	0	0	0	0	0						:
13								0	0	0	0	0	0	0	0						:
14								0	0	0	0	0	0	0	0						:
15								0	0	0	0	0	0	0	0						:
16								0	0	0	0	0	0	0	0						:
17								0	0	0	0	0	0	0	0						:
18								0	0	0	0	0	0	0	0						:
19								0	0	0	0	0	0	0	0						:
20								0	0	0	0	0	0	0	0						:
21								0	0	0	0	0	0	0	0						:
22								0	0	0	0	0	0	0	0						:
23								0	0	0	0	0	0	0	0						:
24								0	0	0	0	0	0	0	0						:
25								0	0	0	0	0	0	0	0						:
26								0	0	0	0	0	0	0	0						:
27								0	0	0	0	0	0	0	0						:
28								0	0	0	0	0	0	0	0						:
29								0	0	0	0	0	0	0	0						:
30								0	0	0	0	0	0	0	0						:
31								0	0	0	0	0	0	0	0						:

DIAGRAM 19-2. THE BID STATUS REPORT

twelve TATs in a year. When we divide $5,000,000 by 12, we get $416,667. We need to have $416,667 worth of bids pending at any one time to meet our sales goal for the year.

If our TAT is two weeks, we would divide $5,000,000 by 26 TAT periods in one year. The result would be that we need $192,308 worth of bids pending at any one time.

a. Seasonal Work

The above example assumes that your work is spread evenly throughout the year. This is hardly ever the case. This is how you would handle seasonal business:

Determine the "SALES GOAL" and then the amount that you need to bid.

Establish monthly sales and amount-pending GOALS for the upcoming season, starting with the month of September of the present season.

b. Target specific goals by certain dates.

Initially, you will not necessarily know how much to shoot for, but this is not important. Just bid as much as you can, track it, and get a feel for how much work you have to bid in September in order to get a certain backlog by March 1st of the following year.

Strive to meet or to exceed your goal. Adjust it as you acquire a feel for the process.

After you monitor this dynamic for 6-12 months, these numbers will make a lot more sense to you, and you will begin to wonder why you did not track them sooner.

(6). "DIFFERENCE"

Subtract the "PENDING TOTAL" from the "GOAL."

If you are below your goal, enter this sum on the "DIFFERENCE" line in **RED** on the board. We want that red number to reach out and slap you in the face to motivate you to get more work.

(7). FY___ SALES GOAL

Enter the yearly sales goal in the "FY___ SALES GOAL" portion of column (3).

(8). CONTRACTS WON

Fill in the "CONTRACTS WON" section and subtract it from the "FY___ SALES GOAL" in order to determine "ADDITIONAL SALES NEEDED." Again, use **RED** if this number is below your goals.

(9). WORK COMPLETED & BILLED

Fill in the "WORK COMPLETED & BILLED" section. Subtract it from the "FY___ SALES GOAL." Enter the difference (in RED, if necessary) in the "ADDITIONAL WORK TO COMPLETE & BILL" section.

D. Motivation

Used properly, the Bid Board can be one of your most powerful motivational tools. It is recommended that:

- You have one per division, if appropriate.
- If you have salespeople, have one board for each person.
- Make sure it is filled in and kept up-to-date **at all times.**

  This allows you to monitor the goals of your salespeople and provides a "report card" for a very crucial part of your business.

  It can greatly facilitate COMMUNICATION and TRAINING. You can help your staff set priorities and focus their energies as you review each board (and progress) throughout the year.

- You (the owner) should be the only person allowed to mark the "hot" items with the red marker on the board.

  This one thing could focus and concentrate your staff on those items that you, the owner, deem important.

- Place the Bid Board in a strategic location.

  Out of sight leads to out of mind. Place the Bid Board in an area that is at eye level. Put it within a few feet

SMITH HUSTON INC.

FROM/CO: _____   **DATE**: _____ / ____ / _____

SUBJECT: MONTHLY AUDIT/DATA **(MAD)** REPORT FOR MONTH OF _____

1. FINANCIAL INFORMATION:

		(MONTH)	(Y-T-D)
A. SALES:	CONST:	_____	_____
	MAINT:	_____	_____
	_____ :	_____	_____

B. PAYABLES: _____   _____   _____   _____   _____

C. RECEIVABLES: _____   _____   _____   _____   _____

           (CUR)   (30)   (60)   (90)   TOTAL

D. FINANCIAL RATIOS:

(1). TOTAL RECEIVABLES + TOTAL PAYABLES   _____ + _____ = _____

(2). CURRENT RECEIVABLES + TOTAL PAYABLES   _____ + _____ = _____

(3). QUICK RATIO: $\dfrac{\text{CASH + RECEIVABLES}}{\text{CURRENT LIABILITIES}}$ = _____ + _____ = _____

2. PRODUCTION (COST-OF-SALES):

		(MO-$)	(MO-HRS)	(YTD-$)	(YTD-HRS)
A. LABOR:	CONST:	_____	_____	_____	_____
	MAINT:	_____	_____	_____	_____
	_____ :	_____	_____	_____	_____
B. MATERIALS:	CONST:	_____	_____	_____	_____
	MAINT:	_____	_____	_____	_____
	_____ :	_____	_____	_____	_____

C. PRODUCTION RATIOS:	(MO-SPH)	(MO-AW)	(YTD-SPH)	(YTD-AW)
CONST:	_____	_____	_____	_____
MAINT:	_____	_____	_____	_____
_____ :	_____	_____	_____	_____

		(MO)	(YTD)
D. MAT/LABOR RATIO:	CONST:	_____.____ :1	_____.____ :1
	MAINT:	_____.____ :1	_____.____ :1
	_____ :	_____.____ :1	_____.____ :1

		BUD	ACT	VAR
E. GPM/HOUR:	CONST:	$____.____	$____.____	$____.____
	MAINT:	$____.____	$____.____	$____.____
	_____ :	$____.____	$____.____	$____.____
F. OVERHEAD/HOUR:	CONST:	$____.____	$____.____	$____.____
	MAINT:	$____.____	$____.____	$____.____
	_____ :	$____.____	$____.____	$____.____

SHI FORM: 08-B(1)

DIAGRAM 19-3A. THE "MAD" REPORT

3. **ESTIMATING**:

	CONST.	MAINT.	OTHER

A. SALES:

    (1). BUDGET FOR YEAR: _____ _____ _____
    (2). CONTRACTS SIGNED Y-T-D: _____ _____ _____
    (3). ADD'L WORK NEEDED: _____ _____ _____
    (4). WORK INSTALLED & BILLED: _____ _____ _____
    (5). WORK INSTALLED & NOT BILLED: _____ _____ _____
    (6). ADD'L WORK TO INSTALL & BILL: _____ _____ _____

B. FIELD LABOR HOURS (BILLABLE):

    (1). BUDGET FOR YEAR: _____ _____ _____
    (2). ACTUAL Y-T-D: _____ _____ _____
    (3). VARIANCE Y-T-D: _____ _____ _____

C. BIDDING:

    (1). # SETS OF PLANS IN OFFICE TO BID/DESIGN ON LAST DAY OF MONTH: _____
    (2). # JOBS BID BUT NOT AWARDED YET: _____
    (3). $ AMOUNT FOR JOBS BID BUT NOT AWARDED: _____
    (4). AVERAGE SIZE OF JOBS BID (3 + 2): _____

4. **MANAGEMENT**:

A. FIVE (5) MAIN CHALLENGES FACING COMPANY NEXT MONTH:

    (1). _____
    (2). _____
    (3). _____
    (4). _____
    (5). _____

B. FIVE (5) MAIN COMPANY GOALS FOR NEXT MONTH:

    (1). _____
    (2). _____
    (3). _____
    (4). _____
    (5). _____

_____
Signature

SHI FORM: 08-B(2)

DIAGRAM 19-3B. THE "MAD" REPORT

of the desk where the person using it will be sitting and looking right at it during the course of the day. It can motivate the dickens out of your people.

It will focus energy and enhance communication by sharing vital information and priorities with your staff.

2. THE BID STATUS REPORT—see Diagram (19-2)

The Bid Status Report is a computerized spreadsheet version of the key elements on the Bid Board.

Steve Smith and I developed the Bid Status Report because we were running out of room on the Bid Board for all of our projects awaiting award. We were weary from making all those continuous changes on the Bid Board and knew that there had to be a better way to track the needed information. Hence, the Bid Status Report was born.

Many of our clients like the Bid Status Report because it displays so much useful information and analysis.

- It is a compact, orderly, and convenient record of every project that you ever bid.
- If kept current and used properly, it is your most dependable means of tracking bids.
- It automatically adjusts to the flow of new information. It will accurately recompute and fill in the appropriate columns (bids pending, lost, won, etc.).

The Bid Status Report is a marvelous tool. If you are interested in purchasing a copy of it on diskette, contact our offices.

3. THE MONTHLY AUDIT DATA (MAD) REPORT—see Diagram (19-3)

This is a monthly, systematic tracking report that charts the process of bringing sales into your business.

This is solid statistics. It is particularly useful for identifying key trends over a long period of time (6-18 months) and for analyzing information in more seasonal locations—identifying and comparing trends and backlogs, comparing bids outstanding to sales, etc.

Many of our clients appreciate its value even more as they encounter periods of economic downturn, because it provides solid numbers, versus anxious feelings, upon which to base business decisions.

The MAD Report is also available on diskette through our offices.

4. THE SALES "BUDGET VS. ACTUAL" GRAPH

Diagram (19-4) depicts your gross sales budget compared to your actual sales and completed work that is billed. You should update this graph weekly. The difference between actual sales and work completed and billed is your work backlog. Used in conjunction with the Work Backlog Graph, it can provide some excellent insight into the results of your sales efforts for the year.

5. THE WORK BACKLOG GRAPH—see Diagram (19-5)

The Work Backlog Graph is a simple chart that you can fill in weekly. It graphs both your backlog goal and actual performance.

CONCLUSION

Establishing sales budgets (goals) and monitoring your progress towards achieving them are essential management tasks. Information properly formatted and collected in a timely manner can provide crucial diagnostic insight into the process of bringing sales into your company. It can also focus management energy and subsequently motivate staff to take corrective action if necessary.

ACTION POINT

Review your methods of tracking your company's sales performance and incorporate the items discussed in this chapter as deemed appropriate.

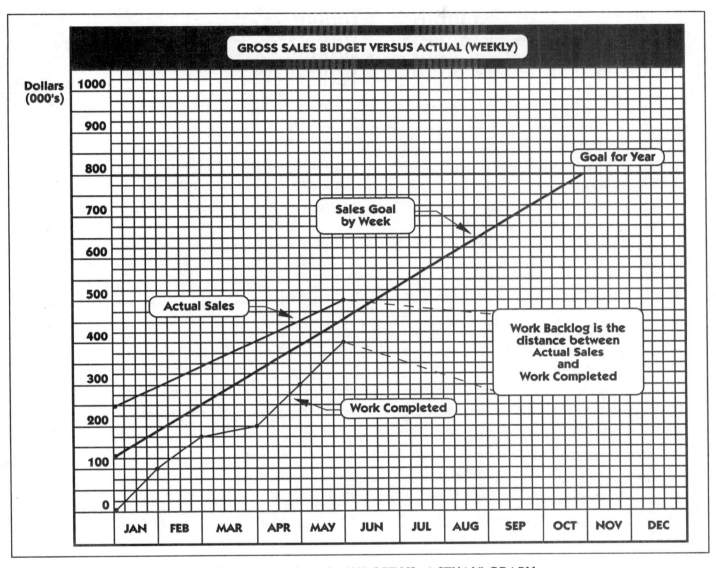

DIAGRAM 19-4. SALES "BUDGET VS. ACTUAL" GRAPH

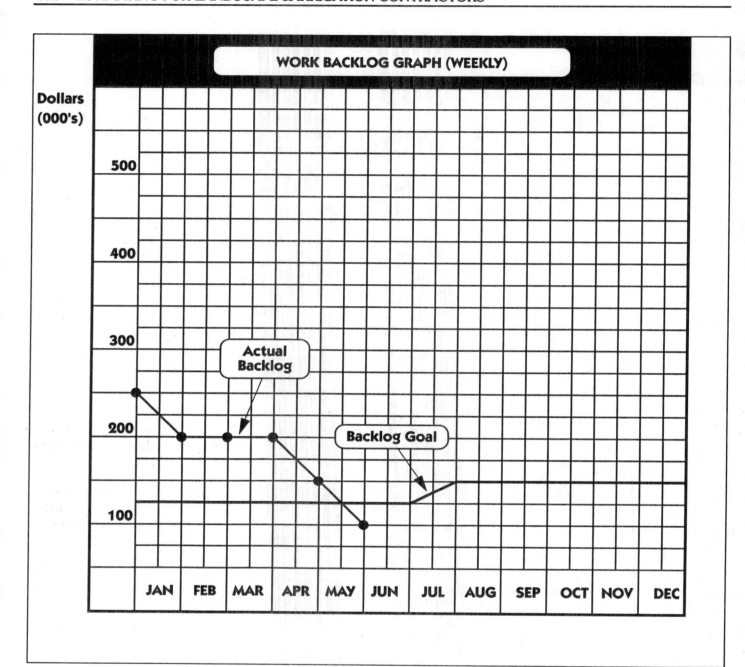

DIAGRAM 19-5. THE WORK BACKLOG GRAPH

CHAPTER 20
JOB COSTING

PURPOSE

To explain the purpose and mechanics of a job-cost system and the steps necessary to start such a system

"I just don't know if I'm making money on this job."

That is the most common complaint that I hear from contractors. Their profit and loss statement may tell them that they are making money, but they usually do not know where. Worse yet, they do not even know where to start to look in order to find an adequate answer for their question.

The most important report for your bookkeeper to produce is a weekly (I prefer daily) job-cost report for each job in progress. Effective job costing (in combination with proper planning and scheduling) can turn jobs that are losers into winners. I have seen it happen time after time: **proper planning** and **preparation produces profitable projects.** Remember those **six Ps.**

If bidding produces your pre-game strategy, it is job costing that keeps score during the game—tells you where you are along the way. Think just how **measured** professional sports have become. Baseball is a game based on data collection, analysis, and comparison. Batting averages, ERAs, RBIs, etc. are all the result of "job costing," if you will.

A professional football game is an even more measured event. Goal lines, sidelines, hash marks, yard markers, goal posts, end zones; 1st, 2nd, 3rd, and 4th downs; offense/defense; quarters, time-outs, two minute warnings and the 40-second clock make up just the playing field. Statistics are kept on everyone, including the coach. There are even guys on the field in "zebra" suits throwing yellow hankies in the air and blowing whistles when someone's play is not up to the accepted standard.

MY POINT IS THIS: professional sports use the same type of information management methodology as should a landscape and irrigation contractor. Standards or performance targets are identified. Performance data is collected and compared to the standard. A variance is determined, and corrections are made to performance in an attempt to improve it. Coaches play the game and adjust their playing strategy based upon data (the scoreboard) that is collected and analyzed. If football coaches played the game of football the way that 99% of the landscape and irrigation contractors ran their companies, you would see 22 guys beating each other's brains out on a 40 acre sea of green. No chalk lines, goal posts, boundaries, referees, time clocks, etc. No standards or data collected and analyzed. Not even a scoreboard—just 22 men going through a lot of pain and wondering in which direction they should go.

Sound familiar?

Job costing is simply the process of measuring and comparing actual field performance to a predetermined standard or sets of standards included in your bid. Just like in professional sports, it can be a very clear-cut, analytical process. If done properly, job costing can provide the objective foundation necessary upon which to build a total quality management (TQM) program in your company. Profits, labor productivity, and management controls will all be maximized when job costing is properly put in place. For a contractor, job costing is the process of comparing direct costs: material costs, field-labor hours (and dollars), equipment hours (and dollars), and subcontractor costs bid into a specific project to the actual hours and dollars expended on these same categories.

Good job costing starts in the bidding process. Unless you establish appropriate, clearly-defined standards in the bid, you have nothing against which to compare actual field performance. There are four specific cost targets that need to be identified and quantified for each job in the bidding process. They are as follows: (1). Material costs to be used on the job (i.e., fertilizers, mulch, etc.). (2). Labor identified in terms of hours estimated to be used on the job as well as the total payroll dollars to pay for those hours. (3). Equipment hours for each piece of equipment to be used on the job (i.e., specific types of mowers, tractors, pickup trucks, string trimmers, etc.) and their corresponding cost per hour (CPH). (4). Subcontractor costs.

The cost targets and/or standards identified in the bidding process are the specific amounts of materials, labor, equipment and subcontractor costs required to *safely* meet the performance standards identified in the plans and specifications for the job being bid. Subsequently, quality control and safety measurement become integrated with job costing. Because we have reduced our performance and safety standards and have translated them into quantifiable, measurable cost standards, the job-costing process is now possible.

Job costing is best accomplished through a computerized accounting system. Otherwise, the appropriate checks and balances and controls are not in place to ensure that field-costing data is accurately allocated to the correct job. Job costing through your accounting system also eliminates duplication of work. I have seen contractors use a computer accounting system for their financial statements and a spreadsheet program for tracking their jobs. The two systems hardly ever agree, and a minimum of twice the labor is required. And all of this effort obtains questionable results. Unfortunately, accounting programs that have adequate job-cost modules are expensive ($5–6,000) and are not easily nor quickly brought on line.

Field payroll should be generated from the data that you collect to produce job-cost reports. If you have two separate systems for collecting this data, payroll will be accurate and job costing will not. Combine these two systems, eliminate redundancy, and improve the accuracy of the data being collected.

TYPES OF DATA TO IDENTIFY

The types of information that a job-costing system should provide and the format for this information are crucial. Whether manual or computer-generated, from an accounting program or a spreadsheet, you need the following direct cost information for each of your jobs in progress at least once a week.

1. MATERIAL COSTS

Job-cost reports should identify all materials used on a job. As much as is practical, vendor invoices should include materials used on one job only, and the name of the job should appear on the invoice to facilitate tracking purposes. Purchase orders or some sort of issue slip should be used for in-house inventory items used on projects. Otherwise, these items (i.e., bulk mulch, sprinkler repair parts, in-house grown and bulk purchase nursery materials, etc.) will not be charged to jobs accurately, and profit margins will appear erroneously high for the jobs in question.

2. FIELD-LABOR COSTS

It is important to identify labor hours and the actual monetary cost of those hours (if possible) and to compare them to your budget. If you do not track and compare both, you might bring the project in on-budget for hours but be over-budget in dollars. This would happen if the crew actually performing the work on the job was comprised of higher-priced labor than that which was used to bid the project.

Many maintenance contractors format their job-cost reports to identify labor-hour targets to be achieved for each crew visit to the site (i.e., not more than 12 hours should be expended per visit, or 168 hours by the 14th visit of the season, for a particular site). Although not always possible at first, labor functions with budget-to-actual targets should eventually be identified on job-cost reports. For the job above using a three-person crew, your budget for each visit might include 45 minutes drive time, 15 minutes to unload/load equipment, three hours for a 48" mower, four hours for a 21" mower, and four hours for trimming and edging. If done this way, you will simultaneously capture equipment as well as labor hours.

3. EQUIPMENT

Identify equipment targets just as we did in the labor category above. Hours accounted for should only include equipment "meter" or running time. The only additional item to be accounted for is the truck used to transport the crew. However, track and combine its meter or running time (drive time) and its curb time when it just sits at the site with its engine turned off. Convert equipment hours into dollars by multiplying the hours charged to a job by your CPH for each piece of equipment.

4. SUBCONTRACTORS

Just like materials, subcontractor costs should be identified and charged to jobs by means of specific invoices for each job. Once the above four categories of information are converted to dollars in your accounting job-cost system, you only need to add your field-labor burden percent (FICA, FUTA, SUTA, WCI, holiday and vacation pay, etc.) to your field labor to identify your actual gross profit margin (GPM). This assumes that sales tax has been included in with material costs. Any good accounting job-cost software package will allow you to accomplish the above and to compare these costs to the same categories in your original bid.

Although simple in concept, effective job costing is more a function of the type of computer software (technology) and the skill levels of both office and field staff (training) needed to collect, assimilate, format, and distribute timely and accurate job-cost reports. You will not be able to obtain timely and accurate reports with all of the information described above unless you have a rather sophisticated computerized accounting job-cost program. Even if you have the software, without properly trained staff, you will be attempting to "push a rope" as far as job costing is concerned. Keep in mind that it takes 6–12 months to implement this type of sophisticated system (assuming that you have the software). However, there is an alternative.

A CONTRACTOR'S BIGGEST CHALLENGE

The biggest challenge faced in contracting and which threatens profitability more than any other factor is field labor. Equipment follows at a distant second place. Tom Peters hits the nail on the head when he says in *Thriving on Chaos,* that, "What gets measured, gets done (controlled)."[7] Contracting is labor. You win or lose depending on how you train and handle (control) people in the field. Materials and subcontractors present minimal risk by comparison.

Consequently, I believe that 90–95% of the benefit of job costing can be realized by obtaining timely and accurate feedback (TAF) concentrating on field-labor and equipment hours only. Don't get me wrong, I would like to have all of the other information that we have discussed collecting. However, we do not always have the ability to do so.

As an interim measure (one that any contractor can implement immediately), simply collect field-labor and equipment hours and compare it to your bid hours in the format outlined below. Preferably, do it in a computer spreadsheet program. If you cannot do so, make a manual form using the format shown. Fill out or print the results daily (if possible) but at a minimum, once a week. Provide each foreman and/or crew leader with a copy of the report as soon as it is made. Get the information (TAF) into the hands of the people that can make the biggest difference in the operation—field crew leaders.

MANUAL CONSTRUCTION JOB-COSTING REPORT

Job: <u>Davis Residence</u>   Date Prepared: <u>2-15-91</u>

Direct Costs:	(1) Budget Hrs/$	(2) Total Used	(2÷1) (3) % Used	(1-2) (4) Balance	(5)* % Complete	(6) Materials to Order	(2+6) (7)** Total Projected Complete	(1-7) (8) Projected Over/ Under
1. Material Costs:								
A. Landscape	2000	1000	50%	1000	NA	950	1950	50
B. Hardscape	1500	1350	90%	150	NA	175	1525	-25
C. Irrigation	1450	1425	98%	25	NA	0	1425	25
D. General Conds.	200	80	40%	120	NA	80	160	40
E._____					NA			
Totals:	5150	3855	75%	1295	NA	1205	5060	90
2. Labor Hours:						Projected Hrs. to go		
A. Landscape	100	60	60%	40	65%	35	95	5
B. Hardscape	40	30	75%	10	75%	10	40	-
C. Irrigation	110	100	91%	10	100%	0	100	10
D. General Conds.	25	20	80%	5	80%	5	25	-
E._____								
Totals:	275	210	76%	65	80%	50	260	15
3. Equipment Hours:								
A. Pick-up	96	80	83%	16	NA	16	96	-
B. 1 Ton	32	24	75%	8	NA	4	28	4
C. Dump Truck	30	24	80%	6	NA	4	28	2
D. Bobcat	24	26	108%	-2	NA	0	26	-2
E. Tractor	0	0	0%	0	NA	0	0	0
*** Totals:	182	154	346%	28	NA	24	178	4
4. Subcontractors Costs								
A. I.R. Masonry	400	400	100%	0	100%	NA	400	0
B._____						NA		
C._____						NA		
D._____						NA		
Totals:	400	400	100%	0	100%	NA	400	0

* = Your estimate of % complete for that portion of project.
** = Projected amount of material costs or hours to complete project.
*** = Equipment totals are somewhat meaningless due to rate variances of equipment.

DIAGRAM 20-1. MANUAL CONSTRUCTION JOB-COST REPORT

CONSTRUCTION JOB-COST REPORTS

1. MANUAL JOB-COST REPORT

Job costing need not be very complicated. You can start with your very next job, if you can pull out labor and equipment hours from your bid, as well as your material and subcontractor costs. Diagram (20-1) is a format for normal job costing. Column (1) is for budget amounts for direct costs (material, labor, labor burden, equipment, and subcontractors). For labor and equipment, hours should be used; dollar amounts should be used for materials and subcontractors. I break down the material and labor into landscape, hardscape, irrigation, and general conditions. Column (2) depicts the total dollars or hours expended to date. Column (3) is the percentage of each category used. Simply divide column (2) by column (1) to obtain it. Subtract column (2) from column (1) to get column (4). Column (5) is your estimate of the completion percent for the job. Column (6) is the projected number of hours or dollars to complete the job and is obtained by dividing column (2) by column (5). Column (7) shows you the projected amount that you will be over- or under-budget. You can make this form more detailed if you like, but do not make it too complex, or you will torpedo your job-costing efforts.

2. COMPUTERIZED JOB-COST REPORT

Diagram (20-2) is a more complex computerized job-cost form. It can be very effective, if used properly.

I developed the sample report format one day out of frustration. It was designed in Lotus 1-2-3 and is only one of many formats that could be used. It is very basic, yet effective, if used properly. However, it is not intended to replace job costing that is integrated with an accounting software program.

Steve Smith and I were trying to bring our accounting software's job costing on line. It was taking too long, and we had a million other things to do. We needed job costing desperately because we had some rather large jobs ahead, and we felt it imperative that we track them closely. In frustration, I closed my office door, got out my laptop computer, and produced Diagram (20-2). Let me explain what I was trying to accomplish.

The job was a medium-size commercial landscape project. Bid information is located primarily on the left third of the form. Supplier information is identified in the middle of the page by vendor. I put labor hour information on the right side of the page, by date. Company-owned equipment is entered under vendors.

We did well on the job since we beat the budget by quite a few dollars and 348 labor hours, some of which were later used to do warranty work. What really made a difference was that we gave the two foremen a copy of this report daily, usually within five minutes after those foremen came into the office with their field report in the afternoon. Not only did we have a plan (a bid), but we provided constant feedback to the crews. The crews were excited because they wanted to beat the budget. We were excited because we wanted to stay in business. Only a few months earlier, this company had tackled a $200,000 project only to lose $20,000, and we could not afford to repeat previous mistakes.

This job (and all future jobs) went smoother, and we did not have to wait until the end of the job to see how it went. We could monitor our progress each and every day. Eventually, we were able to get the accounting job costing up and running, but I must admit that I still like this format better, even though it does have some limitations.

Diagram (20-3) is a "labor only" job-cost report for the Henderson residence project.

MAINTENANCE JOB-COST REPORTS

Few maintenance contractors realize the impact that job costing can have on their operation if the correct data is collected and formatted properly and provided to the right individuals in a timely and accurate manner. This is what TAF (timely, accurate feedback) is all about—empowering people. A format similar to Diagram (20-4) should be used for job costing whether done manually or by your computer accounting system. Frankly, I am amazed at how few owners and managers implement simple yet effective information feedback systems. Tom Peters in his book, *Thriving on Chaos,* states that, "The best leadership is that which insists on visible measures of what is going on in the trenches... simple, visible systems are the (foundation) of success."[8]

	BID	%	ACTUAL	%	VARIANCE
BID PRICE:	150,100	100.0%	150,100	100.0%	0
MATERIAL:	85,252	56.8%	83,696	55.8%	1,557
LABOR:	21,518	14.3%	19,127	12.7%	2,391
BURDEN:	6,886	4.6%	6,121	4.1%	765
EQUIP:	4,351	2.9%	4,234	2.8%	117
RENTALS:	0	0.0%	1,131	0.8%	(1,131)
SUBS:	5,100	3.4%	5,050	3.4%	50
SUB-TOT:	123,108	82.0%	119,359	79.5%	3,749
GEN COND:	0	0.0%	0	0.0%	0
DIR COST:	123,108	82.0%	119,359	79.5%	3,749
GR PROFIT:	26,992	18.0%	30,741	20.5%	(3,749)
OVERHEAD:	12,746	8.5%	11,256	7.5%	1,490
NET PRO:	14,246	9.5%	19,485	13.0%	5,239

BID INFORMATION

OH/FH:	$4.28	(OVERHEAD PER FIELD HOUR)
BURDEN %:	32.0%	
AVE WAGE:	$7.27	(WITHOUT LABOR BURDEN)
SPH:	$50.40	(SALES PER FIELD HOUR)
PPH:	$4.78	(PROFIT PER FIELD HOUR)
MAT/LAB:	4.0	(MATERIAL TO LABOR RATIO)

	BUDGET	ACTUAL	+/-
MATERIAL-IRRIG --->	24,528 −	24,972 =	(444)
MATERIAL-L/S --->	60,724 −	58,724 =	2,000
SUBS --->	5,100 −	5,050 =	50
RENTAL EQUIPMENT --->	0 −	1,131 =	(1,131)
TOTALS: --->	90,352 −	89,876 =	476

VENDORS	TOTAL	IRR	L/S	SUBS	RENT EQ
ACTUAL TOTALS --->	$89,876	24,972	58,724	5,050	1,131
ARTESIA	1,067		1,067		
BERGEN	20,993		20,993		
COAST	530	368	162		
HAULAWAY	102				102
HILLCREST	638		638		
HYDROSCAPE	24,897	24,604	293		
KOR ROCK	273		273		
MACK CONST	700				700
MULLIN LUMBER	10		10		
PETES ROAD SERVICE	329				329
PROGRESSIVE CONCRETE	5,050			5,050	
SANDERS HYD	2,552		2,552		
SOIL & PLANT LAB	140		140		
SOUTHLAND SOD	5,843		5,843		
SUNNY SLOPE	2,975		2,975		
VILLAGE TREES	23,779		23,779		

		RATE/HR	BID HRS	USED	TO GO
EQUIP TOTALS --->	$4,234		372	329	44
1/2 TON PICK-UP	1,068	$3.25	48	44	4
1 TON PICK-UP	191	$4.35	56	67	(11)
BOBCAT	1,005	$15.00	120	112	8
TRENCHER	1,848	$16.50	25	27	(2)
AUGER	122	$4.50			

	BID HRS	USED	TO GO	
TOT --->	2,978	2,630	348	DAYS BID: 46.5
IRR --->	1,033	985	48	DAYS USED: 41.1
L/S --->	1,945	1,645	300	TO GO ---> 5.4

IRR-NOV	L/S-NOV	IRR-DEC	L/S-DEC	IRR-JAN	L/S-JAN
640	0	345	1,027	0	618

Day	%CCT HRS			TO GO	HOLIDAY / RAIN
1	50%				
2	%CCT HRS				
3		SAT	SAT		80
4		SUN	SUN		80
5		40	40		89
6		48	48		
7	6	56	56	8	89
8		24	24	48	
9	40	SAT	SAT	55	89
10		SUN	SUN	43	98
11	SAT	SAT			80
12	SUN	SUN			24
13	48		16	80	NA
14	48		16	72	
15	48		12	84	6
16	48				
17	48				
18	SAT	SAT		54	
19	SUN	SUN		64	
20	40			71	
21	48			72	
22	48			72	
23	HOLIDAY				
24	40	SAT	SAT		EXTRA
25	40	SUN	SUN	72	
26	40	CHAS	CHAS	72	32
27	48		4	80	40
28	NA			80	
29	40		1		
30	48	SAT	SAT		
31	48	SUN	SUN		

DIAGRAM 20-2. COMPUTER CONSTRUCTION JOB-COST REPORT

BENEFITS OF JOB COSTING/TAF

The benefits of timely, accurate job-cost reports are many. Here are some:

(1). They can identify flaws in your estimating methodology. (2). Weaknesses and strengths of your field operation will become apparent. (3). TAF will provide some of the best training that your estimators will ever receive. (4). Job-cost reports are the "scoreboard" for your operation. They will provide field crews and supervisors the feedback that they need to make more timely operational decisions in the field. (5). Field crews will be empowered to take corrective action. Operational problems will be solved faster and more creatively. (6). The organization will become more "bottom-up" vs. "top-down" driven. The burden of field problem-solving will shift from management (bureaucrats) to field people, where it belongs. (7). Customer problems will be resolved faster, so customer relations will improve. (8). TQM will be simultaneously addressed and measured if TAF is established in the manner described. (9). A bonus system can be integrated with this type of TAF system. (10). The number one motivator is timely, accurate feedback. Money and praise do not even come close. Simple job-cost reports in the hands of the right people will maximize motivation. (11). Insisting that "bureaucrats" change their methods and their schedules to provide TAF to the right people will focus all of your bureaucracy, and it will teach non-field (support) personnel what "good" bureaucracy (support) is all about. If this does not excite your bureaucrats, get new ones who do get excited. (12). TAF/job costing applies to many facets of life. It will teach everyone in your company a valuable skill and lesson. (13). Last and most importantly of all, good job costing will improve your bottom line.

MAINTENANCE ILLUSTRATION

To illustrate how TAF can improve your operation, let me address a common problem that I see in the maintenance industry which also applies to construction as well.

Too many maintenance contractors bid their work using a flat hourly rate for projects billing over $500.00 per month. I do not have a problem bidding smaller maintenance projects in this fashion using the package approach covered in Chapter 15. However, if job costing was in place as outlined above, it would identify the following potential problems with this bidding method for larger projects, especially commercial ones.

First, drive time would be identified and charged to jobs. It may have to be charged to jobs as an average per day achieved by dividing total drive time by the number of jobs visited for that day. This process would more accurately identify gross profit margins (GPMs) for each job. By dividing the monthly customer invoice amount for the project by the number of labor hours actually charged to the job (including drive time) you will probably see a dramatic difference between your flat hourly rate "bid" versus the "actual" one realized in the field.

Second, you will identify actual equipment hours used on a job . Because you are charging the same flat rate for every hour that you are on the job, it is a given that you are charging the same amount for each piece of equipment on the job. And unless every piece of equipment is used exactly the same amount of time as every other piece, your equipment cost portion of the bid is inaccurate. Therefore, your original bid was either over- or underpriced. This in turn leads to problem number three.

Third, string trimmers, edgers, and small push mowers cost less per hour to operate than large walk-behinds or ride-on mowers. The **"fatal flaw"** occurs when you bid a flat rate (let's say $25 per man-hour) but you should charge $28 per man-hour for your large ride-on mowers but only $22–23 for the smaller equipment and $19–20 for labor-only functions. The market is not stupid. Over a period of time, it will start to give you more and more of the equipment-intense types of work (because you have underpriced it), but it will not give you the labor-intense jobs (which you have overpriced). Gradually you will see your bottom line deteriorate as your jobs shift towards the underpriced, less profitable equipment-intense ones. This problem becomes more acute in periods of economic downturns.

Fourth, effective job costing will show you that your estimating methods are not flexible enough to accurately identify all of your production costs during the bidding process. It should prompt you to ask questions and to seek a better method.

HENDERSON RESIDENCE

CREW SIZE:------> 4
MAN HRS PER DAY PER MAN:--> 8
DAYS TO GO:-> 0.1

DATE	IRRIG LABOR HRS	SOILPREP LABOR HRS	GRADE LABOR HRS	TREES LABOR HRS	PALMS LABOR HRS	SHRUBS LABOR HRS	SOD LABOR HRS	GRND COVR LABOR HRS	MULCH LABOR HRS	DRIVE & LOAD HRS	WARANTEE LABOR HRS	FOREMAN LABOR HRS	SUPVSR. LABOR HRS	CLN-UP & DETAIL HRS	MISC G.C. HRS	TOTAL LABOR HRS	DAYS	COMP %
TO GO------>	4.1	(0.7)	(1.7)	1.3	1.0	(1.7)	0.9	1.5	1.0	(2.0)	16.0	(2.0)	0.0	1.0	(1.0)	17.6	0.1	3.3%
BUDGET---->	182.1	14.3	6.3	37.3	13.0	69.33	14.9	29.5	12.0	75.0	16.0	15.0	15.0	23.0	8.0	530.6	14.4	100.0%
ACTUAL HRS-->	178.0	15.0	8.0	36.0	12.0	71.0	14.0	28.0	11.0	77.0	0.0	17.0	15.0	22.0	9.0	513.0	14.3	96.7%
JAN 1 (HOLIDAY)																0.0		0.0%
JAN 2																0.0		0.0%
JAN 3 (SUN)																0.0		0.0%
JAN 4	26.0									5.0		2.0	2.0	1.0	1.0	37.0		7.2%
JAN 5	26.0									5.0		2.0	2.0	1.0		36.0		14.2%
JAN 6	26.0									5.0		1.0	1.0	1.0		34.0		20.8%
JAN 7	26.0									5.0		1.0	1.0	1.0		34.0		27.4%
JAN 8	26.0									5.0		1.0		1.0		33.0		33.8%
JAN 9																0.0		33.8%
JAN 10 (SUN)																0.0		33.8%
JAN 11	26.0									5.0		1.0	1.0	1.0	1.0	35.0		40.6%
JAN 12	22.0					4.0				5.0		1.0	1.0	1.0		34.0		47.2%
JAN 13		11.0		10.0		4.0				6.0		1.0	1.0	1.0		34.0		53.8%
JAN 14		4.0		15.0		7.0				5.0		1.0	1.0	1.0		34.0		60.4%
JAN 15				11.0	12.0	3.0				5.0		1.0	1.0	1.0	3.0	37.0		67.6%
JAN 16																0.0		67.6%
JAN 17 (SUN)																0.0		67.6%
JAN 18			7.0			20.0				5.0		1.0			1.0	34.0		74.2%
JAN 19			1.0			6.0		18.0		6.0		1.0	1.0	1.0	2.0	36.0		81.2%
JAN 20						6.0		10.0	10.0	5.0		1.0	1.0	1.0		34.0		87.8%
JAN 21						11.0	14.0		1.0	5.0		1.0	1.0	1.0		34.0		94.4%
JAN 22						10.0				5.0		1.0	1.0	9.0	1.0	27.0		99.7%
JAN 23																0.0		99.7%
JAN 24 (SUN)																0.0		99.7%
JAN 25																0.0		99.7%

DIAGRAM 20-3. HENDERSON RESIDENCE (LABOR ONLY) CONSTRUCTION JOB-COST REPORT

REPORT DATE: MAY 14, 199_
JOB SITE: HENRY RESIDENCE
CREW SIZE: ------> 2
LABOR HOURS PER DAY PER MAN: ------> 8
VISIT NO: ------> 3

VISIT # / ACTUAL DATE	SPRING CLEAN UP MANHRS	DRIVETIME LABOR MANHRS	LOAD/UN-LOAD MANHRS	21" MOW LABOR& EQ HRS	36" MOW LABOR& EQ HRS	52" MOW LABOR& EQ HRS	BLOWER LABOR& EQ HRS	TRIMMER LABOR& EQ HRS	WEED & CLN UP MANHRS	LABOR MANHRS	LABOR MANHRS	FALL CLEAN UP MANHRS	SUPVSR. LABOR MANHRS	TOTALS MANHRS	BUDGET MANHRS TO-DATE	OVER/ UNDER M-HRS	OVER/ UNDER % COMP
TO GO ------>	0.0	7.3	7.5	25.3	25.7	0.0	6.4	12.8	11.8	0.0	0.0	14.0	5.5	116.1	N/A	N/A	0.6%
BUDGET ------>	14.0	9.0	9.0	28.0	28.0	7.0	14.0	14.0	0.0	0.0	14.0	7.0	144.0	144.0	0.0	20.0%	
ACTUAL HRS / %-->	14.0	1.8	1.5	2.8	2.4	0.0	0.7	1.2	2.2	0.0	0.0	0.0	1.5	27.9	28.8	0.9	19.4%
SPRING CLEAN UP(4/22)	14.00	1.00	1.00										1.00	17.0	17.0	0.0	11.8%
VISIT-1 (4/29)		0.25	0.20	1.50	1.25		0.20	0.40	0.20					4.0	3.9	(0.1)	14.5%
VISIT-2 (5/6) SITE VERY WET		0.25	0.10	0.00	0.00		0.20	0.20	2.00					2.8	3.9	1.2	17.3%
VISIT-3 (5/13)		0.25	0.20	1.25	1.10		0.25	0.60					0.50	4.2	3.9	(0.2)	20.0%
VISIT-4														0.0	3.9	3.9	22.7%
VISIT-5														0.0	3.9	3.9	25.4%
VISIT-6														0.0	3.9	3.9	28.2%
VISIT-7														0.0	3.9	3.9	30.9%
VISIT-8														0.0	3.9	3.9	33.6%
VISIT-9														0.0	3.9	3.9	36.4%
VISIT-10														0.0	3.9	3.9	39.1%
VISIT-11														0.0	3.9	3.9	41.8%
VISIT-12														0.0	3.9	3.9	44.5%
VISIT-13														0.0	3.9	3.9	47.3%
VISIT-14														0.0	3.9	3.9	50.0%
VISIT-15														0.0	3.9	3.9	52.7%
VISIT-16														0.0	3.9	3.9	55.5%
VISIT-17														0.0	3.9	3.9	58.2%
VISIT-18														0.0	3.9	3.9	60.9%
VISIT-19														0.0	3.9	3.9	63.6%
VISIT-20														0.0	3.9	3.9	66.4%
VISIT-21														0.0	3.9	3.9	69.1%
VISIT-22														0.0	3.9	3.9	71.8%
VISIT-23														0.0	3.9	3.9	74.6%
VISIT-24														0.0	3.9	3.9	77.3%
VISIT-25														0.0	3.9	3.9	80.0%
VISIT-26														0.0	3.9	3.9	82.7%
VISIT-27														0.0	3.9	3.9	85.5%
VISIT-28														0.0	3.9	3.9	88.2%
FALL CLEAN UP														0.0	17.0	17.0	100.0%

DIAGRAM 20-4. COMPUTER MAINTENANCE JOB-COST REPORT

CONCLUSION

Designing your management systems to provide timely, accurate feedback (TAF) in the form of job-cost reports is one of the most important functions of top management. The process is relatively simple but unfortunately, few contractors understand how to do it. Setting up timely, accurate job-cost reports is very much like turning on the "scoreboard" for your field and staff to see. The benefits are numerous. Operating without the scoreboard keeps everybody in the dark, while turning it on can also "turn on" your field people and staff. Although job costing through your accounting software program is the best method, alternative methods, either manual or computerized, can achieve significant results for your operation. Job costing provides a reality check for your bids. If you bid jobs at a 20% GPM, you should see a 20% GPM on your job-cost reports. If the two differ, job costing will give you the tools to research, identify, and correct this variance. Effective job costing can also provide you with a sound analytical process and the performance standards upon which to build your total quality management (TQM) program.

A CAVEAT

Job costing has limitations which top management needs to understand. It is designed to address only the variance between the gross profit margin (GPM) bid on a specific job versus the GPM actually achieved in the field. It is not meant to address office overhead recovery or allocation, downtime costs for labor, or non-billable equipment costs—to name a few. Job costing only compares "budget-to-actual" performance for direct costs and for specific jobs. To track and monitor these other costs, you need an annual budget and "budget-to-actual" comparisons on your P&L statements to identify targets for overhead recovery, labor downtime, equipment costs, etc.

BASIC STEPS TO START A JOB-COST REPORT SYSTEM

1. Make a master form similar to the sample job-cost report provided. It should contain the majority of the labor functions that your crews perform. Leave a few blank spaces for add-on items. If possible, create it in a personal computer (PC) spreadsheet program. Identify: who in your office will fill it out, how often it will be filled out for each job, and to whom the report will be given.

2. Break down each job into its labor functions. Manually enter this budget data for each job onto a copy of your master job-cost report. Create a similar spreadsheet for each job in your PC software, if possible.

3. Create a manual Field Daily Report (FDR) form similar to the samples provided in Exhibit (9) and (10). It will be used to collect field job site information on a daily basis.

4. Train your crew leader(s) and/or foremen how to fill out the FDR. Notice that the sample FDR is very similar to the format for the job-cost report. This makes feedback to crews more familiar and therefore more meaningful.

5. Collect FDRs on a daily basis from your crew leaders and/or foremen. Emphasize the importance of obtaining them at the end of each day.

6. Complete the job-cost report and distribute it to crew leaders and/or foremen, supervisors and managers on a weekly (preferably daily) basis. Require everyone to review the results (the scoreboard). Build focus and intensity into your company by holding supervisors, foremen, etc. accountable to budget goals.

7. **Make it fun.** Challenge your crews to bring their jobs in on-budget. Reward them with a monetary bonus, pizza, special tee-shirt, etc. if their crew labor hours are on-budget by a set date (e.g., mid-season, end of season, etc.).

ACTION POINT

Review your job-cost procedures and take steps to ensure that, at a minimum, key field personnel and management staff obtain "budget-to-actual" reports for field-labor hours on a daily basis.

CHAPTER 21
BILLABLE FIELD-LABOR HOUR
"BUDGET VS. ACTUAL" PERFORMANCE

PURPOSE
To explain how to focus the company bureaucracy and bureaucrats in support of field operations and how to measure key company performance in relation to it

INTRODUCTION
Everything (overhead, equipment, materials, profit, sales, etc.) should be tracked and compared to billable FLHs . Whenever FLHs are discussed in this chapter, they refer to "billable" hours unless otherwise noted.

RATIONALE
Billable field-labor hours should be the central item to which all others are compared for reasons outlined below. But first, let's define billable FLHs.

DEFINITION: Billable field-labor hours are those hours included in a bid and for which a client is billed. Downtime hours, equipment repair hours, shop time, time spent in the nursery caring for stock, etc. are not considered billable FLHs. However, "T&M" and general condition time are if the client pays for them and/or they are in the bid.

FLHs are:

1. THE COMMON DENOMINATOR
Nothing really good is happening in the company unless your crews are working in the field. The existence of the whole company finds its justification in what happens in the field.

2. EASILY VISUALIZED AND GRAPHED
SPH, PPH, OPH, etc., are figures and ratios that are easily graphed and visualized. Everyone can relate to them quickly and easily.

3. EASILY MONITORED
Keeping track of these figures and ratios is a very simple process.

4. CONNECTED TO JOB COSTING
Focusing and comparing everything to field-labor hours is the most effective way to monitor your field production. If you get into trouble on a job, it is probably because of labor. Hence, you need to be able to focus on and monitor your labor as much as possible.

5. EASILY RELATED TO

Virtually everyone can visualize field labor because labor and its results are a tangible commodity. Monitoring your sales, profit, and overhead will not really mean that much to very many people. However, the monitoring of these areas as they relate to labor will communicate much more effectively to everyone involved.

6. THE MOST INFLUENTIAL ON YOUR COMPANY

Tracking and monitoring your company this way simply works better than anything else. It gets results.

7. ABLE TO FOCUS BUREAUCRATS

Bureaucrats tend to become aloof and to build empires unto themselves. Having them focus on the field helps keep them in touch with reality and with the rest of the company.

TRACKING FLH "BUDGET VS. ACTUAL" PERFORMANCE

It is crucial that you monitor your field-labor hour budget and compare it to actual labor hours used. If your actual field-labor hours consistently EXCEED your projections (and you expect this to hold true throughout the year), you can decrease your OPH and increase your net profit margins, or you could lower your prices and become more competitive. On the other hand, if actual field-labor hours are not going to meet budget projections, you should either increase your OPH (which would make you less competitive) or reduce overhead expenses. This will either bring your OPH back in line with your budget or will lower it and make your bids more competitive.

Field-labor hours must be monitored both on a job-by-job basis (by means of job costing), and overall, by means of a chart similar to Diagram (21-1). Individual jobs coming in on-budget don't do you much good if you are not meeting your yearly budget. Both are important. Both must be constantly tracked and attained or subsequently adjusted.

Diagram (21-2) illustrates a "Weekly Field-Labor Hour Recap Sheet" that you should utilize. It identifies production time as compared to downtime, drive time, etc. Add or delete categories as you wish.

You can also use the "Bid Status Board" and the "MAD Report" (covered in Chapter 19) to track field-labor hours.

DOWNTIME

Field-labor hours paid for by the company but not bid into jobs can become subtle and critical leaks in your company if not monitored closely. Diagrams (21-2) and (21-3) are examples of how you might track them using a computer spreadsheet program. Two specific examples come to mind.

A commercial landscape and irrigation company out West doing about $1M in gross annual sales was showing no net profit on its bottom line. Downtime was identified as one of the problem areas when the two owners described their field labor's daily routine.

Crews (totalling 10 men) would show up at the yard at 6 AM and punch in on a time clock. Next to the yard was a donut shop, so guess where they went next? Crews stood around the yard eating donuts and drinking coffee (and occasionally loading trucks) for 15–30 minutes before driving to job sites. Driving time to job sites averaged 30–45 minutes one way. Crews would work all day, then drive back 30–45 minutes to the yard, punch out on the clock and go home.

I calculated that each man had 1.5 hours per day "loading" trucks in the morning and driving to job sites. This totaled: 10 men x 1.5 hours/day x 220 days/year or 3,300 hours/year. This came to $26,400.00/year in labor dollars at an average wage of $8.00. Adding 30% labor burden would increase the total to $34,320.00 per year.

I'm all for paying reasonable amounts for load and drive time for crews, but my question to the two owners was: "Who was paying the $34,320.00? Was it included in general conditions in the bid?" If it was, the customer was paying for it. If it was not, the company (the two owners) were, and it came out of net profit.

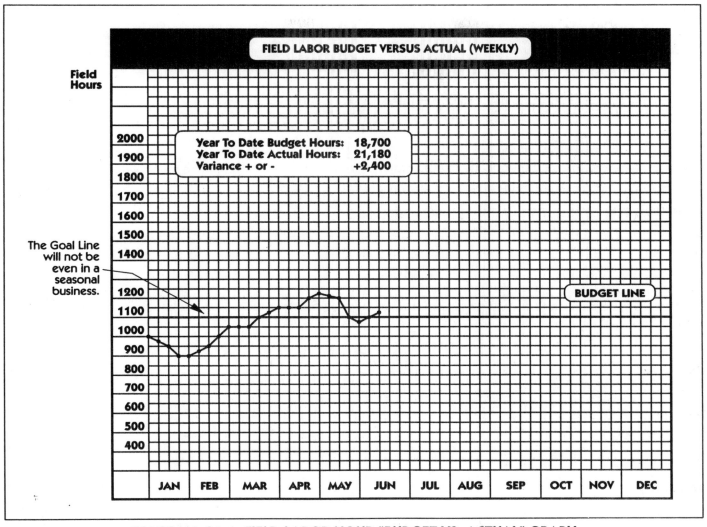

DIAGRAM 21-1. FIELD-LABOR HOUR "BUDGET VS. ACTUAL" GRAPH

FEB 8, 199_

DIV / CREW	LOAD / UNLOAD	YARD AM / PM	EXTRAS	DRIVE TO/FM	MOBIL EQUIP	HAUL MATLS	PRODUCT HRS	MISC	SUBTOTAL	WARANTY	YARD	EQUIP MAINT	SAFETY MEETING	NURSERY	MISC	MISC	SUBTOTAL	TOTALS	%
WK: 16-22 JAN	14.0	10.0	12.0	37.0	6.0	20.0	271.0	11.0	381.0	5.0	5.0	7.0	4.0	2.0	1.0	0.0	24.0	405.0	100.0%
PLANTING	5.0	5.0	0.0	15.0	6.0	8.0	122.0	6.0	167.0	4.0	3.0	2.0	2.0	2.0	0.0	0.0	13.0	180.0	44.4%
IRRIGATION	3.0	3.0	12.0	12.0	0.0	4.0	92.0	3.0	129.0	1.0	0.0	4.0	1.0	0.0	0.0	0.0	6.0	135.0	33.3%
HARDSCAPE	6.0	2.0	0.0	10.0	0.0	8.0	57.0	2.0	85.0	0.0	2.0	1.0	1.0	0.0	1.0	0.0	5.0	90.0	22.2%
	0.0	0.0	0.0	0.0	0.0	0.0	0.0	0.0	0.0	0.0	0.0	0.0	0.0	0.0	0.0	0.0	0.0	0.0	0.0%
	0.0	0.0	0.0	0.0	0.0	0.0	0.0	0.0	0.0	0.0	0.0	0.0	0.0	0.0	0.0	0.0	0.0	0.0	0.0%
	0.0	0.0	0.0	0.0	0.0	0.0	0.0	0.0	0.0	0.0	0.0	0.0	0.0	0.0	0.0	0.0	0.0	0.0	0.0%
	0.0	0.0	0.0	0.0	0.0	0.0	0.0	0.0	0.0	0.0	0.0	0.0	0.0	0.0	0.0	0.0	0.0	0.0	0.0%
	0.0	0.0	0.0	0.0	0.0	0.0	0.0	0.0	0.0	0.0	0.0	0.0	0.0	0.0	0.0	0.0	0.0	0.0	0.0%
	0.0	0.0	0.0	0.0	0.0	0.0	0.0	0.0	0.0	0.0	0.0	0.0	0.0	0.0	0.0	0.0	0.0	0.0	0.0%

<------BILLABLE---HOURS------> <------NON-BILLABLE----HOURS----->

DIAGRAM 21-2. WEEKLY FIELD-LABOR HOUR RECAP SHEET

Simultaneously, the two owners looked at each other. I saw a light go on in both their heads followed by a facial expression indicating a sinking feeling in their stomachs as they realized the problem. They had not passed the 3,300 labor hours on to their clients, and they had "eaten" the $34,320 that it had cost. That's what you call a rather big leak in a company—costs not recaptured in your bids.

An East Coast company identified a similar problem. When its sales dropped from $1.6M to $800K due to the recession in 1989–90, net profits evaporated as sales went down. The two partners liked equipment, and they had been equipment heavy before the drop in sales. Fortunately, most of the equipment was paid for, and it did not have as negative an impact on the company as you might expect. However, it caused another problem that surfaced when we looked at downtime.

Prior to my arrival, the bookkeeper had accounted for 8,000 labor hours which were not included in any of their bids. Approximately 4,000 hours were for two mechanics who were busy working on and repairing equipment. The good news was that the mechanics were busy. The bad news was that the equipment was not. They could have kept four mechanics busy full time because they had so much equipment. Unfortunately, they could not even justify having one full-time mechanic, let alone two.

Approximately 2,500 of the remaining 4,000 hours of downtime hours should have been bid into jobs as drive time, load time, mobilization, etc. Yard time, busy work, and nursery time accounted for the balance.

CONCLUSION

MY POINT IS THIS: Downtime hours can eat up your profits faster than anything else. The one area where you will make the most money or lose the most money is in the way that you handle field-labor hours. If you do not track them, you cannot identify the problem. And you certainly cannot fix a problem that you cannot see.

ACTION POINT

Track and identify the billable vs. non-billable (downtime) field-labor hours in your company and begin to measure company performance in relation to the field where it can make the biggest impact and do the most good.

FEB 8, 199_

TYPE HOURS--->	LOAD / UNLOAD	YARD AM / PM	EXTRAS	DRIVE TO/FM	MOBIL EQUIP	HAUL MATLS	PRODUCT HRS	MISC	SUBTOTAL	WARANTY	YARD	EQUIP MAINT	SAFETY MEETING	NURSERY	MISC	MISC	SUBTOTAL	TOTALS	%
WEEK	37.0	26.0	43.0	98.0	18.0	46.0	864.0	13.0	1,145.0	7.0	13.0	17.0	6.0	7.0	4.0	0.0	54.0	1,199.0	100.0%
1 JAN 1-8	12.0	7.0	18.0	33.0	8.0	12.0	285.0	2.0	377.0	2.0	6.0	8.0	2.0	4.0	2.0	0.0	24.0	401.0	33.4%
2 JAN 9-15	11.0	9.0	13.0	28.0	4.0	14.0	308.0	0.0	387.0	0.0	2.0	2.0	0.0	1.0	1.0	0.0	6.0	393.0	32.8%
3 JAN 16-22	14.0	10.0	12.0	37.0	6.0	20.0	271.0	11.0	381.0	5.0	5.0	7.0	4.0	2.0	1.0	0.0	24.0	405.0	33.8%
4 JAN 23-29	0.0	0.0	0.0	0.0	0.0	0.0	0.0	0.0	0.0	0.0	0.0	0.0	0.0	0.0	0.0	0.0	0.0	0.0	0.0%
5 JAN 30-FEB 5	0.0	0.0	0.0	0.0	0.0	0.0	0.0	0.0	0.0	0.0	0.0	0.0	0.0	0.0	0.0	0.0	0.0	0.0	0.0%
6	0.0	0.0	0.0	0.0	0.0	0.0	0.0	0.0	0.0	0.0	0.0	0.0	0.0	0.0	0.0	0.0	0.0	0.0	0.0%
7	0.0	0.0	0.0	0.0	0.0	0.0	0.0	0.0	0.0	0.0	0.0	0.0	0.0	0.0	0.0	0.0	0.0	0.0	0.0%
8	0.0	0.0	0.0	0.0	0.0	0.0	0.0	0.0	0.0	0.0	0.0	0.0	0.0	0.0	0.0	0.0	0.0	0.0	0.0%
9	0.0	0.0	0.0	0.0	0.0	0.0	0.0	0.0	0.0	0.0	0.0	0.0	0.0	0.0	0.0	0.0	0.0	0.0	0.0%
10	0.0	0.0	0.0	0.0	0.0	0.0	0.0	0.0	0.0	0.0	0.0	0.0	0.0	0.0	0.0	0.0	0.0	0.0	0.0%
11	0.0	0.0	0.0	0.0	0.0	0.0	0.0	0.0	0.0	0.0	0.0	0.0	0.0	0.0	0.0	0.0	0.0	0.0	0.0%
12	0.0	0.0	0.0	0.0	0.0	0.0	0.0	0.0	0.0	0.0	0.0	0.0	0.0	0.0	0.0	0.0	0.0	0.0	0.0%
13	0.0	0.0	0.0	0.0	0.0	0.0	0.0	0.0	0.0	0.0	0.0	0.0	0.0	0.0	0.0	0.0	0.0	0.0	0.0%
14	0.0	0.0	0.0	0.0	0.0	0.0	0.0	0.0	0.0	0.0	0.0	0.0	0.0	0.0	0.0	0.0	0.0	0.0	0.0%
15	0.0	0.0	0.0	0.0	0.0	0.0	0.0	0.0	0.0	0.0	0.0	0.0	0.0	0.0	0.0	0.0	0.0	0.0	0.0%
16	0.0	0.0	0.0	0.0	0.0	0.0	0.0	0.0	0.0	0.0	0.0	0.0	0.0	0.0	0.0	0.0	0.0	0.0	0.0%
17	0.0	0.0	0.0	0.0	0.0	0.0	0.0	0.0	0.0	0.0	0.0	0.0	0.0	0.0	0.0	0.0	0.0	0.0	0.0%
18	0.0	0.0	0.0	0.0	0.0	0.0	0.0	0.0	0.0	0.0	0.0	0.0	0.0	0.0	0.0	0.0	0.0	0.0	0.0%
19	0.0	0.0	0.0	0.0	0.0	0.0	0.0	0.0	0.0	0.0	0.0	0.0	0.0	0.0	0.0	0.0	0.0	0.0	0.0%
20	0.0	0.0	0.0	0.0	0.0	0.0	0.0	0.0	0.0	0.0	0.0	0.0	0.0	0.0	0.0	0.0	0.0	0.0	0.0%
21	0.0	0.0	0.0	0.0	0.0	0.0	0.0	0.0	0.0	0.0	0.0	0.0	0.0	0.0	0.0	0.0	0.0	0.0	0.0%
22	0.0	0.0	0.0	0.0	0.0	0.0	0.0	0.0	0.0	0.0	0.0	0.0	0.0	0.0	0.0	0.0	0.0	0.0	0.0%
23	0.0	0.0	0.0	0.0	0.0	0.0	0.0	0.0	0.0	0.0	0.0	0.0	0.0	0.0	0.0	0.0	0.0	0.0	0.0%
24	0.0	0.0	0.0	0.0	0.0	0.0	0.0	0.0	0.0	0.0	0.0	0.0	0.0	0.0	0.0	0.0	0.0	0.0	0.0%
25	0.0	0.0	0.0	0.0	0.0	0.0	0.0	0.0	0.0	0.0	0.0	0.0	0.0	0.0	0.0	0.0	0.0	0.0	0.0%

Columns under <------BILLABLE---HOURS------> : LOAD/UNLOAD through SUBTOTAL. Columns under <------NON-BILLABLE---HOURS------> : WARANTY through SUBTOTAL.

DIAGRAM 21-3. YEARLY FIELD-LABOR HOUR RECAP SHEET

CHAPTER 22

INDIRECT OR GENERAL AND ADMINISTRATIVE OVERHEAD COSTS:
The October Surprise

PURPOSE
To explain how to monitor G&A costs

Indirect costs or general and administrative (G&A) overhead costs are essential to monitor. Like field-labor downtime, G&A costs can get out of control and eat up your bottom line while individual job-cost reports look great.

I do not recommend that you necessarily attempt to allocate actual overhead expenses to your jobs through job costing. Rather, I prefer to simply compare direct costs bid to actual performance by means of gross profit margin (GPM) on a job-by-job basis. If your GPM turns out better than estimated, great! If not, find the problem and fix it for next time. However, you do need to monitor overhead.

Diagram (22-1) is a sample P&L financial statement format which compares monthly and year-to-date "budget vs. actual" categories. A good accounting software program should allow you to produce such a report. If you do not have that capability, you can produce it manually or enter the data into a computerized spreadsheet program.

BREAK-EVEN POINT (BEP) OR THE "OCTOBER SURPRISE"

A company usually reaches its BEP in the 9th or 10th month of its fiscal year. You reach your BEP when your accumulated GPM (G&A overhead costs plus net profit) equals your G&A budget dollar amount for the year. If you meet your BEP prior to the end of the fiscal year, any dollar amounts on bids above Phases I and II direct costs (plus tax and labor burden) will be profit for that year. That is **IF** the work is completed and billed prior to the end of the fiscal year. Work bid after the BEP is met, which will be completed and billed before the fiscal year end, can be bid with the realization that any amount above direct costs in the bid goes to net profit margin on the bottom line. If your profit markup is 10% and your overhead amounts to another 15%, your profit markup is really 25%, since your overhead is already covered for the year.

Some contractors understand this concept and know that after they meet their BEP, they can bid work cheaper than they normally would at the end of the year without really cutting profit.

(JAN THRU NOV)	NOV BUDGET	NOV ACTUAL	NOV VARIANCE	BUD %	Y-T-D BUDGET	Y-T-D ACTUAL	Y-T-D VARIANCE	BUD %
1. SALES (REVENUES)-------->	195,000	188,500	(6,500)	100.0%	2,340,000	2,080,000	(260,000)	100.0%
2. COST OF SALES (DIRECT COSTS):								
MATERIAL (W/TAX)	64,935	63,008	1,927	33.3%	779,220	726,452	52,768	33.3%
LABOR	35,100	33,467	1,633	18.0%	421,200	389,126	32,074	18.0%
LABOR BURDEN 31.0%	10,881	10,375	506	5.6%	130,572	120,629	9,943	5.6%
SUBCONTRACTORS	18,750	16,000	2,750	9.6%	225,000	155,000	70,000	9.6%
EQUIPMENT	8,775	8,245	530	4.5%	105,300	111,256	(5,956)	4.5%
EQUIPMENT RENTAL	2,917	500	2,417	1.5%	35,004	29,850	5,154	1.5%
MISC.	1,250	659	591	0.6%	15,000	14,525	475	0.6%
TOTAL COST OF SALES---->	142,608	132,254	10,354	73.1%	1,711,296	1,546,838	164,458	73.1%
3. GROSS PROFIT MARGIN----->	52,392	56,246	3,854	26.9%	628,704	533,162	(95,542)	26.9%
4. OVERHEAD (G&A/INDIRECT COSTS):								
ADVERTISING	375	425	(50)	0.2%	4,500	5,025	(525)	0.2%
BAD DEBTS	2,500	0	2,500	1.3%	30,000	45,525	(15,525)	1.3%
COMPUTERS, SOFTWARE	375	375	0	0.2%	4,500	4,678	(178)	0.2%
DONATIONS	42	75	(33)	0.0%	504	600	(96)	0.0%
DOWNTIME LABOR	488	828	(341)	0.3%	5,850	4,378	1,472	0.3%
DOWNTIME LABOR BURDEN	151	257	(106)	0.1%	1,814	1,357	456	0.1%
DUES AND SUBS	108	0	108	0.1%	1,296	898	398	0.1%
INSURANCE (OFFICE)	1,333	1,263	70	0.7%	15,996	16,075	(79)	0.7%
INTEREST AND BANK CHARGES	1,542	1,752	(210)	0.8%	18,504	17,066	1,438	0.8%
LICENSES & SURETY BONDS	46	0	46	0.0%	552	743	(191)	0.0%
OFFICE EQUIPMENT	215	215	0	0.1%	2,580	2,506	74	0.1%
OFFICE SUPPLIES	375	522	(147)	0.2%	4,500	4,200	300	0.2%
PROFESSIONAL FEES	1,375	1,049	326	0.7%	16,500	15,890	610	0.7%
RADIO SYSTEMS	375	375	0	0.2%	4,500	4,769	(269)	0.2%
RENT (OFFICE & YARD)	2,500	2,500	0	1.3%	30,000	30,000	0	1.3%
SALARIES-OFFICE	9,333	9,500	(167)	4.8%	111,996	105,555	6,441	4.8%
SALARIES-OFFICER(S)	6,667	6,667	0	3.4%	80,004	72,000	8,004	3.4%
SALARIES-LABOR BURD.13.0%	2,080	2,102	(22)	1.1%	24,960	23,082	1,878	1.1%
SMALL TOOLS & SUPPLIES	292	256	36	0.1%	3,504	3,756	(252)	0.1%
TAXES (ASSET & MILL TAX)	67	0	67	0.0%	804	1,024	(220)	0.0%
TELEPHONE	400	535	(135)	0.2%	4,800	5,235	(435)	0.2%
TRAINING & EDUCATION	167	225	(58)	0.1%	2,004	3,465	(1,461)	0.1%
TRAVEL & ENTERTAINMENT	183	127	56	0.1%	2,196	1,604	592	0.1%
UNIFORMS/SAFETY EQUIPMENT	100	0	100	0.1%	1,200	1,536	(336)	0.1%
UTILITIES	100	85	15	0.1%	1,200	1,338	(138)	0.1%
VEHICLES, OVERHEAD	1,563	1,478	85	0.8%	18,756	18,027	729	0.8%
YARD EXP./LEASEHOLD IMP.	208	35	173	0.1%	2,496	1,568	928	0.1%
MISCELLANEOUS	42	12	30	0.0%	504	864	(360)	0.0%
TOTAL OVERHEAD---------->	33,002	30,657	2,344	16.9%	396,020	392,764	3,255	16.9%
5. NET PROFIT MARGIN-(NPM)->	19,390	25,589	6,198	9.9%	232,685	140,398	(92,287)	9.9%
6. OVER/(UNDER) BILLING---->	0	0	0	0.0%	0	0	0	0.0%
7. REVISED NPM------------->	19,390	25,589	6,198	9.9%	232,685	140,398	(92,287)	9.9%

DIAGRAM 22-1. P&L STATEMENT BUDGET-TO-ACTUAL EXAMPLE

DETERMINING THE BEP

Use the following formula to calculate a company/division's BEP.

BEP = Annual overhead budget $ ÷ Annual GPM %

For a company whose budget is as follows:

Sales	$1,000,000	100%
G&A Overhead	200,000	20%
Profit	+ 100,000	10%
GPM	$ 300,000	30%

BEP = $200,000 ÷ 30% = $200,000 ÷ .3 = $666,667

The "projected" BEP is calculated to be $666,667.

Remember: This is the "projected" or budgeted BEP. Due to fluctuations in the GPM on individual jobs throughout the year and other factors, the "actual" BEP will in all likelihood be different, as seen on your P&L statement. Therefore, it is necessary to monitor "budgeted-to-actual" BEP by means of the P&L statement .

CONCLUSION

Monitor your G&A costs and "budget-to-actual" BEP by means of a P&L statement similar to the one in Diagram (22-1).

ACTION POINT

Calculate your company/division BEP and monitor it throughout the year as described above.

CHAPTER 23

THE PROFIT AND LOSS (P&L) STATEMENT AND RECAP REPORT

PURPOSE

To explain the P&L statement and the P&L recap report

THE P&L STATEMENT

The P&L statement is a very important report that you need to monitor in order to understand what is (or is not) happening in your company. I think of it as a job-cost report for all of your jobs thrown together. If used properly and in conjunction with other data and reports, it can provide considerable insight into many aspects of your business. If you have multiple divisions with each doing different types of work, provide that division with its own P&L if that division comprises over 15-20% of the business' gross annual sales. You may even want to break a division down into separate subdivisions that may, or may not, do the same types of work. For instance, I know of a landscape maintenance company with over 24 four-man crews in one division. That division is broken down into six subdivisions with four crews and a foreman in each one. Each subdivision is tracked and compared to the others for profitability and cost.

Formatting the P&L is crucial. Diagram (II-1) displays a sample P&L for one division of a corporation (not a sole proprietorship) which we encourage you to use. The sales for the division are at the top of the report, broken down into month and year-to-date (YTD) categories. Cost of sales or direct costs follow next and include everything that can be tied directly to a specific job. You want to put as much as you possibly can into direct costs, including labor burden (i.e., FICA, FUTA workers' compensation insurance, general liability insurance, unemployment insurance, field medical insurance, field holiday and vacation pay, etc.). Depreciation for your field equipment should be identified and included in direct costs. However, depreciation for office/overhead, vehicles and office equipment should be included in overhead.

Overhead is often referred to as indirect or G&A (general and administrative) costs. The point is that you cannot pin G&A costs to a particular field operation or job. Therefore, the term indirect or general is used.

Net profit margin before taxes and non-business-related interest follows overhead. Ten percent NPM is a good target percentage to shoot for after all costs have been included in the expense categories. I like to track NET PROFIT MARGIN "BUDGET VS. ACTUAL" on a graph on a monthly or quarterly basis (see Diagram 23-1). I also prefer that a company be on the accrual basis for its financial statements, because it provides a more consistent and accurate sales and expense picture.

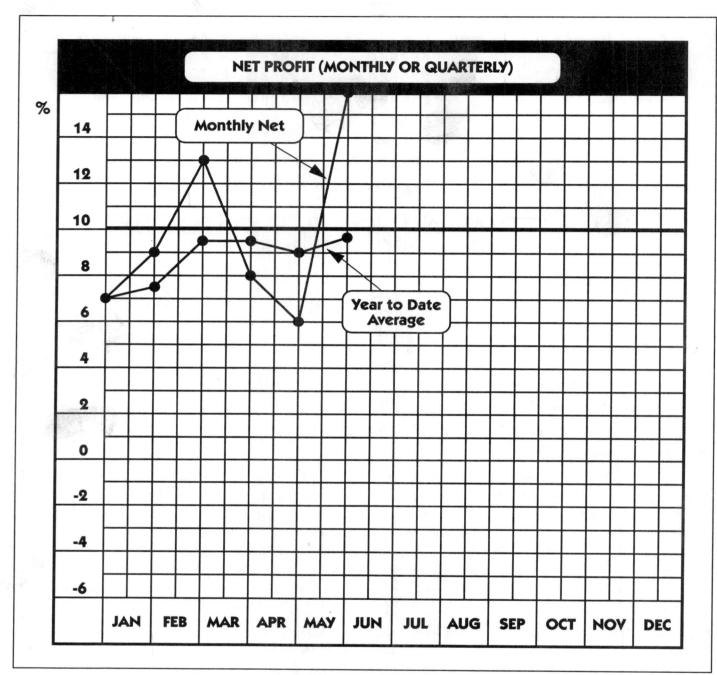

DIAGRAM 23-1. NET PROFIT MARGIN "BUDGET VS. ACTUAL" GRAPH

	MONTH 6	JUN BUDGET	JUN ACTUAL	JUN VARIANCE	JUN ACT %	Y-T-D BUDGET	Y-T-D ACTUAL	Y-T-D VARIANCE	Y-T-D ACT %
1. SALES--------------->		115,905	130,688	14,783	100.0%	695,430	723,089	27,659	100.0%
2. COST OF SALES:									
MATERIAL		18,708	32,345	(13,637)	24.7%	112,250	137,752	(25,502)	19.1%
LABOR		13,274	12,898	376	9.9%	79,646	94,227	(14,581)	13.0%
LABOR BURDEN		4,580	2,656	1,924	2.0%	27,480	27,589	(109)	3.8%
* EQUIPMENT		2,967	698	2,269	0.5%	17,800	6,016	11,784	0.8%
EQUIPMENT RENTALS		1,225	888	337	0.7%	7,350	12,705	(5,355)	1.8%
SUBCONTRACTORS		26,400	25,769	631	19.7%	158,400	141,408	16,992	19.6%
MISC. DIRECT COSTS		347	1,222	(875)	0.9%	2,083	2,500	(418)	0.3%
TOTAL------------>		67,501	76,476	(8,975)	58.5%	405,009	422,197	(17,189)	58.4%
3. GROSS PROFIT------->		48,404	54,212	5,808	41.5%	290,422	300,892	10,471	41.6%
4. OVERHEAD TOTAL----->		29,167	29,219	(52)	22.4%	175,000	206,338	(31,338)	28.5%
A. SALARY OFFICE		3,000	3,300	(300)	2.5%	18,000	18,900	(900)	2.6%
B. SALARY OFFICER		5,000	5,000	0	3.8%	30,000	30,000	0	4.1%
5. NET OP. INCOME----->		19,237	24,993	5,756	19.1%	115,421	94,554	(20,867)	13.1%
6. OTHER INC/(EXP)---->		0	0	0	0.0%		0		0.0%
7. NET INC/(LOSS)----->		19,237	24,993	5,756	19.1%	115,421	94,554	(20,867)	13.1%

8. RATIOS/PER HR INFO:	BUDGET	ACTUAL	VARIANCE		BUDGET	ACTUAL	VARIANCE
A. SALES PER HOUR:	$59.99	$63.72	$3.73		$59.99	$64.97	$4.98
B. DIR COSTS PER HR:	$34.94	$42.04	$7.10		$34.94	$37.93	$3.00
C. GPM PER HOUR:	$25.05	$21.68	($3.38)		$25.05	$27.03	$1.98
D. OVHD PER HOUR:	$15.10	$16.06	$0.97		$15.10	$18.54	$3.44
E. PROFIT PER HOUR:	$9.96	$13.74	$3.78		$9.96	$8.50	($1.46)
F. MAT'L/LAB RATIO:	1.41	2.51	1.10		1.41	1.46	0.05
G. MATERIAL PER HR:	$9.68	$17.78	$8.10		$9.68	$12.38	$2.69
* H. EQUIP/LAB RATIO:	22.3%	5.4%	16.9%		22.3%	6.4%	16.0%
* I. EQUIP PER HOUR:	$1.54	$0.38	$1.15		$1.54	$0.54	$0.99
J. AVERAGE WAGE:	$6.87	$7.09	($0.22)		$6.87	$8.47	($1.60)
K. BILLABLE FLH'S:	1,932	1,819	113		11,593	11,130	463
L. NON-BILL. FLH'S:	20	35	(15)		120	175	(55)
	1,952	1,854	98		11,713	11,305	408

9. ACCOUNTS RECEIVABLE/PAYABLE INFORMATION:

	CURRENT	OVER 30	OVER 60	OVER 90	TOTAL
	66.6%	26.6%	4.0%	2.8%	100.0%
A. ACCTS. RECEV.--->	58,946	23,500	3,500	2,500	88,446
	87.1%	12.0%	0.9%	0.0%	100.0%
B. ACCTS. PAY.----->	42,638	5,863	457	0	48,958

C. CURRENT RECEV./TOTAL PAYABLES-----> 1.20

* -ACTUAL FIGURES DO NOT INCLUDE DEPRECIATION.

DIAGRAM 23-2. P&L STATEMENT RECAP REPORT

Some companies include an over/underbilling category at the bottom of their P&L to reflect profits more accurately. If they are, overbilled or underbilled on their jobs for a given period, they subtract or add said amount for the same and adjust their P&L accordingly.

THE P&L RECAP REPORT

This report provides an encapsulated view of the P&L (see Diagram (23-2)) and compares "budget-to-actual" figures. Each division should have a separate recap sheet. Overhead is condensed into just three categories (I include payables and receivables in order to obtain some percentages at the bottom of the report). The ratios are very important, especially when you graph them over a six- to twelve-month period. Remember, however, that these ratios are meaningless unless taken into consideration within the overall context of the business. As an example, the material-to-labor ratio one month may be very high compared to another, because the materials were bought one month and the labor to install it expended the next. Consequently, you need to view the monthly ratio in context with the yearly one.

CONCLUSION

The P&L statement can provide invaluable insight into your business if it is formatted properly. It can be shortened or recapped to further help you identify key trends and to monitor "budget vs. actual" performance.

ACTION POINT

Review the format of your P&L statement with your accountant or CPA and adjust the format in order to identify direct vs. indirect costs more accurately and thoroughly.

CHAPTER 24
EQUIPMENT

PURPOSE
To explain how to control equipment costs

INTRODUCTION

It isn't enough to ensure that ample field equipment is available and maintained for use in the field. You must monitor equipment use and be sure that equipment is paying for itself. The key to effective equipment planning is to be sure that you are estimating your equipment accurately—and that you are then utilizing your equipment enough to generate sales dollars to cover its costs.

Construction and maintenance companies can get into serious trouble when the income generated by field equipment doesn't cover the cost of that equipment. Often the problem gets buried. No one in the company is the wiser until the bottom line begins to deteriorate. Management then has to investigate, diverting valuable time and resources, when the entire problem could have been avoided from the beginning, or at least identified much sooner.

It is important that you compare your equipment "budget-to-actual" costs, for individual jobs as well as for your company as a whole.

EQUIPMENT COSTS BID INTO JOBS

Equipment costs should be included in bids as the job requires. Equipment is bid much the same as labor. Actual usage in hours on a particular job should be multiplied by a predetermined cost per hour figure. Monitor equipment usage through job costing. Use the job-cost format in Diagram (20-2) to track and compare equipment "budget-to-actual" costs for individual jobs. Track the use of equipment on each separate job in order to be sure that equipment bid into separate jobs is very close to equipment actually used.

EQUIPMENT BUDGETS AND MONITORING COMPANY-OWNED EQUIPMENT

Concerning equipment, the primary function of accounting, other than processing accounts payable invoices and loan payments for equipment, is to **"keep score"** by comparing your equipment budgets to their actual cost.

One of three methods, or a combination of these methods, can be utilized to compare budgeted costs to actual costs for the entire company.

1. ESTABLISH AN EQUIPMENT CHECKBOOK.

As you would calculate payroll for labor (based on actual hours multiplied by a labor rate), do the same for each piece of equipment (hours used multiplied by its CPH). Total this amount for all equipment. Deposit the equipment "paycheck" into the equipment checkbook. All equipment expenses for fuel, repairs, monthly payments, insurance, etc., should be made from this account.

A. If this account always has too little money in it to cover all your equipment expenses:

   (1). You have more equipment than you can justify owning, **or...**

   (2). Your calculated cost per hour is too low, **or...**

   (3). A combination of both (1) and (2).

B. If you consistently have more money in this checking account than you need, and your job costing shows that your equipment bid vs. actually used is in line, accumulate this money for planned replacement of equipment that wears out.

2. ESTABLISH AN EQUIPMENT DIVISION ON PAPER.

You do this by establishing an equipment division P&L statement in your accounting software.

A. Direct costs consist of materials and labor expended on vehicle upkeep and repair, labor burden, equipment used for repairs and preventive maintenance, and subcontractor costs for repairs done outside the company.

Overhead consists of indirect G&A expenses or a portion of them which is used to support the equipment division.

B. Sales are generated by charging field production divisions a rental fee (at the CPH rate) for equipment used. This fee is then charged to the respective division as a direct cost for equipment used.

The main purpose of the equipment division is to format equipment costs in such a way as to indicate promptly whether equipment is paying for itself or not.

The goal of the equipment division is to break even (at a minimum). Division "profits" should be set aside in a separate bank account for future replacement costs of equipment.

You should seriously consider establishing an equipment division on paper once your company reaches about $1.5M in gross annual sales.

3. FORM AN ACTUAL DIVISION (WITH A DIVISION MANAGER) OR COMPANY.

Many companies who establish a "paper" equipment division eventually create an actual division with a designated manager in charge.

The manager's primary responsibility is to run this division and to service all of the corporate equipment needed while keeping costs to a minimum. The equipment division's P&L statement serves as the manager's report card.

For tax purposes, this division can eventually be split off as a separate company. If you desire to pursue this, be sure to consult with your CPA.

JUSTIFYING EQUIPMENT PURCHASES (THE 50% RULE)

As a general rule, you need to be able to bill out a piece of equipment approximately **50%** of the time.

For example, a tractor that has a CPH of $15 should be included in bids, operated, and billed to jobs at least twenty hours each week. This would generate $300 per week in revenue (before profit and overhead markups) to cover its costs.

If you cannot bill out a piece of equipment 50% of the time, you will probably not generate enough revenue from it to cover its costs. Purchase of specialty pieces of equipment, and/or ones that are difficult to rent, may be justified if you believe that the convenience of owning them offsets the lack of billable hours generated by them.

KEEPING SCORE

1. Keeping score means that you are constantly verifying that the revenue generated from equipment bid into jobs and installed in the field is sufficient to cover all of your equipment expenses.

2. By making a few variations to the job-cost format in Diagram (20-2) as I did on Diagram (24-1), you can use it to combine all of your costs and compare them to your equipment budget for the year.

CHARGE TO JOBS--->

CHARGE TO JOBS--->	BUDGET	%	ACTUAL	%	VARIANCE
	90,160	100.0%	47,456	52.6%	42,704
MATERIAL:	10,000	11.1%	6,366	7.1%	3,634
LABOR:	19,280	21.4%	10,930	12.1%	8,350
BURDEN:	6,170	6.8%	3,498	3.9%	2,672
SHOP P/UP:	7,280	8.1%	2,835	3.1%	4,445
RENTALS:	500	0.6%	35	0.0%	465
SUBS/BLADES:	2,500	2.8%	1,477	1.6%	1,023
PAYMENTS:	24,324	27.0%	14,189	15.7%	10,135
FUEL/POL:	12,000	13.3%	11,756	13.0%	244
		0.0%		0.0%	0
		0.0%		0.0%	0
DIR COST:	82,054	91.0%	51,086	56.7%	30,968
GR PROFIT:	8,106	9.0%	(3,630)	-4.0%	11,736
OVERHEAD:	4,820	5.3%	2,733	3.0%	2,088
NET PRO:	3,286	3.6%	(6,362)	-7.1%	(9,649)

BUDGET INFORMATION (MONTH 7)

OH/HR:	$2.50	(OVERHEAD PER FIELD HOUR)
BURDEN %:	32.0%	(WITHOUT LABOR BURDEN)
AVE WAGE:	$10.00	(SALES PER FIELD HOUR)
SPH:	$46.76	(SALES PER FIELD HOUR)
PPH:	$1.70	(PROFIT PER FIELD HOUR)
MAT/LAB:	0.5	(MATERIAL TO LABOR RATIO)

MONTHLY EQUIPMENT PAYMENTS/LEASE PAYMENTS

ITEM	PMT	PAID Y-T-D
P/UP # 2	$189	$1,323
P/UP # 3	$205	$1,435
1 TON P/UP	$276	$1,932
DUMPTRUCK	$354	$2,478
TRACTOR/BOBCAT	$414	$2,898
MOWERS COMBINED	$589	$4,123
TOTALS--->	$2,027	$14,189

	BUDGET	ACTUAL	+/-
MATERIAL-PARTS--->	18,000 -	6,366 =	11,634
FUEL/OIL/LUBRICANTS--->	22,000 -	11,756 =	10,244
SUBS--->	2,500 -	1,477 =	1,023
RENTAL EQUIPMENT--->	500 -	35 =	465
TOTALS:--->	43,000 -	19,634 =	23,366

VENDORS	TOTAL	PARTS	FUEL/POL	SUBS	RENT EQ
ACTUAL TOTALS--->	$19,634	6,366	11,756	1,477	35
A-1 AUTO	3,345	3,345			
CHIEF AUTO PARTS	783	783			
HENLEY OIL	10,038		10,038		
JERRY'S MUFFLER	278			278	
JIM'S TIRE & WHEEL	1,034	1,034			
LUBE-N-TUNE	389		389		
READY EQUIP. RENTAL	35				35
SMALL ENGINE REPAIR CO.	465			465	
SO. SHORE BOBCAT	1,549	1,204		345	
TEXACO	1,053		1,053		
SHELL OIL	665		665		
	0				
	0				
	0				
	0				

		COST/HR	BUD HRS/$	CHARGED	TO GO
SHOP PICK UP--->	2,835	$3.50	2,080	810	1,270
EQUIP CHARGED TO JOBS->	$47,456	COST/HR	BUD HRS/$	CHARGED	TO GO
P/UP # 1	3,637	$3.00	2,080	1,212	868
P/UP # 2	3,940	$3.25	2,080	1,212	868
P/UP # 3	3,940	$3.25	2,080	1,212	868
1 TON P/UP	4,720	$5.00	2,080	944	1,136
DUMPTRUCK	9,990	$18.00	1,200	555	645
TRACTOR	7,290	$15.00	1,200	486	714
BOBCAT	8,636	$17.00	1,200	508	692
RENTAL EQUIPMENT	5,302	N/A	$10,000	5,302	4,698

LABOR HRS:	BUD	USED	% TO GO	%	% COMP	VAR
TOT--->	1,928	1,093	57%	835	43%	58%	2%
PM/REP>	1,668	977	59%	691	41%	58%	0%
ADMIN->	260	116	45%	144	55%	58%	14%

PAY PERIOD	HOURS--PM&REPAIR	HOURS----ADMIN
A=1-15 , B=16-31	977	116
1 JAN A	78	8
2 JAN B	75	9
3 FEB A	74	12
4 FEB B	62	7
5 MAR A	71	9
6 MAR B	73	9
7 APR A	65	10
8 APR B	72	8
9 MAY A	76	6
10 MAY B	74	9
11 JUN A	72	9
12 JUN B	77	7
13 JUL A	66	8
14 JUL B OFF 1 WK	42	5
15 AUG A		
16 AUG B		
17 SEP A		
18 SEP B		
19 OCT A		
20 OCT B		
21 NOV A		
22 NOV B		
23 DEC A		
24 DEC B		

DIAGRAM 24-1. EQUIPMENT BUDGET-TO-ACTUAL REPORT FOR THE WHOLE CO/DIV.

It is best to have accounting software compare equipment budget-to-actual costs. However, so few contractors have accounting software up and running (let alone providing budget-to-actual expense comparisons) that accomplishing the same, manually as described above, may be necessary as an interim measure until you can do so in an accounting software system.

3. Always ask yourself these all-important questions for each piece of equipment:

   A. Is this piece of equipment paying for itself?

   B. Is it justifying its existence and expense?

Let me attempt to make what I have said more real with an illustration from Chapter 18 of what I have observed in businesses ranging in size from $100,000 to $10,000,000 plus:

ABC Landscape & Irrigation Company plans to employ enough labor in the field next year to total approximately 50,000 field-labor hours. Total equipment costs are estimated at about $450,000 for the year. Indirect costs (overhead) are estimated to be another $300,000. For bidding purposes, direct costs (material, sales tax, labor, labor burden and subcontractors) are included in the bid at cost. Equipment costs are combined with overhead and added to the bid at $15 per estimated field-labor hour in the bid ($450,000 equipment costs plus $300,000 overhead, both divided by the 50,000 projected field-labor hours).

Sounds simple enough, but there is a **fatal flaw** in the process. Equipment costs are not bid based on what is needed on that particular job. Rather, all equipment is averaged and spread on jobs regardless of how much or how little equipment is required in the field for that specific job. Jobs requiring nothing more than pickup trucks and wheelbarrows are charged the same ($15 per field-labor hour) as ones needing bobcats, trenchers, and backhoes (you could get away with this if every job required the same amount of labor and equipment). The consequences are subtle and eventually disastrous. Labor-intense jobs (the ones requiring only pickup trucks and wheelbarrows) are estimated with too much equipment costs included to cover the equipment. Subsequently, you do not get these jobs in a competitive market because your price for the job is too high. Jobs which are extremely equipment-intense are charged too little for equipment costs; but because the bids are underpriced, you get these jobs—and you keep getting these underpriced projects. The result is that you are using all of your equipment, but charging your customers for only a fraction of its actual cost. The bottom line evaporates, and until you correct your estimating method, you will lose more and more money.

The problem is further compounded when a very common solution is offered to remedy the problem: "Let's bid more jobs and increase sales volume." This is like putting more chickens into a chicken coop with a fox in it. Until you get rid of the fox, you have not dealt with the real problem.

Two things need to happen to correct this scenario. First, equipment should be included in bids as the job requires. Actual usage in hours should be multiplied by a predetermined cost per hour figure. In this manner, every job is bid with equipment much as it is bid with labor: estimated hours times a cost per hour. Second, equipment usage should be monitored through job costing on a job-by-job basis to ensure that equipment bid into specific jobs is very close to equipment actually used on the job.

CONCLUSION

Whether you simply have an equipment checking account, an equipment division/profit center with a manager, or an altogether separate equipment leasing company, remember: keeping score means that you are constantly verifying that the revenue generated from your equipment bid into jobs is sufficient to cover all of your equipment expenses. Too few contractors track this data, when to do so is actually fairly simple.

ACTION POINT

Review your company's ability to "keep score" of equipment costs and ensure that all costs are included in your bids and covered by revenue generated by your jobs completed in the field.

CHAPTER 25

TQM RE-EVALUATED:
The Good, the Bad, and Those Stuck in the Middle

PURPOSE
To expose the pros and cons of the current TQM movement

TQM (Total Quality Management) is the latest "buzzword" echoing throughout the corporate world. National, state, and local trade associations have picked up the banner and have added to the fervor.

The construction and maintenance services industry has joined the parade in no small way. The American Society for Quality Control (ASQC) is addressing quality issues concerning the construction industry in its regional operations centers and conferences. The Association for Quality and Participation (AQP) is forming a Construction Industry Division. The Construction Industry Institute (CII) said in August of 1989, "Companies which do not implement Total Quality Management in their firms will not be competitive in the national and international markets within the next 5-10 years!"[9] Ernst & Young's March/April 1992 issue of Contractor Briefing echoed CII when it stated, "To remain competitive, contractors must use (the) total quality management technique."[10]

However, do the results justify the hundreds and thousands of dollars (and hours) being expended on TQM? Is it just another fad that stressed-out and overworked contractors latch onto in desperation as they attempt to control the chaos surrounding them? Or is the TQM movement really the panacea that some portend that it is?

The answer to these questions is neither a resounding "yes" nor a definite "no." Rather, the answer is probably most analogous to the advent of the personal computer (PC) and its arrival onto the construction and maintenance services industry scene in the mid-1980's. Finally, after seven or eight years of both horror stories and rave reviews, the dust has begun to settle. Some companies have made significant progress using PCs. Others improved neither their management methods nor their bottom line even after spending thousands of dollars and hundreds of labor hours on personal computers.

The results obtained from TQM have not been unlike those achieved with personal computers. Fortunately, there is hard data to help us ferret out the good from the bad.

There are basically two types of TQM programs. The first targets incremental **results-driven** improvements by focusing on specific goals and the management processes involved to obtain them. The second is **activity-centered** and focuses on employees doing the right types of activities. These activities will, in turn, improve performance as a natural by-product of the activity.

These two methods have come under much scrutiny. Robert H. Schaffer and Harvey A. Thomson conclude in the January/February 1992 issue of the *Harvard Business Review* that, "Most corporate change programs mistake means for ends, process for outcome. The solution: focus on results, not on activities." They further state that, "The performance efforts of many companies have as much impact on operational and financial results as a ceremonial rain dance has on the weather. While (results-driven) companies constantly improve measurable performance, managers in (activity-centered ones) continue to dance round and round the campfire—exuding faith and dissipating energy."

"The 'rain dance' is the ardent pursuit of activities that sound good, look good, and allow managers to feel good—but which in fact contribute little or nothing to bottom line performance."

They go on to cite a 1991 survey which covered over 300 electronics firms. Findings indicated that 73% of the companies had TQM programs but 63% had improved quality defects by less than 10%. These are dismal results for efforts expended.[11]

Tracy E. Benson of *Industry Week* magazine remarked in its October 5, 1992 issue that, "The large number of quality activities in U.S. firms 'that aren't making a difference' is alarming....Remember *'The Emperor's New Clothes,'* the fairy tale about the Emperor whose robes were made of fabric so opulent that it could not be seen by those who were unfit for their jobs or impossibly stupid? The Emperor paraded daily before the townspeople until one day a young boy could contain himself no longer. 'The Emperor has no clothes,' he cried. So it is with quality."[12]

In a May 3, 1993 *IW* article, Benson wrote, "Employee-based programs that are designed to boost morale but are not tied directly to performance can actually frustrate and alienate workers instead." She goes on to quote the authors of *Quality on Trial* (West Publishing Co., 1992): Roger J. Howe, Dee Gaeddert, and Maynard A. Howe. They wrote "When the primary focus...is measuring participation in activities (e.g., number of hours of training, number of team meetings, number of suggestions, and the like) rather than on the impact of those activities on business results, the corporation's time and money are wasted."[13]

VERDICT ON TQM

Change and quality improvement programs offer great hope to the contractor facing seemingly overwhelming problems who is desperate for solutions. TQM has the potential not only to offer many real solutions while improving the bottom line, but it also has the ability to revolutionize a company and the management structure in it. However, **buyer beware! TQM is not a "magic wand."** Like the dollars and labor hours spent on personal computers during the last eight years, many contractors improved management performance and profits. Unfortunately, many other contractors wasted a lot of money and a lot of labor hours on PCs because precise expectations were not established and compared to measurable-targeted results. Contractors who make similar mistakes with TQM and who do not target specific incremental results-driven improvements will probably be very disappointed with the return in their investment and disheartened with the very concept of TQM. This is unfortunate because many others will obtain the results and see the benefits of the tried and proven methods of TQM when it is used properly.

TQM RE-EVALUATED

Philip B. Crosby, in *Quality is Free*, tackles two key myths regarding TQM and quality improvement. He stated, "The first erroneous assumption is that quality means goodness, or luxury, or shininess, or weight. The word 'quality' is used to signify the relative worth of things in such phrases as 'good quality,' 'bad quality,' and that brave new statement 'quality of life'...It is a situation in which individuals talk dreamily about something without ever bothering to define it...That is precisely the reason we must define quality as 'conformance to standards' if we are to manage it."[14]

Crosby goes on to say "In fact, quality is precisely measurable by the oldest and most respected of measurements—cold hard cash." [15]

Tom Peters builds on Crosby's basic premises when he challenges us to "MEASURE WHAT'S IMPORTANT."[16] And to remember that in today's marketplace, "What gets measured gets done has never been so powerful a truth."[17]

TQM is simply the threefold process of (1). establishing a performance standard, (2). collecting actual performance data, and (3). comparing the two to obtain a variance (or deviation) between them. As pointed out in the Introduction, this is a similar methodology as the one used by science. For the green industry, the first step is a budgeting or estimating one which this book addresses. The second and third are data collection and job-costing which are really information management issues.

The typical contractor has problems with the very concept of TQM for the following reasons. First, he or she simply does not understand the methodology (or the very concept) of TQM as described above. The process remains too ethereal and non-quantifiable. The "rubber never meets the road." Second, most contractors do not understand that you "de-mystify" TQM by means of the budgeting and estimating process and thereby make it an integral part of a business. Third, the typical contractor either does not have key staff trained to collect the right types of data and to compare it to the original performance standards (budgets and estimates) or does not have an accounting software system in place to do so in a cost-effective manner, or both.

Annual company/division budgets and individual job estimates provide the majority of the "standards" or targets required to run a landscape and irrigation company. It is the monthly profit and loss statement (for your company and/or division) and individual job-cost reports which provide the means of comparing "estimated" or budgeted standards to actual performance. It is this very process of identifying standards, collecting data, and comparing the two that comprises the "guts" of any TQM system. However, if you do not establish quantifiable measurable standards in the bidding and budgeting process, you can neither direct and control your jobs nor your company. You cannot effectively job cost a job bid by means of material-times-two or one whose price is calculated by the Market-driven Unit Pricing method. You have nothing to which to compare actual performance data.

This is why a good estimating system provides you with not only an accurate **PRICE,** but also a **PLAN** by which to run your job (and your company), plus a **PROCESS** by which to self-adjust the whole process.

Crosby hits the nail on the head when he states, "In business,...requirements must be clearly stated so that they cannot be misunderstood. Measurements are then taken continually to determine conformance to those requirements. The non-conformance detected is the absence of quality. Quality problems become non-conformance problems and quality

**Corporate Vital Signs
(The TQM Process)**

becomes definable. All through (his) book, whenever you see the word 'quality,' (it means) 'conformance to requirements' (standards)."[18]

CONCLUSION

Total quality management can significantly improve your operation. However, in order to use the TQM process effectively, you have to analyze and measure things against predetermined standards (budgets, goals, bids, etc.). This book, in essence, shows you how to establish the standards upon which to build an effective TQM program. However, do not be tricked into thinking that only budgeting and bidding fall under the TQM realm. Chapter 19 covers the Bid Board and the MAD Report which help you monitor the nebulous process of bringing business into the company. They help you to "grab the handle," as we say, by measuring the process of selling jobs and comparing targeted standards to actual performance. Just about anything can be measured and brought into the TQM umbrella. It simply is not necessary to measure everything, only that which is important. Analysis can lead to paralysis if not properly monitored and data is simply collected and never used. One of the key roles of the owner/manager is to collect and focus upon the right data at the right time and to understand the process in doing so.

TQM can improve the whole company and the systems/processes in it if it is understood and used properly. Otherwise, "magic wand" management and management by "mysticism" prevail as owners and/or managers employ "rain dance" methods in their attempts to control and direct their companies.

CHAPTER 26
COMPUTER ESTIMATING

PURPOSE
To outline the pros, cons, and pitfalls of computer estimating

Estimating with a personal computer (PC) holds great potential for you and your company **IF** it is approached wisely and done correctly. PCs can "move" (calculate and reformat) huge quantities of numbers, letters, and symbols in a very short period of time. PCs can do to data what backhoes do to dirt, so to speak.

POTENTIAL BENEFITS OF PC ESTIMATING

1. Increases accuracy and consistency.
2. Increases your ability to control the bidding process.
3. Increases confidence (yours and your clients).
4. Increases bid volume per bidding labor hour (up to 100% increase).
5. Decreases overhead per field-labor hour (20-30%).
6. Increases your ability to plan and run jobs.
7. Provides data for job costing.
8. Can be used as a sales tool.
9. Provides a packaged system for teaching other employees how to bid.
10. Eliminates errors in arithmetic.
11. Thoroughly analyzes job prices.
12. Produces unit and lump sum prices simultaneously.
13. Familiarizes your employees with computers (and their other applications within your company).
14. Clearly defines quality standards.
15. Produces tools for quick analysis and review of bids.

WHAT BACKHOES DO TO DIRT...

16. Permits large amounts of data to be manipulated accurately by means of task data libraries.

17. Excites/motivates your people by improving their professional skill level.

18. Increases bidding flexibility and versatility.

DRAWBACKS OF PC ESTIMATING

Some people become entranced by these little white boxes called computers, thinking that they are a panacea for their corporate ills. You must combine the right hardware and software. Your estimating software must reflect correct business philosophies, methodologies, systems, and strategies. If not, **all you are doing is making mistakes at the speed of light.**

LESSONS LEARNED

Steve Smith and I have computerized the estimating of numerous construction and maintenance service contractors throughout North America. Here are some of the lessons that we have learned:

1. You must have a stable staff in place who have the aptitude and time to get your estimating software up and running.

2. Implementation should take thirty (30) days, as long as you work with the system two to three hours per day producing bids. In other words, you should feel comfortable producing bids after about one month if you have consistently used the system.

3. During the implementation period, qualified support and training for your staff is crucial.

4. You should plan to have a qualified estimating trainer return to your office every six months in order to review your bidding system (tune it up) and to provide additional training.

TOOLS PROVIDED BY COMPUTER ESTIMATING:

Estimating software should provide the following tools and items that will help you to run your company and to run those jobs that were bid (see Exhibits (11) through (13) also).

1. Unit and lump sum prices.
2. Material lists (help in ordering material).
3. Capacity to produce proposals.
4. Overhead costs per field-labor hour (OPH).
5. Sales per field-labor hour (SPH).
6. Profit per field-labor hour (PPH).
7. Unit pricing by the square foot, yard, etc.
8. Equipment-to-labor ratio analysis.
9. Material-to-labor ratio analysis.
10. Equipment costs analysis.
11. Labor tables by type labor and crews.
12. Prevailing wage tables.
13. Equipment tables for specific equipment rates.
14. Overhead recovery.
15. Field-labor hours per job per function.
16. Detail analysis as required.

THE BIG MISTAKE

A good estimating system should provide you with a tool that does one more very important thing for you. It should help to prevent you from getting jobs that you would be wise not to take. I know numerous contractors who have taken jobs that they should not have taken—jobs that they would not have taken if they had had the right information in front of them at the time decisions were being made when the bid was being submitted. Eliminating big mistakes by eliminating little mistakes that add up to big ones is one of advantages of a computerized estimating system. It should help you to recap and review a bid quickly once the bid is finalized.

Making Mistakes at the Speed of Light

Reviewing a bid properly can identify production rate errors; erroneous units; mathematical formula errors; inaccurate costs; wrong labor rates, burdens, average wage; etc. When I am reviewing bids, it is not uncommon for me to find mistakes that total 2-10% of the total bid price. That is a lot of percentage points to come out of the net profit margin (where else is it to come from?), especially when you are only bidding ±10% net profit to begin with.

Certainly, you want your computer estimating system to get you work by bidding it more accurately and quickly. However, in so doing, you also want it to provide a tool that will help you review bids and accurately calculate **ALL** costs associated with a particular job. This is your **BEP.** You then know with confidence how low you can go on a bid before you start cutting into meat and bone. Identifying this number properly (the BEP) can save you a lot of lost dollars, because it will help prevent you from getting into the mud with "low-ballers" and winning bids and getting jobs that you do not want. It can help prevent you from making that Big Mistake.

SUMMARY

Computerized estimating provides significant potential to improve your bottom line. By developing the right philosophies, methodologies, systems, and strategies **(PMSS),** you can learn to bid with confidence and to develop tools that will help you understand your bidding, and your competitors' bidding much better. Computerized estimating offers many ways to improve profits by providing the tools to recap and review bids easily and accurately, and ways to prevent big mistakes that can take you years from which to recover. But you have to be careful. If your estimating software does not reflect the correct methods, **all you will be doing is making mistakes at the speed of light.**

ACTION POINT

Review the sample construction and maintenance bid examples in Exhibits (11) through (13) and take note of how they display all five of the estimating methods previously discussed as well as other key data, ratios, and standards.

CHAPTER 27
ORDERING AND CONTROLLING MATERIALS

PURPOSE
**To explain the process of ordering and controlling materials,
and key tools used to do so**

INTRODUCTION

The ability to produce accurate bids means little if you cannot control your material inventory, resources, and tools. The following is offered to help you maintain accountability of some of the "nuts and bolts" of your operation.

USE PURCHASE ORDERS

Purchase orders can cut material order processing time while increasing your ability to job cost quickly and accurately.

Further ordering time can be saved, and you will have greater control over the amount of material you order, if you have computerized estimating that produces a materials list for the job. This list of items and quantities can then be annotated onto purchase orders and subsequently faxed to vendors.

1. BENEFITS OF USING PURCHASE ORDERS

A. Job costing accuracy increases.

All orders can be identified to a particular job and to a specific individual ordering the material.

B. Reduced labor is required to job cost.

Providing purchase order numbers to vendors for specific deliveries to specific jobs speeds up the verifying and matching of purchase orders to delivery tickets and invoices.

C. Accountability, or an audit trail, is established.

This allows you to verify how much of a particular material was ordered for a particular job and to identify and stop potential theft.

D. Controlling inventory is possible.

By putting "stock" in the supplier/vendor portion of the purchase order, purchase orders can help you track the use of inventory or stock on hand.

E. Verification of bid quotes is also possible.

Purchase orders allow you to match and compare actual invoice prices to quotes received when you were bidding the job. This allows you (or your bookkeeper) to request credit for overcharges.

F. The foreman's work is simplified.

Purchase orders can help your foremen organize themselves by providing a simple, yet effective, tool that clearly identifies the material ordering process.

2. PROCEDURE FOR USING PURCHASE ORDERS

The individual using and/or installing the materials should be the person ordering them. This eliminates finger-pointing and excuses. Your foreman should realize that part of being a foreman is the scheduling and ordering of materials and the processing of necessary paperwork.

Specific individuals allowed to order materials should be identified, in writing, to vendors. Only these individuals should be allowed to order materials and ONLY if they provide a purchase order number when ordering.

No single purchase order should be used to order materials for more than one job. This will greatly simplify job costing.

Vendors should be asked to enter your purchase order numbers on all delivery tickets and invoices.

It is not uncommon for contractors to fax copies of their purchase orders to vendors thus eliminating the need to dictate items ordered over the phone.

Some contractors use a different size purchase order (e.g, 5" x 8" vs. 8" x 11") to help identify and document change orders.

3. SOURCES FOR PURCHASE ORDERS

Have purchase orders printed with your company name and address on them. Buy blank two-part NCR paper and make your own purchase orders on your computer's printer. Or buy generic purchase orders (available from an office supply store, fifty to the pad in two-part NCR format).

4. FORMAT (SEE DIAGRAM (27-1))

Each purchase order should be sequentially numbered and should have a place for the following:
- Vendor's name and salesperson's name
- Job name
- Date materials required
- Date ordered
- Unit
- Quantity
- Item description
- Signature of person ordering material

INVENTORY CONTROL OF STOCK AND MATERIAL UPON RECEIPT

1. MAKE IT A RULE TO HAVE PERSONNEL INVENTORY OR COUNT A DELIVERY BEFORE SIGNING FOR IT.

No exceptions. This may take a little more time, but it can be time well spent.

2. USE PURCHASE ORDERS WHEN ITEMS ARE TAKEN OUT OF COMPANY STOCK.

Put "stock" or "inventory" in the vendor section of the purchase order. Use fair market value for these items when you job cost them.

3. INVENTORY STOCK ON A REGULAR BASIS.

Conduct an in-depth inventory at least every six months. Do a random spot check on 5-10% of your line items during the months between inventories. If you discover a discrepancy, investigate more fully.

4. PROVIDE SECURE INVENTORY LOCATIONS.

REMEMBER: "a lock is not meant to keep a thief out; it is meant to keep an honest man honest." Do not provide

| | | | PURCHASE ORDER | | | | | ORDER NO. 0496 |

PURCHASE ORDER

ORDER NO.
0496

TO _____   SHIP TO _____

ADDRESS _____   ADDRESS _____

CITY _____   CITY _____

REQ. NO.	FOR		DATE REQUIRED	TERMS	HOW SHIP		DATE
QUANTITY			**PLEASE SUPPLY ITEMS LISTED BELOW**			**PRICE**	**UNIT**
ORDERED	**RECEIVED**						
1							
2							
3							
4							
5							
6							
7							
8							
9							
10							
11							
12							
13							
14							
15							
16							
17							
18							
19							
20							
21							
22							

IMPORTANT

OUR ORDER NUMBER MUST APPEAR ON ALL IN-
VOICES, PACKAGES, ETC.

PLEASE NOTIFY US IMMEDIATELY IF YOU ARE UN-
ABLE TO SHIP COMPLETE ORDER BY DATE SPECIFIED.

PLEASE SEND   COPIES OF YOUR INVOICE WITH ORIGINAL BILL OF LADING

PURCHASING AGENT

1H 146 REDIFORM.   **ORIGINAL**

DIAGRAM 27-1. SAMPLE PURCHASE ORDER

temptation or opportunity for your people to steal.

Without locked and secure facilities, it is not a question of *if* people will steal, but ***when.*** If you allow for the opportunity, ***it will happen.***

5. INSIST ON TIMELY JOB COSTING.

One of your best methods of preventing unnecessary loss or theft is to job cost invoices quickly. Identify overages A.S.A.P. Show everyone that there is a system in place that will hold people accountable.

TOOLS

1. USE A PURCHASE ORDER WHEN ORDERING ALL TOOLS. AGAIN, NO EXCEPTIONS.

2. KEEP YOUR TOOLS:

A. Paint hand tools an obnoxious color. They will stick out like a sore thumb, and fewer people will want to take them home.

B. Assign specific tools, or a basic issue of tools, to each individual.

- Provide a specific place where people can store their tools (e.g., a locker or bin with the employee's name on it).
- Have foremen inventory these tools at the end of each day.
- If legal in your locality, hold individuals responsible for replacing "lost" tools.

C. Subtract dollars spent on "lost" tools, or tools broken due to negligence, from the bonus pool. Be certain to identify to all concerned in the pool the amount deducted for tools.

D. Buy high quality.

Good tools last longer. The question is whether your company or a thief is going to be the benefactor of this extended life.

GASOLINE AND FUEL

The rule is that checks and balances prevent theft. Without checks and balances, theft is going to happen. The key to controlling gasoline and fuel purchases and preventing abuse is two-fold:

1. Have a system that tracks purchases to specific individuals and/or vehicles.

Whether you use commercial credit cards or a bulk fuel outlet, be certain that you review your monthly statements and/or receipts. Ensure that you are able to identify who made the purchase.

2. As you review statements and receipts, look for excess quantities or the fueling of non-company vehicles.

Let people know that you are monitoring fuel purchases by asking questions of the appropriate people. Simply asking questions subtly lets people know that someone is watching this area, and that fact alone may very well reduce temptation and possibly prevent a theft.

PAYING INVOICES/BILLS

The method by which you process and pay your invoices can help to save thousands of dollars by identifying overcharges, missed credits, and possible misappropriation. Here are some tips:

1. INVOICES

A. Process invoices quickly.

B. Job cost invoices within two to three days of receiving them.

Memories tend to fade the longer invoices sit around unprocessed. It becomes more difficult to solve any problems or to answer with accuracy any questions that may arise.

C. Match purchase orders to delivery tickets and invoices.

Again, this matching is done as soon as the invoice arrives, not when you are ready to submit it to the owner and/or field superintendent for final approval before payment.

2. PAYMENT

A. Only one person (possibly two) should be authorized to sign checks. Bookkeeping and accounting personnel should NOT be allowed to sign checks.

B. Checks should not be signed unless all supporting data (i.e., purchase orders, delivery tickets, invoices) is attached to them prior to signing. Since you (hopefully) have already pre-approved these items, the whole process should be very simple and completed very quickly.

C. Staple a copy of the check to the company's copy of the invoice, or annotate the check number on the invoice copy.

D. Attach the purchase order and delivery tickets.

E. File this supporting documentation in a vendor file for future reference.

3. BANK STATEMENTS AND CANCELED CHECKS

The owner or CEO should be the only person authorized to open bank statements arriving in the mail that contain canceled checks. This is done to prevent someone from doctoring checks or removing fraudulent payments.

CONCLUSION

Accurate and insightful bidding won't mean a thing if "leaks," misappropriation, or theft drain your business.

Good people do bad things when checks and balances are not in place. This may seem mean and harsh, but I have seen case after case of theft by people I have known and trusted. I have testified at courts martial and have seen military officers lose their careers and their futures over a couple thousand dollars worth of merchandise.

Not long ago, a contractor I know well told me that three of his employees had embezzled over $40,000 from his company. This shocked me, as I knew all three of the employees.

I have seen large companies that were out of control lose $20,000 a month in inventory. It just disappeared.

You may have a family business and think that you are exempt. I know of an individual who was embezzling thousands of dollars from his own brothers by running invoices for "side jobs" through the company. I know of a man and wife who were doing the same thing in a business owned by the man's father.

Believe me, **if you provide the opportunity, theft is going to happen.** The key to ordering and processing materials is to implement a system of checks and balances that will remove the possibility of theft or misappropriation.

ACTION POINT

Review your procedures and policies for ordering and controlling materials and establish sound checks and balances where needed.

CHAPTER 28
NOTES FOR A CPA/ACCOUNTANT

PURPOSE
To explain my role as a consultant to the CPA/outside accountant and how we can best work together as a team for the benefit of the contractor

The pace of operations both in the field and in the office of a landscape or irrigation company makes it essential that outside resources associated with the company work together and are all pulling in the same direction. Otherwise, the contractor receives conflicting information and is often confused.

When I begin to work with a new client, I like to meet or at least talk with his CPA and/or outside accountant as early as possible in the process. I want to be more than just a name or a face (who in some cases may be perceived as a threat) to these key individuals who impact decision making and who help implement systems in the organization. To this end, I encourage clients to bring their CPAs or outside accountants (at no charge) to the first day of my two-day estimating workshops. The better that we understand what the other is attempting to accomplish, the more we can supplement each other and implement information management systems that allow the client to direct and control his operation more effectively and more profitably. It is almost axiomatic that good business results from good relationships. Therefore, as a member of a cohesive team, I can better meet the expectations of the client, implement sound business systems, and improve the directing and controlling of the company.

MY CONSULTING OBJECTIVES FOR THE CONTRACTOR
There is no magic wand to wave or fairy dust to throw that implements the systems, methodologies, and strategies overnight. It takes two to three seasons to collect and to refine the data and to thoroughly train everyone as to how to use the systems necessary to run a company. Even when good systems are already in place, it takes this long for two reasons. First, the fast pace and operational necessity simply do not allow you to stop everything and to fix what is broken. It is like changing a flat tire on a car that is still moving down the road—pretty difficult. Second, it requires two seasons of data collection and analysis to diagnose all (or most all) of what is happening within a company. Therefore, it is not until year three that the fine details can be effectively addressed. Consequently, my desire is to develop a long-term relationship with a client and to bring about change in this manner. **"Fixing" a contracting company requires a marathon approach, not a 100-yard dash one.**

It will take a contractor four to five years to master and implement all of the products and services that I have to offer, whether it be budgeting and estimating, strategic planning, computer systems, etc. Meanwhile, I am busy adding new products and services that my client base will need in the years ahead. Lifetime clients, repeat business, and referrals are at the very center of my business plan. My long-term objective for the client is not only to implement sound information management systems as quickly as possible but also to provide him or her and the related staff with an MBA (Masters Degree in Business Administration) education of sorts, concerning their business during the course of our relationship. Therefore, the team approach becomes even more critical when viewed over the entire life cycle of the business.

MY CONSULTING STYLE

FUNDAMENTAL CONSULTING PRINCIPLES

There are fundamental principles upon which I build a consulting relationship. They are as follows:

1. Business is a **process.** You do not "go" into business, you "grow" into business. The "quick fix" approach simply is not effective.

2. All of the systems, methods, and strategies that I employ in a business are based on objective historical data as much as is possible. This approach provides a logical audit trail that others can understand and relate to. This is not "management by mysticism."

3. **Lifetime clients,** repeat business, referrals, and "partnering" (where clients aid in the development and refinement of products and services) are the center of my business plan. Customers are in a position to best critique all of my services. Therefore, I do not only continually attempt to qualify customers, but my customers, as part of a partnering relationship, also continually qualify and improve me—the consultant.

4. Business is often **paradoxical.** Consequently, my "style" of consulting is often:

   A. Immediate-results-driven but over a long period.

   B. Market-/big-picture-oriented while focused on minute details of the business.

   C. Market-/big-picture-oriented while focused on the personal development of individuals within the context of a business and by means of "mentoring."

   D. Seriously humorous (or is it "humorously serious?") in an attempt to break through the tension, mental blocks, anxiety, and tunnel vision that develop in the heat of operational battles.

   E. Revenue- and bottom-line-driven within the constraints of budgets, quality standards, safety, and schedules.

   F. Entrepreneurial yet socially responsible.

   G. Analytical yet intuitive.

   Analysis must lead to action; otherwise, it results in paralysis. Decisions must be made at the "gut" level but based upon, not contrary to, sound analysis. Otherwise, decision making becomes arbitrary and incommunicative to other members of the organization.

MY OBSERVATIONS REGARDING MISCONCEPTIONS CONTRACTORS HAVE ABOUT CPAs

Landscape and irrigation contractors quite often do not understand the area of expertise of a CPA/accountant and the limitations of the profession. I often hear a contractor complain that his CPA does not understand his business nor give him the advice that he needs to run his business. Upon further questioning, I usually discover that the contractor is expecting operational guidance which is not a part of the CPA's field of expertise.

I tell the contractor to look at the situation a little differently. Actually, asking your CPA or accountant for detailed operational guidance is unfair. Asking for this type of information is like an NFL football coach asking the team's sports medicine doctor for strategy as to how to play the next Sunday's game. The doctor can tell him all about a specialized area (e.g., training regimens, injury prevention, nutrition, etc.) concerning the players but that specialty does not include "field operations" (playing the game).

I hear two other common complaints from a contractor. The first is that his accountant's business has grown and that he does not get the personal attention that he once did. The second is that tax planning occurs too late in the fiscal year. Subsequently, actions cannot be taken to reduce the tax liability at year end. This often costs a contractor thousands of dollars and an accountant his client.

HOW WE (ACCOUNTANTS AND SHI) CAN BEST HELP A CONTRACTOR

I believe that the best way for both of us to help a contractor is for us to understand each other's discipline and field of expertise and where the two overlap. However, we both need to understand some important differences between us. In this manner, we can supplement each other without duplicating each other's work. I see the key areas of overlap or "juncture points" as follows:

1. BUDGETING

Three types of budgets were discussed in the Introduction. They are the tax planning budget, the cash flow budget, and the estimating budget. The first falls under the cognizance of an accountant. The third falls under my domain while the second could be either. It is important that we realize that although the three types of budgets do overlap, that there are important differences which are spelled out in the Introduction.

It is extremely helpful to a contractor to have all three types of budgets in-place and in similar formats. Doing so will help the contractor to relate to the budgeting process much better and it will eliminate confusion.

2. FORMATTING THE PROFIT AND LOSS STATEMENT

The P&L statement should have a format similar to that of the budgets above. Identifying direct versus indirect costs and the resultant gross profit margin (GPM) is very important in order for a contractor to properly identify job related costs from administrative ones. Splitting and allocating overhead item costs to divisions is not nearly as important as correctly identifying the GPM for that division. A sample format of a P&L is contained in Diagram (II-1) and discussed in Chapters 22 and 23.

3. JOB COSTING

Timely, accurate feedback (TAF) on a job-by-job basis is the number one need of any contractor. Whether job-cost reports are produced manually or by an accounting software program, getting this information in front of owners, division managers, crew foremen, etc. is the number one job for any information management system (IMS). Accurately identifying actual field performance GPMs for each job (to include equipment costs) can improve not only the corporate bottom line, it can also focus management attention on the right problems at the right time.

It is important for both the CPA and myself to strive to help the contractor implement an IMS that provides this type of information easily, accurately and regularly. Which leads to my next point.

4. ACCOUNTING SOFTWARE

If job costing is the number one need of a contractor, a timely, accurate P&L statement follows as a close second. Identifying the GPM by division on a monthly basis is a very important management tool for a contractor. While job cost reports provide a focused job-by-job scoreboard, the monthly division P&L statement looks at the big picture. These two used on a consistent basis and in the hands of the right people can empower key decision makers and improve production dramatically.

It does not matter if monthly statements are produced in-house or by an outside accountant. The important thing is that they are accurate, and produced and used on a timely, regular basis.

Because job costing is so critical to the directing and controlling of a contracting operation, it should be one of the primary considerations when purchasing accounting software. Chapter 20 identifies the key information that a job-cost program should be able to provide.

Because an accounting software program touches two critical yet different disciplines (accounting/tax planning and operations in the form of job costing and the P&L statement), it is imperative that both the CPA and I work in harmony concerning this crucial information management system for the company.

5. AUDITED STATEMENTS FOR BONDING

It is important that both the accountant and I are on the same information management systems if we are to maximize our assistance in helping a client obtain bonding. This is especially important if the bonding company requires an audited financial statement from the CPA.

6. LINE OF CREDIT (LOC)

A contractor with timely, accurate financial statements, thorough budgets (all three types), a strategic plan, backed by two professional consultants just a telephone call away, who are reinforcing one another's position and backing the client presents a formidable force when the client presents himself in front of a banker in order to negotiate a LOC or obtain other financing.

CONCLUSION

Let's talk. Other items could be added to the above list. The important thing is that you (the accountant) and I develop the list together in support of "our" client. We prosper only when the client prospers.

Please do not hesitate to call my 24-hour toll-free telephone number to discuss matters of mutual interest or to provide feedback concerning what I have written in these pages. As in my two-day estimating seminars, there will be no charge to CPAs or outside accountants for such calls. The more closely that you and I work together, the better it is for our clients and thus the better it is for us. As one of my favorite sayings goes, "The right hand washes the left, the left hand washes the right, and both hands wash the face."

CHAPTER 29
CONCLUDING REMARKS

DON'T PANIC!

This book covers a lot of material. My laptop computer screen shows that I started this writing project three years ago and that I have hundreds of hours into it to date. I have spent thousands of hours in the "trenches" teaching hundreds of

contractors and their staffs how to apply the material contained in these pages. Believe me when I say that I understand if you feel somewhat overwhelmed right now. But don't panic. Numerous contractors have successfully implemented the methods and systems described in this book. There also is plenty of help available to help you to do the same.

THINK LONG-TERM

It takes 2-3 years to implement all of the systems contained in this book. Once implemented, you still must constantly monitor and refine them. There is no magic wand or fairy dust to throw at your business problems to make them go away. Study your business. Keep peeling (analyzing) its layers away like the proverbial onion, until there is nothing left to peel. Forget about the quick fix. We're talking about hard work, long hours, bulldog tenacity, and

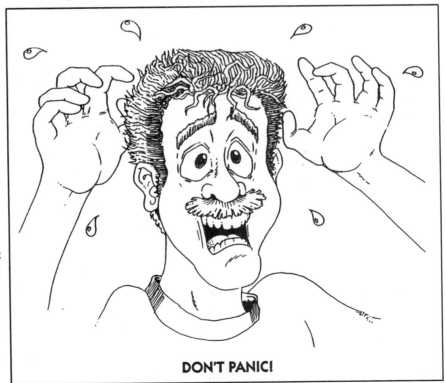

DON'T PANIC!

good old American ingenuity. You will see significant progress but it takes time because, as I said in the Introduction, fixing your business is like fixing a tire on a car that is still moving down the road—extremely difficult.

THINK SMART

It is important that you see and understand the big picture (the puzzle boxtop) for your company. Once you do, the various pieces of it begin to fall into place almost automatically. You can then begin to plan success into your company using methods similar to those of science to establish goals, targets, and budgets (standards) and to compare actual performance to them. It is this process of measuring actual performance against predetermined standards that produces a variance or deviation and which allows you to gain control over the organization as you attempt to reduce the variance to zero. Remember, a good estimating system not only produces an accurate **price** but also a **plan** and a self-correcting **process** as well.

Use this book as a reference tool in your company. Train estimators how to apply these principles at the bid table. Share the financial statement formats, definitions, and report formats with your accountant and CPA. However, keep in mind that you cannot expect to see significant results if you merely dabble with the methods described here. You must make a commitment and implement the whole system. Otherwise, you will not format and collect the data that you will need to adjust and refine the whole process. That is like owning a race car with only three tires: you do not own a race car, just a pile of very expensive junk that is not going anywhere.

Compare the prices obtained from the OPPH (overhead and profit per hour) method to the prices produced by your present method of estimating. They should be reasonably close. If they are not, examine why not and attempt to explain the differences. The OPPH method will produce a safe price if you have done your homework. If you still have questions, call our offices to set up a telephone appointment to further examine your bid.

Peeling an Onion

Keep in mind as you price a particular job that the key question is not how the GPM (gross profit margin) on that job compares to your annual budget. Rather, ask whether the job being bid at a certain price will help you realize your budgeted amounts of gross and net profit dollars (not percentages) for the year. Percentages are too misleading and mathematically incorrect when comparing yearly budgeted gross and net profit margins to individual jobs. It is dollars of gross and net profit that we desire, not percentages.

This leads to the next point: there is no right or wrong way to charge for or recover overhead. If there was, overhead would be a direct cost. Consequently, the goal is not to determine the "right" way to charge for overhead but rather to effectively measure, allocate, and control overhead costs throughout the company. The OPPH method facilitates this process better than any other.

STICK TO AND CONSTANTLY REVIEW THE BASICS

Since page one, we have discussed pricing (estimating) in conjunction with planning and within the context of the overall business process. They all must work together. You take care of the business (the systems) and the business (the systems) will take care of you. It is the responsibility of top management (officers, owners, and managers) to ensure that the correct systems are in place and used properly.

An estimating system, whether manual or computerized, must reflect the correct philosophies, methodologies, systems, and strategies. Otherwise it may contain serious and even fatal flaws. The primary function of an estimating system is to identify two numbers. The first is the **BEP** (break-even point) or **how low** you can go on a bid and still cover all of your costs. The second is the maximum price that the market will bear or **how high** you can go and still get the job. It is this bidding "envelope" that you are attempting to identify in the bidding process. Successful companies consistently price their jobs within this envelope. In this light, four of the five most common methods of pricing used in the market today contain serious shortcomings as they do not accurately or easily identify this bidding envelope.

The OPPH method of estimating compensates for the shortfalls of the other four methods better than any other. Use it, compare prices produced by it to your present pricing methods, and collect job-cost data to compare bid-to-actual performance. Make adjustments as necessary.

Fill out a BAR (bid analysis review) worksheet for jobs that you have already bid using your present system and compare the ratios, percentages, and per hour standards to ones for the same jobs bid using the OPPH method. This exercise should give you some significant insight into your market as well as your present bidding methods.

GET HELP

Our books, computer estimating software, audio and video tapes, workshops, seminars, and personal consultations are all designed to help you implement sound business management systems as quickly and as efficiently as possible. However, in order to optimize this process as we attempt to drive, direct, change, and control the whole company towards increased profitability, let me offer some suggestions that many of our clients have used successfully.

Talk to other contractors about the systems and methods covered in this book. Obtain a list of references either from us or from a client or a state or local association with whom we work. Read our other books and attend our two-day estimating workshop or a brainstorming seminar and meet other contractors using these methods. Show this book to your CPA or accountant and discuss it with him.

Clients who take advantage of the full range of services and products that we offer realize the most improvement in their companies. Just like you, they were once skeptical. They were not about to waste valuable time and hard-earned dollars on unproven methods and overnight-wonder consultants. However, they did the smart thing. They obtained additional information, did their homework, and made an informed decision. Once that decision was made and the process begun, most found that the best way to get everybody (estimators, bookkeepers, owners, etc.) trained and marching in the same direction, was to have one of our consultants come into their offices for one day every six months or so. This approach maximizes training while shortening the learning curve. Such 5 to 6 month increments of training allow individuals to use the information presented and to absorb it at their own pace. Questions between visits are usually easily handled by telephone.

I tell contractors to think of us in terms of that regular doctor's examination that we all should get. Have us come in and monitor the company's "vital signs" and train everyone how to get and stay healthy. It usually costs less than the amount that a contractor pays his accountant and lawyer each year. More often than not, the money you will save on one or two bids that we will help you with during the season will cover the cost.

Do not fall into the trap that some contractors fall into. They like the systems and the methods taught in this book. However, they do not want their competition to discover them. This is incomplete thinking. Think of it this way. If contractors in a given geographical area all had accurate estimating methods, they would reinforce one another's prices more often than not. Costs are costs and with few exceptions, everyone's costs are about the same. The problem occurs when low-ballers bid work and ruin the pricing in a market because they do not identify and fail to include all of their costs during the bidding process. They keep grinding down and diluting prices until finally they go out of business. Unfortunately, it takes them five years to do so and for every one low-baller that goes under, two more spring up and take his place. It's like killing one fly and having 50 more come to the funeral.

Contractors with accurate estimating systems should and will reinforce one another's pricing. Therefore, the more contractors who use accurate estimating methods in a given market, the better. Encourage your local, state, and national associations to provide estimating training materials and workshops in your local market. It will help all concerned.

TRAIN YOUR PEOPLE

Successful contractors consistently use and teach their people to use good habits. They coach and train their people by providing a "protected" environment of sorts. Call it a business greenhouse, if you will. Within this environment are all the ingredients necessary for success: systems, goals, budgets, methods, etc. It is within this protected environment that people are constantly pursuing goals while exchanging ideas and information. It is a beehive of networking activity.

Leaders are readers. They never stop learning. They teach their people to do the same. Leaders constantly take notes, ask questions, and seek out new technologies and ideas. Today's leading small business contractors see the benefits of the management and information revolution which personal computers are bringing to the industry and they are constantly training their staffs to take advantage of them.

CONCLUSION

I trust that this book has helped you deal with the "numbers" side of your business more effectively. If you love numbers (like the Count on Sesame Street), great! If you do not, I hope that I have made using them more meaningful and a little less painful.

Please make a copy of and fill out the survey form in the back of the book and tell me what you thought of *Estimating for Landscape & Irrigation Contractors*. Better yet, get on board that train that is revolutionizing contracting management as it heads toward the year 2000 and come join me at a seminar. The trip won't be easy. Nor will it be a stroll through Hollywood. However, it should be fun and it should help us make a little money.

Come join the revolution!

Just another day in Hollywood

APPENDIX

A

PRACTICAL EXERCISES

INTRODUCTION: The following Practical Exercises assume normal site conditions. Unless otherwise provided, the following items apply.

Crew...................4 men, 8 man-hours each per day
CAW....................$8.33 per hour
OTF....................10.0% (company average for season)
RF.....................10.0%
CAW (w/OTF & RF)....$10.00 (rounded down for ease of math)
Labor burden.........30.0%
Sales tax.............6.0%
OPH...................$10.00

Answers to Practical Exercises are found in the back of Appendix A. Some computer calculations on the Bid Analysis Review (BAR) Worksheets differ from manual calculations due to rounding by the computer. These differences are insignificant.

Exhibits 11-13 provide additional computerized construction, commercial maintenance and residential maintenance bid examples, respectively. They were bid using the Smith Huston Estimating Program. The bid in Exhibit 11 is for the Henderson residence project outlined below. Review these bids to gain additional insight into the various ways of formatting your bids and the data contained in them.

CHAPTER 11

1. Mr. Barkley wants two 2-2.5" caliper Red Maples and 10 assorted five gallon shrubs planted in his front year. He also wants 600 square feet of sod installed with the appropriate amount of soil preparation included. Mr. Barkley's home is approximately 45 minutes from the yard. The desired net profit margin for the job is 15%. Wholesale material costs, equipment costs and production rates are as follows:

Amendments...............$18.00 per cubic yard
Labor....................1 cubic yard per man-hour to spread
Rototiller CPH...........$5.00
Red Maples...............$125 each or $131 with a 5% loss factor
Labor....................1 man-hour per tree
Stake kits...............$8.00 each
Labor....................2 stake kits per man-hour
Plant tabs...............$ .10 each, 10 per tree
Five gallon shrubs.......$11.50 each or $12 with a 5% loss factor
Labor....................5 shrubs per man-hour
Plant tabs...............$ .10 each, 2 per shrub
Sod......................$ .30 per square foot
Labor....................200 square feet per man-hour
Crew pickup truck CPH....$4.00

A. Fill out the Recap Bid Worksheet for the project.

B. Fill out the Planting Bid Worksheet for the project.

C. Fill in the material and labor portions of the Planting Bid Worksheet.

D. Fill in the rototiller portion of the Planting Bid Worksheet.

2. Prepare and fill in all of the Bid Worksheets for the Henderson project as outlined below. Do not fill in the General Conditions or Markup portions of the worksheets.

A. Materials list and production rates:

	Item	Qty	Unit	Unit cost	Qty/MHR
1.	Point of connection	2	HR	CAW	N/A
2.	Controller w/hardware	1	EA	370.00	.5
3.	Wire UF # 14	2,000	LF	.04	500
4.	Backflow preventer	1	EA	347.00	.25
5.	Quick coupler .75"	4	EA	34.00	1
6.	ML schedule 40 1.25"	240	LF	.32	25
7.	Sleeve schedule 40 2.5"	120	LF	.82	25
8.	Laterals class 200 1"	520	LF	.08	25
9.	Laterals class 200 .5"	1,200	LF	.05	25
10.	Valves w/box & fittings	14	EA	82.50	.5
11.	Heads, 6" pop-up	170	EA	8.50	5
12.	Heads, 4" pop-up	60	EA	8.50	5
13.	Fine tune system labor	8	HR	CAW	N/A
14.	Soil prep	6,300	SF	N/A	1,500
15.	Trees 24" box	6	EA	95.00	.5
16.	Trees 15 gl	15	EA	26.00	1
17.	Stake kits (2/tree)	21	EA	9.00	3
18.	Palm (Wash.Rob.) 10' BTF	3	EA	250.00	.25
19.	Grade	6,300	SF	N/A	1,000
20.	Shrubs 5 gl	215	EA	6.00	5
21.	Shrubs 5 gl espalier	3	EA	19.50	4
22.	Shrubs 1 gl	283	EA	3.00	12
23.	Vine 1 gl	24	EA	3.50	12
24.	Grnd cover (8"OC 64/flt)	23	FLT	9.00	2
25.	Grnd cover (12"OC 64/flt)	36	FLT	9.00	2
26.	Sod	3,345	SF	.31	225
27.	Mulch (1" deep 2,955 SF)	9	CY	18	1
28.	Maint. 90 days w/1 fert.	3	MO	N/A	N/A

B. Miscellaneous notes:

1.	Pipe fittings	25% of pipe cost			
2.	Nitro sawdust (4CY/M)	25	CY	16.00	1
3.	Fertilizer (125#/M)	16	BAG	7.50	15
4.	Gypsum (120#/M)	10	BAG	8.00	15
5.	As-builts	1	LS	100.00	N/A
6.	Soil test	1	LS	75.00	N/A
7.	Plant tabs	96	EA	.10	N/A

C. Equipment:

		CPH
Qty/HR		
1.	Tractor/Bobcat..........................18.00	
2.	Rototiller...............................6.00	1,500
3.	Const. crew 1/2 ton pickup..............3.25	
4.	Trencher, walk-behind...................8.00	100
5.	Supervisor's 3/4 ton pickup............4.00	
6.	21" mower...............................1.25	7,500
7.	Edger & trimmer.........................2.00	
8.	Maintenance crew pickup................3.25	
9.	Materials/mobilization truck..........10.00	

CHAPTER 12

1. Fill in the General Conditions Bid Worksheet for the Barkley project in Chapter 11.

2. Fill in the General Conditions Bid Worksheet as outlined below for the Henderson project in Chapter 11.

```
Foreman admin time............1 man-hour per day
Crew truck....................8 hours per day
Mobilization-labor............2 man-hours at $9.00 CAW
Mobilization-truck............2 hours at $10.00 CPH
Haul materials-labor..........6 man-hours at $9.00 CAW
Haul materials-truck..........6 hours at $10.00 CPH
Clean up at end of each day...15 minutes per man per day
Waranty-labor.................16 man-hours at $10.00 CAW
Waranty-truck/equipment.......8 hours at $3.25 CPH
Supervisor....................1 man-hour per day at $15 CAW
Supervisor's truck............1 hour per day at $4.00 CPH
Detail job at end of job......8 man-hours $10.00 CAW
As-builts.....................$100.00
Soil test.....................$75.00
Crew drive time...............1 man-hour per man per day
load time.....................1 man-hour per day
```

CHAPTER 13

1. Calculate YOUR company/division annual budget for:

A. Total billable field-labor hours: _____

B. Total estimating overhead budget: $_____

C. Labor & burden overhead markup percent _____ %
   using the MORS Method.

D. Annual budget average SPH: $_____

E. Annual budget average OPH: $_____

F. Annual budget average PPH: $_____

2. Complete the following exercise:

Calculate a price for the Jones' residence project using the data
provided, the OPPH overhead recovery method and a BAR Worksheet.

          Materials...........$6,000
          Field-labor CAW.....$10.00 (with RF & OTF)
          Labor burden........30.0%
          Sales tax...........6.0%
          Equipment...........10 days for a one-ton pickup at $5 CPH;
                              32 hours of tractor time at $15 CPH
          Subcontractors......None
          Job duration........10 crew-days at 10 hours per man per day
                              of which 60 hours are general conditions
          Crew................3 men
          OPH.................$10.00
          PPH.................$5.00

3. Add the appropriate Markups from your company and a net profit
margin of 15% to the Barkley project outlined in Chapters 11 and
12.

4. Add the appropriate Markups from your company and a net profit
margin of 15% to the Henderson project in Chapters 11 and 12.

5. Fill out a BAR Worksheet for the Barkley and Henderson
projects.

CHAPTER 14

1. Calculate the GPM (overhead and profit) markup percentages
for the the Henderson project to be used for calculating unit
prices. Utilize an OPH of $10.00 and a net profit margin of 15%.
Do not include maintenance in the unit price calculations.

2. Calculate the unit and lump sum prices for the irrigation
items in the Henderson project.

3. Calculate the unit and lump sum prices for the landscape
items in the Henderson project.

CHAPTER 15

1. Use a BAR Worksheet to recalculate the curb-time crew rate
for the example in Diagram (15-3) after changing it as follows:

A. Increase daily drive time from 2 to 4 hours.

B. Decrease labor burden from 45% to 30%.

C. Decrease man-hours per day per man from 9 to 8, 5 days per
week. Also decrease daily production labor hours from 16 to 12.

2. Calculate the following:

A. Curb time rate per man-hour and per crew-hour for the following project using the Bid Worksheet.

        Crew...........................3 men
        CAW............................$8.00
        OTF (40 hrs/wk, 8 hrs/day)....0.0%
        RF............................10.0%
        CAW loaded....................$8.80
        Labor burden..................25.0%
        Materials.....................None
        Subcontractors................None
        Pickup & trailer CPH..........$4.00 (8 hours/day)
        48" ride-on mower CPH.........$3.75 (6 hours/day)
        21" push mower CPH............$1.25 (5 hours/day)
        Edger CPH.....................$2.00 (4 hours/day)
        Blower CPH....................$2.00 (2 hours/day)
        Drive & load time.............1 hour per man per day
        OPH...........................$5.00
        Net profit margin.............10.0%

B. Fill out a BAR worksheet for exercise 2A.

Note: The above exercises were compiled utilizing numerous bidding scenarios found throughout North America. They may differ somewhat from the practices found in your specific geographical region. However, they are still valid for instructional purposes.

BID WORKSHEET

PROJECT: __BARKLEY RESIDENCE__ PAGE (1) OF (3)

LOCATION: __789 MCKINLEY ST. LOS ANGELES__ DATE PREPARED: __7__ / __25__ / __9___

P.O.C./G.C.: __MR. BARKLEY__ ESTIMATOR: __M. THRUE__

DUE: __8__ / __1__ / __19___ PH.: __619 726-2185__ FAX: __N/A__ CREW SIZE: __3__

TYPE WORK: __RECAP__ HOURS/DAY: __8__ CAW: $ __10.00__

REMARKS:

DESCRIPTION	QTY	UNIT	U/C	MAT'L	LABOR	EQUIP	SUBS
- PLANTING							
- GENERAL CONDITIONS							

SHI FORM 04-A

BID WORKSHEET

PROJECT: __BARKLEY RESIDENCE__ PAGE (2) OF (3)

LOCATION: _____

P.O.C./G.C.: _____ DATE PREPARED: ____ / ____ / ____

DUE: ___ / ___ / ___ PH.: _____ FAX: _____ ESTIMATOR: _____

TYPE WORK: __PLANTING__ CREW SIZE: __3__

HOURS/DAY: __8__ CAW: $ __10.00__

REMARKS:

DESCRIPTION	QTY	UNIT	U/C	MAT'L	LABOR	EQUIP	SUBS
– SOIL PREP AMEND.<5CY/M>	3	CY	18 –	54			
– SPREAD AMEND.<1CY/MHR>	3	HR	10 –		30		
– ROTOTILL LABOR<1MHR>	1	HR	10 –		10		
– ROTOTILLER<1HR MIN>	1	HR	5 –			5	
– RED MAPLES 2-2.5"	2	EA	131 –	262			
– LABOR <1/MHR>	2	HR	10 –		20		
– STAKE KITS	2	EA	8 –	16			
– LABOR <2/MHR>	1	HR	10 –		10		
– PLANT TABS <10/TREE>	20	EA	.10	2			
– SHRUBS 5 GALLON	10	EA	12 –	120			
– LABOR <5/MHR>	2	HR	10		20		
– PLANT TABS <2 EA>	20	EA	.10	2			
– FINE GRADE	1	HR	10		10		
– SOD	600	SF	.30	180			
– LABOR <200 SF/MHR>	3	HR	10 –		30		
– ROLL SOD	1	HR	10 –		10		
				636	140	5	∅

SHI FORM 04-A

PHASE I. IRRIGATION COSTS:

BID ITEM	QTY	UNIT	UNIT COST	MATL'S IRRIG	MATL'S PLANTING	LABOR $$	EQUIP $$	SUBS	
TOTALS				$4,402	$0	$1,820	$156	$0	$6,378
POC LABOR	2	HR	10.00			20			
BACKFLOW	1	EA	279.00	279					
FITTINGS	1	LS	68.00	68					
LABOR (2MHR/EA)	4	HR	10.00			40			
MAIN LINE 1.25" SCH 40	260	LF	0.32	83					
FITTINGS 25%	260	LF	0.08	21					
LABOR (4MHR/100')	10	HR	10.00			100			
TRENCHER (100'/HR)	2.5	HR	8.00				20		
SLEEVES 2.5" SCH 40	140	LF	0.82	115					
LABOR (4MHR/100')	5	HR	10.00			50			
VALVES	14	EA	64.00	896					
BOX	14	EA	10.00	140					
FITTINGS(7/VLV/$1EA)	14	EA	7.00	98					
PENTITES(3/VLV/$.5EA)	14	EA	1.50	21					
LABOR (2MHR/EA)	28	EA	10.00			280			
WIRE UP # 14	2000	LF	0.04	80					
LABOR (500'/MHR)	4	HR	10.00			40			
QUICK COUPLERS .75"	4	EA	20.00	80					
BOX	4	EA	10.00	40					
FITTINGS	4	EA	4.00	16					
LABOR (1MHR/EA)	4	HR	10.00			40			
CONTROLLER, WALL MOUNT	1	EA	370.00	370					
LABOR (2MHR/EA)	2	HR	10.00			20			
LATERALS 1" CL 200	580	LF	0.08	46					
FITTINGS 25%	580	LF	0.02	12					
LABOR (4MHR/100')	21	HR	10.00			210			
TRENCHER (100'/HR)	5	HR	8.00				40		
LATERALS 1/2" CL 315	1320	LF	0.05	66					
FITTINGS 25%	1320	LF	0.013	17					
LABOR (4MHR/100')	48	HR	10.00			480			
TRENCHER (100'/HR)	12	HR	8.00				96		
HEADS, 6" POP-UP	170	EA	6.00	1,020					
SWING JOINT	170	EA	2.50	425					
LABOR (5/MHR)	34	HR	10.00			340			
HEADS, 4" POP-UP	60	EA	6.00	360					
SWING JOINT	60	EA	2.50	150					
LABOR (5/MHR)	12	HR	10.00			120			
FINE TUNE	8	HR	10.00			80			

> A computerized version of The Bid Worksheet.

PHASE I. SOIL PREP & SOD COSTS:

BID ITEM	QTY	UNIT	UNIT COST	MATL'S IRRIG	MATL'S PLANTING	LABOR $$	EQUIP $$	SUBS	TOTALS
TOTALS				$0	$1,627	$370	$114	$0	$2,111
SOIL PREP	6300	SF							
NITRO SAWDUST (4CY/M)	25	CY	16.00		400				
TRACTOR (5CY/HR)	5	HR	18.00				90		
OPERATOR	5	HR	10.00			50			
GYPSUM (120#/M-80#/BAG)	10	BG	7.00		70				
FERTILIZER (125#/M-50#)	16	BG	7.50		120				
LABOR (1 MAN W/TRACTOR)	5	HR	10.00			50			
ROTOTILLER (1500/HR)	4	HR	6.00				24		
OPERATOR	4	HR	10.00			40			
GRADE	6,300	SF							
LABOR (1,000/MHR)	6	HR	10.00			60			
SOD	3,345	SF	0.31		1,037				
LABOR (225/MHR)	15	HR	10.00			150			
LABOR TO ROLL SOD	2	HR	10.00			20			

PHASE I. PLANTING COSTS

BID ITEM	QTY	UNIT	UNIT COST	MATL'S IRRIG	MATL'S PLANTING	LABOR $$	EQUIP $$	SUBS	TOTALS
TOTALS				$0	$5,167	$1,600	$81	$0	$6,848
TREES 24" BOX or 2" CAL	6	EA	104.50		627				
PLANT TABS (6/TREE)	36	EA	0.10		4				
LABOR (2MHR/EA)	12	EA	10.00			120			
STAKE KITS (2/TREE)	6	KIT	9.00		54				
LABOR (3/MHR)	2	EA	10.00			20			
TRACTOR (.25HR/EA)	1.5	HR	18.00				27		
OPERATOR	1.5	HR	10.00			15			
TREES 15 GL or 1.5" CAL	15	EA	27.30		410				
PLANT TABS (4/TREE)	60	EA	0.10		6				
LABOR (1/MHR)	15	HR	10.00			150			
STAKE KITS (2/TREE)	15	KIT	9.00		135				
LABOR (3/MHR)	5	HR	10.00			50			
PALM TREE 10' BTF	3	EA	275.00		825				
LABOR (4MHR/EA)	12	HR	10.00			120			
TRACTOR (.33HR/EA)	1	HR	18.00				18		
OPERATOR	1	HR	10.00			10			
SHRUBS 5 GL	215	EA	6.30		1,355				
LABOR (5/MHR)	43	HR	10.00			430			
SHRUBS 5 GL (ESPALIER)	3	EA	20.48		61				
LABOR (4/MHR)	1	HR	10.00			10			
SHRUBS 1 GL	283	EA	3.15		891				
LABOR (12/HR)	24	HR	10.00			240			
VINES 1 GL	24	EA	3.68		88				
LABOR (12/HR)	2	HR	10.00			20			
GROUND COVER (8"OC)	655	SF							
MATERIALS (64 CNT/FLAT)	23	FLT	9.00		207				
LABOR (2/MHR)	11.5	HR	10.00			115			
GROUND COVER (12"OC)	2,300	SF							
MATERIALS (64 CNT/FLAT)	36	FLT	9.00		324				
LABOR (2/MHR)	18	HR	10.00			180			
MULCH (1" DEEP)	2,955	SF							
MULCH	10	CY	18.00		180				
LABOR (1CY/MHR)	10	HR	10.00			100			
TRACTOR (5CY/HR)	2	HR	18.00				36		
OPERATOR	2	HR	10.00			20			

PHASE I. MAINTENANCE COSTS:

MAINT. CREW AVE WAGE-----> $7.50   (3 MAN CREW)

BID ITEM	QTY	UNIT	UNIT COST	MATL'S IRRIG	MATL'S PLANTING	LABOR $$	EQUIP $$	SUBS	TOTALS
TOTALS				$0	$15	$292	$58	$0	$365
MAINTENANCE (3 MAN CREW: 1 VISIT PER WK/1 CREW-HOUR PER VISIT)									
LABOR (WEED, MOW, ETC.)	39	HR	7.50			292			
MOWING (3,345 SF/VISIT)	43,151	SF							
21" MOWER (7500/MHR)	6	HR	1.25				8		
EDGE/TRIM	4	HR	2.00				8		
FERTILIZE (10#/M)	2	BG	7.50		15				
PICKUP TRUCK	13	HR	3.25				42		

```
                                                             $365.03
SALES TAX----------------------------------------------->       1.00
LABOR BURDEN------------------------------------------->       88.00
                                                           =========
                                           SUBTOTAL --------> 454.03
OVERHEAD--------------> 39.0 MHRS $10.00 OPH-------------->    390.00
                                                           =========
                                                            $844.03
```

MAINT. (BREAK-EVEN POINT) MAN-HOUR RATE---> $21.64 ROUNDUP-> $22.00 X 3 MEN   3 $66.00

OR

YOU COULD USE A MAINTENANCE PACKAGE APPROACH:

	HITS	RATE		
MAINT PACKAGE HOURS---->	13 CR-HR	66.00	---------------------->	$858.00
FERTILIZE (10#/M)------>	2 BG	7.50	---------------------->	15.00

TOTAL MAINTENANCE PRICE ADDED TO BID AS A SUBCONTRACTOR COST-------------------> $873.00

P.E. 12-1
BID WORKSHEET

$$CREW\text{-}DAYS = \frac{20.75}{24} = .87$$

PROJECT: BARKLEY RESIDENCE

PAGE (3) OF (3)

LOCATION: _____

DATE PREPARED: _____ / _____ / _____

P.O.C./G.C.: _____

ESTIMATOR: _____

DUE: ___ / ___ / ___ PH.: _____ FAX: _____

CREW SIZE: 3

TYPE WORK: GEN CONDS

HOURS/DAY: 8   CAW: $ 10.00

REMARKS:

DESCRIPTION	QTY	UNIT	U/C	MAT'L	LABOR	EQUIP	SUBS
— FOREMAN ADMIN	1	HR	15 —		15		
— LOAD/UNLOAD <.5 HR/MAN>	1.5	HR	10 —		15		
— DRIVETIME <.75 HR/MAN>	2.25	HR	10 —		23		
— CREW PICKUP	8	HR	4 —			32	
— WATER IN PLANTS/CLEANUP	1	HR	10 —		10		
— WARRANTY LABOR	4	HR	10 —		40		
— WARRANTY EQUIP/PICKUP	4	HR	4 —			16	
— DETAIL JOBSITE	1	HR	10 —		10		
				Ø	113	48	Ø

SHI FORM 04-A

-224-

PHASE II. GENERAL CONDITIONS: CREW DAYS = 379/(32-6) =   14.6 CREW DAYS

BID ITEM	QTY	UNIT	UNIT COST	MATL'S IRRIG	MATL'S PLANTING	LABOR $$	EQUIP $$	SUBS	TOTALS
TOTALS				$0	$175	$1,617	$556	$0	$2,348
FOREMAN (1MHR/DAY)	15	HR	12.00			180			
CREW TRUCK (8HR/DY)	120	HR	3.25				390		
MOBILIZATION (LABOR)	2	HR	9.00			18			
MOBILIZATION (EQUIP)	2	HR	10.00				20		
HAUL MATERIALS (LABOR)	6	HR	9.00			54			
HAUL MATERIALS (EQUIP)	6	HR	10.00				60		
CLEAN UP (.25HR/MAN/DAY)	15	HR	10.00			150			
WARANTY (LABOR)	16	HR	10.00			160			
WARANTY (EQUIP)	8	HR	3.25				26		
SUPERVISOR (1HR/DAY)	15	HR	15.00			225			
SUPERVISOR'S TRUCK	15	HR	4.00				60		
DETAIL JOB	8	HR	10.00			80			
AS-BUILTS	1	LS	100.00		100				
SOIL TEST	1	LS	75.00		75				
CREW DR.TIME(1HR/M/DY)	60	HR	10.00			600			
LOAD TIME(1 HR/DAY)	15	HR	10.00			150			

OPH------>	$10.00	CAW/AVE WAGE	$8.33	OTF------>	10.00%
PPH------>	$5.00	PROFIT---->	10.00%	RF------>	10.00%
		TAX------>	6.00%	CAW-LOADED>	$10.00
NO. UNITS->	1.0	LABOR BURDEN	30.00%	CREW TRUCK>	$5.00

===

	MAT	LABOR	EQUIP	SUBS
	=======	=======	=======	=======

I. PRODUCTION OF FINISHED PRODUCT:

240 HRS

	MAT	LABOR	EQUIP	SUBS
	6,000	2,400	480	0

II. GENERAL CONDITIONS:

60 HRS   80 HRS

	0	600	400	0
SUBTOTALS:	6,000	3,000	880	0

III. MARKUPS:

A. SALESTAX   360

B. LABOR BURDEN------>   900

SUBTOT:	6,360	3,900	880	0

TOTAL DIRECT COSTS-----------------------------------> 11,140   71.23%

C. OVERHEAD RECOVERY:

300 (NUMBER OF HOURS X OPH) $10.00 -->   3,000   19.18%

"BEP" SUBTOTAL (DIRECT COSTS + OVERHEAD)----> 14,140   90.41%

D. CONTINGENCY FACTOR (IF DESIRED)-------------> 0   0.00%

E. PROFIT:

300 HOURS X PPH $5.00 --> $1,500 -->   1,500   9.59%

F. TOTAL PRICE FOR THE JOB---------------------> $15,640   100.00%

IV. ANALYSIS:

	$	%			$/RATIO	%
A. SPH:	$52.13	100.00%	J. MAT/LAB:		2.00 :1	
B. DCPH:	$37.13	71.23%	K. MPH:		$20.00	38.36%
C. OPH:	$10.00	19.18%	L. EQ/LAB:			29.33%
D. PPH:	$5.00	9.59%	M. EQPH:		$2.93	5.63%
E. BEP:	$14,140.00	90.41%	N. GC:		$1,000.00	6.39%
F. OVHD:	$3,000.00	19.18%	O. GCPH:		$3.33	6.39%
G. PROF:	$1,500.00	9.59%	P. GCH/TH:			20.00%
H. GPM:	$4,500.00	28.77%	Q. FACTOR:		2.61 X	MAT'L
I. GPMPH:	$15.00	28.77%	R. UNIT PRICE-------->		$15,640	

V. MORS PERCENTAGE MARKUP ON LABOR & BURDEN-----------------> 54.97%
   (ASSUMING A MARKUP OF 10/25/5% ON MAT'L., EQUIP., & SUBS.)

BID WORKSHEET

PROJECT: __BARKLEY RESIDENCE__ PAGE (1) OF (3)

LOCATION: __789 McKINLEY ST LOS ANGELES__ DATE PREPARED: __7__ / __25__ / __19—__

P.O.C./G.C.: __MR. BARKLEY__ ESTIMATOR: __M. THRUE__

DUE: __8__ / __1__ / __19—__ PH.: __619 726-2185__ FAX: __N/A__ CREW SIZE: __3__

TYPE WORK: __RECAP__ HOURS/DAY: __8__ CAW: $ __10.00__

REMARKS:

DESCRIPTION	QTY	UNIT	U/C	MAT'L	LABOR	EQUIP	SUBS
- PLANTING				636	140	5	Ø
- GEN CONDS				Ø	113	48	Ø
				636	253	53	Ø
- SALES TAX	6	%		38			
- LABOR BURDEN	30	%			76		
				674	329	53	Ø
					674		
					53		
					Ø		
SUBTOTAL					1056		
- OVERHEAD ⟨OPH = $10.00⟩	25	HR	10 —		250		
SUBTOTAL					1306		
- CONTINGENCY FACTOR					Ø		
SUBTOTAL					1306		
- PROFIT ⟨1306 ÷ ⟨1 - .15⟩⟩	15	%			230		
- FINAL PRICE					1536		

BEP = $1306 85%

OVHD = $ 250 16%

NET PM = $ 230 15%

GPM = 480 31%

SPH = 1536 ÷ 25 = $61.44

SHI FORM 04-A

P.E. 13-4

BID WORKSHEET

PROJECT: HENDERSON RES.
LOCATION: < NO SUBCONTRACTORS >
P.O.C./G.C.: _____
DUE: __/__/__ PH.: _____ FAX: _____
TYPE WORK: RECAP

PAGE (1) OF (7)
DATE PREPARED: __/__/__
ESTIMATOR: _____
CREW SIZE: 4
HOURS/DAY: 8 CAW: $ 10.00

REMARKS:

DESCRIPTION	QTY	UNIT	U/C	MAT'L	LABOR	EQUIP	SUBS
– IRRIGATION				4402	1820	156	
– PLANTING				5167	1600	81	
– SOIL PREP				590	200	114	
– SOD				1037	170		
– MAINTENANCE <90 DAYS>				15	292	58	
– GEN. CONDS.				175	1617	556	
				11386	5699	965	0
– SALES TAX				683			
– LABOR BURDEN					1710		
				12069	7409	965	0
					12069		
					965		
					0		
					20443		
– OVERHEAD	570	HR	10 –		5700		
					26143		
– CONTINGENCY FACTOR					0		
					26143		
– PROFIT	15	%			4613		
– FINAL PRICE					30756		

SHI FORM 04-A

– 229 –

BID WORKSHEET

PROJECT: _HENDERSON RES._ PAGE (_1_) OF (_7_)

LOCATION: _< MAINTENANCE AS SUBCONTRACTOR >_ DATE PREPARED: ____ / ____ / ____

P.O.C./G.C.: _____ ESTIMATOR: _____

DUE: __ / __ / __ PH.: _____ FAX: _____ CREW SIZE: _4_

TYPE WORK: _RECAP_ HOURS/DAY: _8_ CAW: $ _10.00_

REMARKS:

DESCRIPTION	QTY	UNIT	U/C	MAT'L	LABOR	EQUIP	SUBS
- IRRIGATION				4402	1820	156	
- PLANTING				5167	1600	81	
- SOIL PREP				590	200	114	
- SOD				1037	170		
- MAINTENANCE							873
- GEN CONDS				175	1617	556	
				11371	5407	907	873
- SALES TAX	6	%		682			
- LABOR BURDEN	30	%			1622		
				12053	7029	907	873
					12053		
					907		
					873		
					20862		
- OVERHEAD	531	HR	10 —		5310		
					26172		
- CONTINGENCY FACTOR					0		
					26172		
- PROFIT	15	%			4619		
- FINAL PRICE					30791		

SHI FORM 04-A

P.E. 13-5. BAR WORKSHEET FOR THE BARKLEY RESIDENCE

OPH------->	$10.00	CAW/AVE WAGE	$8.33	OTF------->	10.00%
PPH------->	$5.00	PROFIT----->	15.00%	RF-------->	10.00%
		TAX-------->	6.00%	CAW-LOADED>	$10.00
NO. UNITS->	1.0	LABOR BURDEN	30.00%	CREW TRUCK>	$4.00

==

	MAT	LABOR	EQUIP	SUBS
	=======	=======	=======	=======

I. PRODUCTION OF FINISHED PRODUCT:

		14 HRS		
	636	140	5	0

II. GENERAL CONDITIONS:

		11 HRS	12 HRS	
	0	113	48	0
	===========	=========	=========	==========
SUBTOTALS:	636	253	53	0

III. MARKUPS:

A. SALESTAX   38

B. LABOR BURDEN------>   76

	===========	=========	=========	==========
SUBTOT:	674	329	53	0

TOTAL DIRECT COSTS----------------------------------> 1,056   68.76%

C. OVERHEAD RECOVERY:

        25 (NUMBER OF HOURS X OPH) $10.00 -->   250   16.24%
                                                       ===========
        "BEP" SUBTOTAL (DIRECT COSTS + OVERHEAD)----> 1,306   85.00%

D. CONTINGENCY FACTOR (IF DESIRED)------------->   0   0.00%

E. PROFIT:
        15.00%X BEP-------------------> $230   230   15.00%
        24.75 HOURS X PPH $5.00 --> $124 -->   0   0.00%
                                                       ===========
F. TOTAL PRICE FOR THE JOB--------------------> $1,536 100.00%

IV. ANALYSIS:

	$	%			$/RATIO	%
A. SPH:	$62.06	100.00%	J.	MAT/LAB:	2.51 :1	
B. DCPH:	$42.67	68.76%	K.	MPH:	$25.70	41.41%
C. OPH:	$10.00	16.11%	L.	EQ/LAB:		20.95%
D. PPH:	$9.31	15.00%	M.	EQPH:	$2.14	3.45%
E. BEP:	$1,305.56	85.00%	N.	GC:	$161.00	10.48%
F. OVHD:	$249.50	16.24%	O.	GCPH:	$6.51	10.48%
G. PROF:	$230.39	15.00%	P.	GCH/TH:		43.43%
H. GPM:	$479.89	31.24%	Q.	FACTOR:	2.42 X	MAT'L
I. GPMPH:	$19.39	31.24%	R.	UNIT PRICE------->		$1,536

V. MORS PERCENTAGE MARKUP ON LABOR & BURDEN---------------->   51.33%
   (ASSUMING A MARKUP OF 10/25/5% ON MAT'L., EQUIP., & SUBS.)

P.E. 13-5. BAR WORKSHEET FOR THE HENDERSON PROJECT (NO SUBS)

OPH------>	$10.00	CAW/AVE WAGE	$8.33	OTF------>	10.00%
PPH------>	$5.00	PROFIT----->	15.00%	RF------->	10.00%
		TAX------->	6.00%	CAW-LOADED>	$10.00
NO. UNITS->	6,300.0	LABOR BURDEN	30.00%	CREW TRUCK>	$3.25

===

	MAT	LABOR	EQUIP	SUBS
	=======	=======	=======	=======

I. PRODUCTION OF FINISHED PRODUCT:

418 HRS

	MAT	LABOR	EQUIP	SUBS
	11,211	4,082	409	0

II. GENERAL CONDITIONS:

		152 HRS	120 HRS	
	175	1,617	556	0
SUBTOTALS:	11,386	5,699	965	0

III. MARKUPS:

A. SALESTAX   683

B. LABOR BURDEN------>   1,710

	MAT	LABOR	EQUIP	SUBS
SUBTOTAL:	12,069	7,409	965	0

TOTAL DIRECT COSTS----------------------------->   20,443   66.47%

C. OVERHEAD RECOVERY:

570 (NUMBER OF HOURS X OPH) $10.00 -->   5,700   18.53%

"BEP" SUBTOTAL (DIRECT COSTS + OVERHEAD)---->   26,143   85.00%

D. CONTINGENCY FACTOR (IF DESIRED)------------->   0   0.00%
E. PROFIT:

15.00%--->   4,613   15.00%

F. TOTAL PRICE FOR THE JOB--------------------->   30,756   100.00%

IV. ANALYSIS:

	$	%		$/RATIO	%
A. SPH:	$53.96	100.00%	J. MAT/LAB:	2.00 :1	
B. DCPH:	$35.86	66.47%	K. MPH:	$19.98	37.02%
C. OPH:	$10.00	18.53%	L. EQ/LAB:		16.93%
D. PPH:	$8.09	15.00%	M. EQPH:	$1.69	3.14%
E. BEP:	$26,142.86	85.00%	N. GC:	$2,348.00	7.63%
F. OVHD:	$5,700.00	18.53%	O. GCPH:	$4.12	7.63%
G. PROF:	$4,613.45	15.00%	P. GCH/TH:		26.67%
H. GPM:	$10,313.45	33.53%	Q. FACTOR:	2.70 X	MAT'L
I. GPMPH:	$18.09	33.53%	R. UNIT PRICE------->		$4.88

V. MORS PERCENTAGE MARKUP ON LABOR & BURDEN-----------------> 57.39%
(ASSUMING A MARKUP OF 10/25/5% ON MAT'L., EQUIP., & SUBS.)

P.E. 13-5. BAR WORKSHEET FOR THE HENDERSON PROJECT (MAINT. AS SUB)

OPH------->	$10.00	CAW/AVE WAGE	$8.33	OTF------->	10.00%
PPH------->	$5.00	PROFIT----->	15.00%	RF-------->	10.00%
		TAX------->	6.00%	CAW-LOADED>	$10.00
NO. UNITS->	6,300.0	LABOR BURDEN	30.00%	CREW TRUCK>	$3.25

==

	MAT	LABOR	EQUIP	SUBS
	=======	=======	=======	=======

I. PRODUCTION OF FINISHED PRODUCT:

379 HRS

MAT	LABOR	EQUIP	SUBS
11,196	3,790	351	873

II. GENERAL CONDITIONS:

152 HRS   120 HRS

MAT	LABOR	EQUIP	SUBS
175	1,617	556	0
==========	==========	==========	==========
SUBTOTALS: 11,371	5,407	907	873

III. MARKUPS:

A. SALESTAX   682

B. LABOR BURDEN------>   1,622

MAT	LABOR	EQUIP	SUBS
============	==========	==========	==========
SUBTOTAL: 12,053	7,029	907	873

TOTAL DIRECT COSTS--------------------------->   20,862   67.75%

C. OVERHEAD RECOVERY:

531 (NUMBER OF HOURS X OPH) $10.00 -->   5,310   17.25%
                                             ==========

"BEP" SUBTOTAL (DIRECT COSTS + OVERHEAD)---->   26,172   85.00%

D. CONTINGENCY FACTOR (IF DESIRED)------------->   0   0.00%
E. PROFIT: ----------

15.00%-------------------------------------->   4,619   15.00%
                                                 ==========

F. TOTAL PRICE FOR THE JOB-------------------->   30,791   100.00%

IV. ANALYSIS:	$	%			$/RATIO	%
A. SPH:	$57.99	100.00%	J.	MAT/LAB:	2.10 :1	
B. DCPH:	$39.29	67.75%	K.	MPH:	$21.41	36.93%
C. OPH:	$10.00	17.25%	L.	EQ/LAB:		16.77%
D. PPH:	$8.70	15.00%	M.	EQPH:	$1.71	2.95%
E. BEP:	$26,172.36	85.00%	N.	GC:	$2,348.00	7.63%
F. OVHD:	$5,310.00	17.25%	O.	GCPH:	$4.42	7.63%
G. PROF:	$4,618.65	15.00%	P.	GCH/TH:		28.63%
H. GPM:	$9,928.65	32.25%	Q.	FACTOR:	2.71 X	MAT'L
I. GPMPH:	$18.70	32.25%	R.	UNIT PRICE------>		$4.89

V. MORS PERCENTAGE MARKUP ON LABOR & BURDEN----------------->   54.55%
(ASSUMING A MARKUP OF 10/25/5% ON MAT'L., EQUIP., & SUBS.)

HENDERSON RESIDENCE   UNIT PRICE BID

UNIT PRICE WORKSHEET   SMITH HUSTON, INC.-16-A

ITEM NO.	DESCRIPTION	QUANTITY	UNIT	MATERIALS + SALES TAX	LABOR + BURDEN	EQUIPMENT	TOTALS	MARKUP % 57	UNIT PRICE	UNIT PRICE ROUNDED	TOTAL PRICES
1	MOBILIZATION	1	LS								783
2	BACKFLOW	1	EA	368	78		446	783	783	783	783
3	VALVES (w/FINE TUNE +WIRE)	14	EA	1309	520		1829	3209	22921	230	3220
4	QUICK COUPLERS	4	EA	144	52	20	196	344	86	86	344
5	MAIN LINE 1.25"	240	LF	110	130		260	456	190	190	456
6	SLEEVES 2.5"	120	LF	122	65		187	326	273	275	330
7	CONTROLLER	1	EA	992	26	40	418	733	733	735	735
8	LATERAL LINE 1"	520	LF	61	273	96	374	536	126	130	676
9	LATERAL LINE 1/2"	1200	LF	89	624		806	1418	118	120	1440
10	HEADS-6" POP-UP	170	EA	1532	442		1974	3463	2037	2050	3485
11	HEADS-4" POP-UP	60	EA	541	156		597	1223	2038	2050	1230
12	SOIL PREP	6300	SF	635	182	114	921	616	26	26	1638
13	TREES 24" BOX / 2" CAL	6	EA	724	202	27	955	1675	27917	280	1680
14	TREES 15 GL	15	EA	584	260		844	1481	9873	99	1485
15	PALMS (WASH ROB) 10'BTF	3	EA	815	169	18	1042	1863	621	625	1875
16	SHRUBS 5 GL	215	EA	1434	557		1795	3500	1628	1650	3547
17	SHRUBS 5 GL ESPALIER	3	EA	65	13		72	137	46	46	138
18	SHRUBS 1 GL	283	EA	944	312		1256	2204	779	8	2264
19	VINES 1 GL	24	EA	93	26		119	209	871	9	216
20	GROUND COVER	57	FLT	563	384		947	1641	2815	29	1711
21	SOD (w/GRADING)	3345	SF	1099	299	76	1398	2453	73	75	2509
22	MULCH 1" DEEP	2985	SF	191	156	351	383	672	23	25	739
	SUBTOTAL (#1)			11868	4928	55%	17110				35501
	GENERAL CONDITIONS (#1)	15%		186	2102	907	2844				
	SUBTOTAL (#2)			12054	7030		17991				
	OVERHEAD (O.P.H. = $10.00)	531	HR				5310				
	SUBTOTAL (#3)						25301				
	PROFIT & CONT. FACTOR	15%					4765				
	TOTAL PRICE						29766				

ADD SUBCONTRACTORS (WITH ALL MARKUPS) HERE→ $1027

(DOUBLE-CHECK #1)
(DOUBLE-CHECK #2)
(DOUBLE-CHECK #3)

$ 735.—
# $ 735.—

DIFFERENCE DUE TO ROUNDING DECIMAL

$ 31528

1   $29,766 − 17,147 = 12,619

2   12,619 ÷ 29,766 = .4239

3   .4239 = .43

4   1.0 − .43 = .57 ≈ 57%

873
$873   .85
<1 − .15> = $1027

$31528

UNIT PRICING MODEL

SALES TAX:	6.00%	PRICE:	$30,141	OVHD:	17.84%
AVERAGE WAGE:	$10.00	SPH:	$56.76	PROFIT:	15.00%
OFH:	$10.00	HRS:	531.0	PPH:	$8.41
DESIRED MINIMUM PPH:	$5.00	CREW:	4	GPM:	32.84%
LABOR BURDEN:	30.00%	HRS/DAY:	8	GC:	9.55%
PROFIT AS A %	15.00%				

SET-----> 14.22% 20.99% 15.00%

14.22% 20.99% 15.00%

#	DESCRIPTION	QTY	UNIT	MATL+TAX	LABOR+LB	EQUIP	TOTALS	U/C	+ G.C.	+OH (BEP)	+ PROFIT	U/P	ROUNDED	TOTAL BID	VARIANCE
		1	2	3	5	7	8	CALCS	14.22%	20.99%	15.00%	10	11	12	
1	MOBILIZATION	1	LS				0	0.00	0	0	0	0.00	0.00	0.00	0.00
2	BACKFLOW	1	EA	368	78		446	445.82	520	658	774	773.90	775.00	775.00	1.10
3	VALVES(F.TUNE & WIRE)	14	EA	1,309	520		1,829	130.65	2,132	2,699	3,175	226.80	230.00	3,220.00	44.86
4	QUICK COUPLERS	4	EA	144	52		196	49.04	229	289	341	85.13	85.00	340.00	(0.51)
5	MAIN LINE 1.25"	240	LF	110	130	20.00	260	1.08	303	384	452	1.88	1.90	456.00	4.25
6	SLEEVES 2.5"	120	LF	122	65		187	1.56	218	276	324	2.70	2.70	324.00	(0.44)
7	CONTROLLER	1	EA	392	26		418	418.20	488	617	726	725.95	730.00	730.00	4.05
8	LATERAL PIPE 1"	520	LF	61	273	40.00	374	0.72	437	553	650	1.25	1.25	650.00	(0.06)
9	LATERAL PIPE 1/2"	1200	LF	88	624	96.00	808	0.67	942	1,192	1,403	1.17	1.20	1,440.00	37.43
10	RB 1806 SAM HEADS	170	EA	1,532	442		1,974	11.61	2,301	2,912	3,426	20.15	20.25	3,442.50	16.35
11	RB 1804 SAM HEADS	60	EA	541	156		697	11.61	812	1,028	1,209	20.15	20.25	1,215.00	5.77
12	SOIL PREP	6300	SF	625	182	114.00	921	0.15	1,074	1,360	1,599	0.25	0.26	1,606.50	7.04
13	TREES 1 1/2" CAL	6	EA	726	202	27.00	955	159.10	1,113	1,409	1,657	276.18	280.00	1,680.00	22.91
14	TREES 15 GL	15	EA	584	260		844	56.27	984	1,245	1,465	97.68	99.00	1,485.00	19.79
15	PALMS (WASH ROB) 10'	3	EA	875	169	18.00	1,062	353.83	1,238	1,566	1,843	614.22	615.00	1,845.00	2.34
16	SHRUBS 5 GL	215	EA	1,436	559		1,995	9.28	2,326	2,944	3,464	16.11	16.25	3,493.75	30.10
17	SHRUBS 5 GL ESPALIER	3	EA	65	13		78	25.89	91	115	135	44.94	45.00	135.00	0.19
18	SHRUBS 1 GL	283	EA	944	312		1,256	4.44	1,465	1,854	2,181	7.71	7.75	2,193.25	12.16
19	VINES 1 GL	24	EA	93	26		119	4.97	139	176	207	8.63	8.75	210.00	2.94
20	GROUND COVER	59	FLT	563	384		946	16.04	1,103	1,396	1,643	27.84	28.00	1,652.00	9.21
21	SOD (+GRADE)	3345	SF	1,099	299		1,398	0.42	1,630	2,063	2,427	0.73	0.75	2,508.75	81.58
22	MULCH 1" DEEP	2955	SF	191	156	36.00	383	0.13	446	565	665	0.22	0.25	738.75	74.25

SUBTOTAL:				11,869	4,927	351	17,147	57.6%	19,990	25,300	29,765	XXXXXXXX	XXXXXXXX	30,141	
GEN. CONDS.				186	2,102	556	2,844	14.22%							
SUBTOTALS:				12,054	7,029	907	19,990	67.16%	$0.00						
									0.00%						
OVERHEAD:	531.0 HOURS X OPH ---------------------->						5,310	20.99%	(DBL CK-1)						
SUBTOTALS:							25,300	-------->	-------->	$0.00					
PROFIT:	15.00%						4,465	15.00%		0.00%					
										(DBL CK-2)	V			V	
							29,765	100.00%	-------->	-------->	$0.00	-------->	-------->	VARIANCE	
											0.00%			$375.30	
											(DBL CK-3)			1.26%	
														(DBL CK-4)	

This is a computerized version of P.E. 14-1/2/3. Total bid amounts differ due to the rounding of the markup decimal on the manual worksheet.

P.E. 15-1. BAR WORKSHEET CURB TIME MAN-HOUR RATE

OPH------->	$10.00	CAW/AVE WAGE	$9.00	OTF------->	0.00%
PPH------->	$3.00	PROFIT----->	10.00%	RF-------->	10.00%
		TAX-------->	0.00%	CAW-LOADED>	$9.90
NO. UNITS->	12.0	LABOR BURDEN	30.00%	CREW TRUCK>	$20.00

===

	MAT	LABOR	EQUIP	SUBS
	=======	=======	=======	=======

I. PRODUCTION OF FINISHED PRODUCT:

12 HRS

	0	119	63	_____

II. GENERAL CONDITIONS:

4 HRS   4 HRS

	_____	40	80	_____
	==========	=========	=========	==========
SUBTOTALS:	0	158	143	0

III. MARKUPS:

A. SALESTAX   0

B. LABOR BURDEN------>   48

	==========	=========	=========	==========
SUBTOT:	0	206	143	0

TOTAL DIRECT COSTS------------------------------->   349

C. OVERHEAD RECOVERY:

16 (NUMBER OF HOURS X OPH) $10.00 -->   160

==========

"BEP" SUBTOTAL (DIRECT COSTS + OVERHEAD)---->   509

D. CONTINGENCY FACTOR (IF DESIRED)------------->   0

E. PROFIT:

10.00%-------------------------------> $56.55 -->   57

16 HOURS X PPH $3.00 --> $48.00 -->   0

==========

F. TOTAL PRICE FOR THE JOB-------------------> $565 $565

IV. ANALYSIS:

	$	%		$/RATIO	%
A. SPH:	$35.34	100.00%	J. MAT/LAB:	0.00 :1	
B. DCPH:	$21.81	61.70%	K. MPH:	$0.00	0.00%
C. OPH:	$10.00	28.30%	L. EQ/LAB:		90.28%
D. PPH:	$3.53	10.00%	M. EQPH:	$8.94	25.29%
E. BEP:	$508.92	90.00%	N. GC:	$119.60	21.15%
F. OVHD:	$160.00	28.30%	O. GCPH:	$7.48	21.15%
G. PROF:	$56.55	10.00%	P. GCH/TH:		25.00%
H. GPM:	$216.55	38.30%	Q. FACTOR:	0.00 X MAT'L	
I. GPMPH:	$13.53	38.30%	R. CURB MAN-HOUR RATE->	$47.12	

V. MORS PERCENTAGE MARKUP ON LABOR & BURDEN-------------------> 60.34%

(ASSUMING A MARKUP OF 10/25/5% ON MAT'L., EQUIP., & SUBS.)

P.E. 15-1. BAR WORKSHEET FOR CURB TIME CREW-HOUR RATE

```
OPH------->      $10.00   CAW/AVE WAGE    $9.00    OTF------->      0.00%
PPH------->       $3.00   PROFIT----->   10.00%    RF-------->     10.00%
                          TAX-------->    0.00%    CAW-LOADED>      $9.90
NO. UNITS->         6.0   LABOR BURDEN   30.00%    CREW TRUCK>     $20.00
```
==

	MAT	LABOR	EQUIP	SUBS
	=======	=======	=======	=======

I. PRODUCTION OF FINISHED PRODUCT:

		12 HRS		
	0	119	63	_____

II. GENERAL CONDITIONS:

		4 HRS	4 HRS	
	_____	40	80	_____
	==========	=========	=========	==========
SUBTOTALS:	0	158	143	0

III. MARKUPS:

A. SALESTAX   0

B. LABOR BURDEN------>   48

	==========	=========	=========	==========
SUBTOT:	0	206	143	0

TOTAL DIRECT COSTS------------------------------------>   349

C. OVERHEAD RECOVERY:

        16 (NUMBER OF HOURS X OPH) $10.00 -->   160
                                                    ===========
    "BEP" SUBTOTAL (DIRECT COSTS + OVERHEAD)---->   509

D. CONTINGENCY FACTOR (IF DESIRED)------------->   0

E. PROFIT:
        10.00%------------------------> $56.55 -->   57
            16 HOURS X PPH $3.00 --> $48.00 -->   0
                                                    ===========
F. TOTAL PRICE FOR THE JOB--------------------> $565 $565

IV. ANALYSIS:

	$	%			$/RATIO	%
A. SPH:	$35.34	100.00%	J. MAT/LAB:		0.00 :1	
B. DCPH:	$21.81	61.70%	K. MPH:		$0.00	0.00%
C. OPH:	$10.00	28.30%	L. EQ/LAB:			90.28%
D. PPH:	$3.53	10.00%	M. EQPH:		$8.94	25.29%
E. BEP:	$508.92	90.00%	N. GC:		$119.60	21.15%
F. OVHD:	$160.00	28.30%	O. GCPH:		$7.48	21.15%
G. PROF:	$56.55	10.00%	P. GCH/TH:			25.00%
H. GPM:	$216.55	38.30%	Q. FACTOR:		0.00 X	MAT'L
I. GPMPH:	$13.53	38.30%	R. CURB CREW-HOUR RATE>		$94.24	

V. MORS PERCENTAGE MARKUP ON LABOR & BURDEN------------------>   60.34%
(ASSUMING A MARKUP OF 10/25/5% ON MAT'L., EQUIP., & SUBS.)

BID WORKSHEET

PROJECT: _PRACTICAL EXERCISE 15-2 **A**_ PAGE (1) OF (1)

LOCATION: _____

DATE PREPARED: _____ / _____ / _____

P.O.C./G.C.: _____

ESTIMATOR: _____

DUE: ___ / ___ / ___ PH.: _____ FAX: _____ CREW SIZE: _3_

TYPE WORK: _MAINTENANCE CREW_ HOURS/DAY: _8_ CAW: $ _8.80_

REMARKS:

DESCRIPTION	QTY	UNIT	U/C	MAT'L	LABOR	EQUIP	SUBS
I. PRODUCTION OF SERVICE							
- LABOR	21	HR	8 80		185		
- 48" MOWER	6	HR	3 75			23	
- 21" MOWER	5	HR	1 25			6	
- EDGER	4	HR	2 —			8	
- BLOWER	2	HR	2 -			4	
				Ø	185	41	Ø
II. GEN. CONDS.							
- DRIVE & LOAD TIME	3	HR	8 80		27		
- CREW TRUCK	8	HR	4 —			32	
				Ø	212	73	Ø
III. MARKUPS							
- SALES TAX	6	%		Ø			
- LABOR BURDEN	25	%			53		
				Ø	265	73	Ø
					Ø		
					73		
					Ø		
					338		
- OVERHEAD	24	HR	5 —		120		
					458		
- PROFIT & CONT. FACTOR	10	%			51		
- FINAL PRICE					509		

CREW CURB TIME RATE = $509 / 21 HRS = $24.24 × 3 MEN = $72.72

MAN-HOUR CURB TIME RATE = $509 / 21 HRS = $24.24

SHI FORM 04-A

P.E. 15-2 (B). BAR WORKSHEET CURB TIME MAN-HOUR RATE

OPH------>	$5.00	CAW/AVE WAGE	$8.00	OTF------>	0.00%
PPH------>	$0.00	PROFIT----->	10.00%	RF------->	10.00%
		TAX-------->	6.00%	CAW-LOADED>	$8.80
NO. UNITS->	21.0	LABOR BURDEN	25.00%	CREW TRUCK>	$4.00

	MAT	LABOR	EQUIP	SUBS

I. PRODUCTION OF FINISHED PRODUCT:

21 HRS

	0	185	41	0

II. GENERAL CONDITIONS:

3 HRS   8 HRS

	0	26	32	0
SUBTOTALS:	0	211	73	0

III. MARKUPS:

   A. SALESTAX   0

   B. LABOR BURDEN------>   53

SUBTOT:	0	264	73	0

     TOTAL DIRECT COSTS------------------------------->   337

   C. OVERHEAD RECOVERY:

        24 (NUMBER OF HOURS X OPH) $5.00 -->   120

     "BEP" SUBTOTAL (DIRECT COSTS + OVERHEAD)---->   457

   D. CONTINGENCY FACTOR (IF DESIRED)------------->   0

   E. PROFIT:

       10.00%---------------------------> $50.75 -->   51

            24 HOURS X PPH $0.00 --> $0.00 -->   0

   F. TOTAL PRICE FOR THE JOB--------------------> $508 $508

IV. ANALYSIS:

	$	%			$/RATIO	%
A. SPH:	$21.15	100.00%	J.	MAT/LAB:	0.00 :1	
B. DCPH:	$14.03	66.35%	K.	MPH:	$0.00	0.00%
C. OPH:	$5.00	23.65%	L.	EQ/LAB:		34.45%
D. PPH:	$2.11	10.00%	M.	EQPH:	$3.03	14.33%
E. BEP:	$456.75	90.00%	N.	GC:	$58.40	11.51%
F. OVHD:	$120.00	23.65%	O.	GCPH:	$2.43	11.51%
G. PROF:	$50.75	10.00%	P.	GCH/TH:		12.50%
H. GPM:	$170.75	33.65%	Q.	FACTOR:	0.00 X	**MAT'L**
I. GPMPH:	$7.11	33.65%	R.	CURB MAN-HOUR RATE>	$24.17	

V. MORS PERCENTAGE MARKUP ON LABOR & BURDEN------------------>   38.57%
    (ASSUMING A MARKUP OF 10/25/5% ON MAT'L., EQUIP., & SUBS.)

P.E. 15-2 (B). BAR WORKSHEET CURB TIME CREW-HOUR RATE

```
OPH------->      $5.00   CAW/AVE WAGE    $8.00   OTF------->    0.00%
PPH------->      $0.00   PROFIT----->   10.00%   RF-------->   10.00%
                         TAX------->     6.00%   CAW-LOADED>    $8.80
NO. UNITS->       7.0    LABOR BURDEN   25.00%   CREW TRUCK>    $4.00
```
==

	MAT	LABOR	EQUIP	SUBS
I. PRODUCTION OF FINISHED PRODUCT:		21 HRS		
	0	185	41	0
II. GENERAL CONDITIONS:		3 HRS	8 HRS	
	0	26	32	0
SUBTOTALS:	0	211	73	0

III. MARKUPS:

	MAT	LABOR	EQUIP	SUBS
A. SALESTAX	0			
B. LABOR BURDEN------>		53		
SUBTOT:	0	264	73	0

```
   TOTAL DIRECT COSTS---------------------------------->    337
```

C. OVERHEAD RECOVERY:

```
      24 (NUMBER OF HOURS X OPH)       $5.00 -->     120
                                                 ==========
   "BEP" SUBTOTAL (DIRECT COSTS + OVERHEAD)---->     457
                                                 ----------
D. CONTINGENCY FACTOR (IF DESIRED)------------->      0
                                                 ----------
E. PROFIT:
      10.00%-------------------------->   $50.75 -->    51
      24 HOURS X PPH    $0.00 -->         $0.00 -->     0
                                                 ==========
F. TOTAL PRICE FOR THE JOB------------------->      $508    $508
```

IV. ANALYSIS:

		$	%			$/RATIO	%
A.	SPH:	$21.15	100.00%	J.	MAT/LAB:	0.00 :1	
B.	DCPH:	$14.03	66.35%	K.	MPH:	$0.00	0.00%
C.	OPH:	$5.00	23.65%	L.	EQ/LAB:		34.45%
D.	PPH:	$2.11	10.00%	M.	EQPH:	$3.03	14.33%
E.	BEP:	$456.75	90.00%	N.	GC:	$58.40	11.51%
F.	OVHD:	$120.00	23.65%	O.	GCPH:	$2.43	11.51%
G.	PROF:	$50.75	10.00%	P.	GCH/TH:		12.50%
H.	GPM:	$170.75	33.65%	Q.	FACTOR:	0.00 X	MAT'L
I.	GPMPH:	$7.11	33.65%	R.	CURB CREW-HOUR RATE>	$72.50	

```
V. MORS PERCENTAGE MARKUP ON LABOR & BURDEN------------------->   38.57%
   (ASSUMING A MARKUP OF 10/25/5% ON MAT'L., EQUIP., & SUBS.)
```

APPENDIX

B

LABOR PRODUCTION RATES

I. IRRIGATION

ITEM	UNIT	QTY/ MHR	TIME/ UNIT	COMMENTS
1. Point of connection (POC)	LS	.25-.5	2-4 HRS	
2. Automatic controllers	EA	.5-1.0	1-2 HRS	
3.* Trench/clean/fittings/ backfill (6"X12-18")	LF	20-30	2.4 MIN	walk-behind trencher
4.* Trench/clean/fittings/ backfill (6"X12-18")	LF	25-35	2.0 MIN	ride-on trencher
5.* Trench/clean/fittings/ backfill (6"X24")	LF	20-30	2.4 MIN	ride-on trencher
6.* Pull/fittings/backfill 10-16" depth	LF	20-30	2.4 MIN	
7. Trench only	LF	65-100	.9-.6 MIN	walk-behind
8. Trench only	LF	100-150	.6-.4 MIN	ride-on trencher
9. Install pipe (only)	LF	100-125	.5-.6 MIN	
10. Backfill (only)	LF	100-125	.5-.6 MIN	w/tractor
11. Backflow preventors	EA	.5-1.0	1-2 HRS	
12. Remote control valves	EA	2-3.0	15-20 MIN	
13. Quick coupler valves	EA	2-4.0	15-30 MIN	
14. Ball/gate valves	EA	2-3.0	20-30 MIN	
15. Wire	LF	500-600	6-7.2 SEC	
16. Shrub heads	EA	5-6	10-12 MIN	
17. Turf spray heads	EA	4-5	12-15 MIN	
18. Turf rotor heads	EA	2-3	20-30 MIN	
19. Drip emmitters	EA	20-30	2-3 MIN	
20. Drip pipe tubing	LF	100-125	.5-.6 MIN	
21. Sleeving	LF	20-30	2.5 MIN	

* Trenching and pulling includes all functions (e.g., trenching, cleaning, installing fittings and backfilling trench).

II. SOIL PREP

	ITEM	UNIT	QTY/ MHR	TIME/ UNIT	COMMENTS
1.	Rough grade to +/-.1 ft	SF	4,000-6,000	10-15 MIN/M **	w/tractor
2.	Rototill (4" depth)	SF	1,500-2,500	30 MIN /M	w/walk behind rototiller
3.	Rototill (6" depth)	SF	1,000-2,000	40 MIN /M	"
4.	Rototill (4" depth)	SF	3,500-4,500	15 MIN /M	w/tractor tiller
5.	Rototill (6" depth)	SF	2,500-3,500	20 MIN /M	"
6.	Finish/fine grade	SF	1,000	60 MIN /M	
7.	Finish/fine grade	SF	2,500	24 MIN /M	w/tractor

III. PLANTING (rates include spotting, digging the hole, amending, cutting container, planting, and basin preparation).

1. Trees (box):

	UNIT	QTY/MHR	TIME/UNIT	COMMENTS
72" box	EA	.125	8 HRS	1 HR for backhoe
60" box	EA	.167	6 HRS	.8 HR for backhoe
48" box	EA	.25	4 HRS	.5 hr for backhoe
36" box	EA	.33	3 HRS	.3 HR for backhoe
24" box	EA	.5	2 HRS	.25 HR for b-hoe
15 gallon	EA	1.0	1 HR	
5 gallon	EA	5.0	12 MIN	
1 gallon	EA	12-15	4-5 MIN	

2. Trees (B&B):

	UNIT	QTY/MHR	TIME/UNIT	COMMENTS
5" cal/60" B&B	EA	.167	6 HRS	.8 HR for backhoe
4" cal/48" B&B	EA	.25	4 HRS	.5 HR for backhoe
3" cal/36" B&B	EA	.33	3 HRS	.3 HR for backhoe
2.5" cal/30" B&B	EA	.4	2.5 HRS	.3 HR for backhoe
2" cal/24" B&B	EA	.5	2 HRS	.25 HR for b-hoe
1.5" cal/18" B&B	EA	.75	1.5 HRS	

2. Trees (B&B) contd.	UNIT	QTY/ MHR	TIME/ UNIT
1" cal/12" B&B	EA	1	1 HR
Stake kit (1 stk/tree)	KIT	4	15 MIN
Stake kit (2 stk/tree)	KIT	3	20 MIN
Stake kit (3 stk/tree)	KIT	2	30 MIN
Guying kit (3 pt guy)	kit	2	20 MIN
3. Shrubs:			
3.0'	EA	.75	1.3 HRS
2.5'	EA	1.0	1.0 HRS
24"	EA	1.2	.8 HRS
18"	EA	2-2.5	.33-.5 HRS
12"	EA	3-4.0	15-20 MIN
15 gallon	EA	1.0	1.0 HRS
5 gallon	EA	4-5.0	12-15 MIN
1 gallon	EA	12-15	4-5 MIN
4. Ground cover:			
1 gallon	FL	12-15	4-5 MIN
Flat (2 1/4" pots)	FL	1.5-2	30-45 MIN
Flat (4" pots)	FL	2-2.5	24-30 MIN
Flat (64 cnt)	FL	2-2.5	24-30 MIN
Flat (81 cnt)	FL	1.5-2	30-45 MIN
Flat (100 cnt)	FL	1-1.5	45-60 MIN
5. Miscellaneous			
Mulch	CY	1.5-2	30-45 MIN
Sod	SF	225-300	20-27 MIN /100

IV. MAINTNENACE

ITEM	UNIT	QTY/MHR
1. Mowing:		
18-21" rotary	SF	7,500-10,000
36" rotary	SF	15,000-20,000
48" rotary	SF	30,000-35,000
60" rotary	SF	40,000-50,000
72" rotary	SF	45,000-55,000
72" reel	SF	35,000-45,000
12' reel	SF	55,000-65,000
_____	____	_____
_____	____	_____
_____	____	_____
2. Power edging	LF	300-500
3. String trimming	LF	300-500

** M = 1,000

NOTE: The above rates are approximations and based on normal/flat site conditions. They are provided for purposes of comparisons only. Your production rates will vary depending upon site conditions, experience of labor force, types of equipment used, etc.

APPENDIX
C

EQUIPMENT COST PER HOUR (CPH) RATES

EQUIPMENT COSTS

I. TRUCKS:

COST PER---->	HOUR	DAY	WEEK	MONTH
MINI-PICKUP	2.75	22	110	473
1/2 TON PICKUP	3.25	26	130	559
3/4 TON PICKUP	4.00	32	160	688
1 TON	5.00	40	200	860
2 TON	10.00	80	350	1,100
5 CY DUMP	15.00	120	525	1,650
8 CY DUMP	17.00	136	595	1,870
10 CY DUMP	20.00	160	700	2,200
SEMI & TRAILER	20.00	160	700	2,200
BOOM TRUCK	25.00	200	875	2,750
TANK TRUCK (500 GL)	16.00	128	560	1,760
TANK TRUCK (1,000 GL)	20.00	160	700	2,200
HYDRO-SEEDER (500 GL)	22.00	176	770	2,420
HYDRO-SEEDER (1000 GL)	23.00	184	805	2,530
* BRONCO II	4.00	32	160	688
* BRONCO (FULLSIZE)	5.00	40	200	860
* SUBURBAN	6.00	48	240	1,032
_____	__.__	__.__	__.__	__.__
_____	__.__	__.__	__.__	__.__

II. AUTOMOBILES / OVERHEAD VEHICLES:

COST PER---->	HOUR	DAY	WEEK	MONTH
SUB-COMPACT	2.50	20	100	430
COMPACT	3.00	24	120	516
MID-SIZE	3.50	28	140	602
FULL SIZE	4.25	34	170	731
_____	__.__	__.__	__.__	__.__
_____	__.__	__.__	__.__	__.__

EQUIPMENT COSTS

III. TRACTORS COST PER--->	HOUR	DAY	WEEK	MONTH
1/8 CY - 12-15 HP	12.00	96	420	1,320
1/4 CY - 15-20 HP	14.00	112	490	1,540
1/2 CY - 25-35 HP	16.00	128	560	1,760
1 CY - 45-55 HP	18.00	144	630	1,980
3 CY - 55-75 HP	24.00	192	840	2,640
WHEEL / UNI-LOADER	25.00	200	875	2,750
BACKHOE - SMALL	25.00	200	875	2,750
BACKHOE - LARGE	30.00	240	1,050	3,300
D-4 DOZER	25.00	200	875	2,750
D-7 DOZER	45.00	360	1,575	4,950
_____	__.__	__.__	__.__	__.__
_____	__.__	__.__	__.__	__.__

IV. TRENCHERS/VIBRATORY PLOWS: COST PER--->	HOUR	DAY	WEEK	MONTH
WALK-BEHIND	5.00	40	175	550
TRENCHER (SMALL)	15.00	120	525	1,650
TRENCHER (LARGE)	20.00	160	700	2,200
_____	__.__	__.__	__.__	__.__
PULLER (SMALL)	15.00	120	525	1,650
PULLER (MED)	20.00	160	700	2,200
PULLER (LARGE)	23.00	184	805	2,530
_____	__.__	__.__	__.__	__.__
_____	__.__	__.__	__.__	__.__
_____	__.__	__.__	__.__	__.__
_____	__.__	__.__	__.__	__.__

EQUIPMENT COSTS

V. MOWERS:

COST PER--->	HOUR	DAY	WEEK	MONTH
21" PUSH	1.00	8	35	110
21" SELF-PROPELLED	1.50	12	53	165
36" ROTARY	2.75	22	96	303
48" ROTARY	4.00	32	140	440
51" ROTARY	4.15	33	145	457
60" ROTARY	4.25	34	149	468
72" ROTARY	4.75	38	166	523
15' ROTARY	9.50	76	333	1,045
15' REEL	9.75	78	341	1,073
** MAINT PACKAGE (2 MEN)	5.00	40	175	550
*** MAINT PACKAGE (3 MEN)	6.00	48	210	660
_____	___.__	___.__	___.__	___.__
_____	___.__	___.__	___.__	___.__
_____	___.__	___.__	___.__	___.__
_____	___.__	___.__	___.__	___.__

VI. MISCELLANEOUS CONSTRUCTION EQUIPMENT:

COST PER--->	HOUR	DAY	WEEK	MONTH
JACK HAMMER (ELECTRIC)	4.00	32	140	440
CHAINSAW	2.50	20	88	275
AUGAR	3.00	24	105	330
SOD CUTTER	5.00	40	175	550
AERATOR	6.00	48	210	660
TAMPER / COMPACTOR	5.00	40	175	550
ROTOTILLER	5.00	40	175	550
_____	_____	_____	_____	_____
_____	_____	_____	_____	_____

EQUIPMENT COSTS

VII. MISCELLANEOUS MAINTENANCE EQUIPMENT:

COST PER---->	HOUR	DAY	WEEK	MONTH
CHIPPER	9.00	72	315	990
TRIMMER, HEDGE	2.00	16	70	220
WEED-EATER	1.25	10	44	138
EDGER	2.00	16	70	220
TRAILER (1 AXLE)	1.00	8	35	110
TRAILER (2 AXLE)	1.50	12	53	165
TRAILER, LOWBOY	2.50	20	88	275
**** HYDRO SEEDER (500 GL)	6.00	48	210	660
**** HYDRO SEEDER (1000 GL)	8.00	64	280	880
**** HYDRO SEEDER (1500 GL)	10.00	80	350	1,100
BRICK SAW (W/O BLADES)	3.00	24	105	330
CONCRETE MIXER	4.00	32	140	440
STUMP GRINDER	6.00	48	210	660
_____	__.__	__.__	__.__	__.__
_____	__.__	__.__	__.__	__.__
_____	__.__	__.__	__.__	__.__
_____	__.__	__.__	__.__	__.__
_____	__.__	__.__	__.__	__.__
_____	__.__	__.__	__.__	__.__

```
    * OR EQUIVALENT.
   ** CONSISTS OF: A 1 TON TRUCK, (1-2) 36-48" & (1) 21" MOWER,
            TRIMMER, BLOWER, WEED-EATER AND HAND TOOLS.
  *** CONSISTS OF: A 1 TON TRUCK, (2-3) 36-48" & (2) 21" MOWERS,
            TRIMMER, BLOWER, WEED-EATER AND HAND TOOLS.
 **** MOUNTED ON A TRAILER. DOES NOT INCLUDE TRUCK.
```

NOTE: THE EQUIPMENT COST PER HOUR RATES ABOVE ARE AVERAGES AND
ARE FOR COMMARISON USE ONLY. YOUR CPH RATES WILL VARY
DEPENDING UPON ENGINE SIZE, MAINTENANCE PRACTICES,
UTILIZATION PRACTICES, ETC.

APPENDIX

D

ABBREVIATIONS

```
ACT...............actual
AIMS..............advanced information management system
ASAP..............as soon as possible/practical
AW................average wage
B&B...............balled & burlapped
BAR...............bid analysis review worksheet
BEP...............break-even point
BF................board foot
BL................bale
BUD LITE..........unrealistically low budget
BUD...............budget
C.................hundred
CAD...............computer aided design
CAL...............caliper
CAW...............crew average wage
CF................cubic foot
CF................contingency factor
CI................cubic inch
CIP...............constant improvement process
CO/DIV............company/division
CO................company
COB...............close of business
COGS..............cost of goods sold (same as COS)
CON or CONST......construction
COPH..............corporate overhead per hour
COS...............cost of sales (same as COGS)
CPA...............certified public accountant
CPH...............cost per hour
CY................cubic yard
DCPH..............direct costs per hour
DOPH..............division overhead per hour
EA................each
EQ/L RATIO........equipment-to-labor ratio
EQ or EQUIP.......equipment
EQPH..............equipment per hour
FDR...............field daily report
FF................fudge factor
FICA / F.I.C.A....Federal Insurance Contributions Act
FLH...............field-labor hour
FMV...............fair market value
FT................foot
FUTA / F.U.T.A....Federal Unemployment Tax Act
FY................fiscal year
FYTD..............fiscal year-to-date
G&A...............general & administrative
GA................gauge
GAS...............gross annual sales
GC................general conditions
GCH/TH............general condition hours divided by total hours
GL................gallon
GLI...............general liabiltiy insurance
GPM...............gross profit margin
```

```
GPM................gallons per minute
GPMH or GPMPH.....gross profit margin per hour
GVW................gross vehicular weight
HD.................head
HP.................horsepower
HR(S).............hour(s)
IMS...............information management system
IRR or IRRIG......irrigation
K or M............thousand
KT................kit
LB, #.............pound
LB................labor burden
LF................linear foot
LS................lump sum
LSC...............landscape
M/L/E/S or MLES...material, labor, equipment, subcontractors
M/L RATIO.........material-to-labor ratio
M or K............thousand
MAT/LAB RATIO.....material-to-labor ratio
MAT'L.............material
MDAY..............man day
MDUP..............market-driven unit pricing
MHR or MH.........manhour
MMT...............meaningful management term(s)
MO................month
MORS..............multiple overhead recovery system
MOY...............monthly
MPH...............material per hour
MSF...............thousand square feet
MTBF..............mean-time between failures
NO, #, NR.........number
NPH or NPMPH......net profit margin per hour, same as PPH
NPM...............net profit margin
NWC...............net working capital
OC................on center
OH or OVHD........overhead
OPH...............overhead per hour
OPPH..............overhead & profit per hour
OT................overtime
OTF...............overtime factor
P&L...............profit & loss
PATT..............place, activity, time, tools
PC................personal computer
PKG...............package
PMSS..............philosophies, methodologies, systems & strategies
POC...............point of connection
POC...............point (person) of contact
POL...............petroleum, oil, lubricants
PPH...............profit per hour, same as NPH
PSI...............pounds per square inch
PVC...............polyvinyl chloride
PW................prevailing wage
```

```
QA..................quality assurance
QCP.................quality control process
QTR.................quarter
QTRLY...............quarterly
QTY.................quantity
RF..................risk factor
RL..................roll
SDI.................state disability insurance
SF..................square foot
SI..................square inch
SORS................single overhead recovery system
SPH.................sales per hour
SUB(S)..............subcontractors
SUTA / S.U.T.A......State Unemployment Tax Act
SY..................square yard
T/O.................takeoff
T&M.................time and materials
TAF.................timely, accurate feedback
TAT.................turn around time
TN..................ton
TOPH................total overhead per hour
TQM.................total quality management
U/P or UP...........unit price
U/C or UC...........unit cost
W/S or WS...........worksheet
WCI.................workers' compensation insurance
Y-T-D or YTD........year-to-date
YR..................year
```

FOOTNOTES

1. Phillip B. Crosby, *Quality is Free*, (New York: Mentor, 1979); p. 66.

2. James R. Huston, *Strategic Planning for Landscape & Irrigation Contractors*, (Oceanside, CA: Smith Huston, Inc., 1990); p. 73.

3. Crosby, Ibid.

4. *Yard & Garden Remodeling*, The Garden Council, Plant a Little Paradise ™ brochure (Chicago, 1991).

5. Tom Peters, *Thriving on Chaos*, (New York: Knopf, 1987); p. 490.

6. Ibid, p. 91.

7. Ibid, p. 490.

8. Ibid, pp. 484-5.

9. *Construction Industry Institute* (1989), p. 1.

10. *Ernst & Young's Contractor Briefing*, Vol. XI, No.2, March/April 1992 (Washington), p. 1.

11. Robert H. Schaeffer and Harvey A. Thomson, "Successful Change Programs Begin with Results," *Harvard Business Review*, (Jan/Feb 1992), pp. 80-81.

12. Tracy E. Benson, "Quality Is Not What You Think It Is", *Industry Week*, (October 5, 1992), p. 22.

13. Benson, "A Business Strategy Comes of Age", (May 3, 1993), p. 40.

14. Crosby, pp. 14-15.

15. Ibid.

16. Peters, p. 481.

17. Ibid. p. 490.

18. Crosby, p. 15.

DIAGRAMS

EXHIBITS

1. BUDGET WORKSHEET

EXHIBIT 1

DIVISION: _____

PREPARED BY: _____

BUDGET PERIOD: _____

DATE PREPARED: ___/___/___

	LAST YEAR	%	NEW BUD	%

I. SALES: _____ ____ _____ ____

II. COST OF SALES (DIRECT COSTS):

1. MATERIALS & SUPPLIES (W/TAX) _____ ____ _____ ____
2. DIRECT LABOR _____ ____ _____ ____
3. LABOR BURDEN _____ ____ _____ ____
4. SUBCONTRACTORS _____ ____ _____ ____
5. EQUIPMENT _____ ____ _____ ____
6. RENTAL EQUIPMENT _____ ____ _____ ____
7. COMMISSIONS _____ ____ _____ ____
8. MISC._____ _____ ____ _____ ____

   TOTAL COST OF SALES-------------> _____ ____ _____ ____

III. GROSS PROFIT MARGIN (GPM) (I-II)---> _____ ____ _____ ____

IV. OVERHEAD OR INDIRECT G&A COSTS:

1. ADVERTISING _____ ____ _____ ____
2. BAD DEBTS _____ ____ _____ ____
3. COMPUTERS & SOFTWARE _____ ____ _____ ____
4. DONATIONS _____ ____ _____ ____
5. DOWNTIME LABOR _____ ____ _____ ____
6. DOWNTIME LABOR BURDEN ____% _____ ____ _____ ____
7. DUES & SUBSCRIPTIONS _____ ____ _____ ____
8. INSURANCE: OFFICE/MEDICAL/KEYMAN _____ ____ _____ ____
9. INTEREST & BANK CHARGES _____ ____ _____ ____
10. LICENSES & SURETY BONDS _____ ____ _____ ____
11. OFFICE EQUIPMENT _____ ____ _____ ____
12. OFFICE SUPPLIES _____ ____ _____ ____
13. PROFESSIONAL FEES _____ ____ _____ ____
14. RADIOS/BEEPERS/CAR PHONES _____ ____ _____ ____
15. RENT (OFFICE & YARD) _____ ____ _____ ____
16. SALARIES - OFFICE STAFF ____% _____ ____ _____ ____
17. SALARIES - OFFICER(S) ____% _____ ____ _____ ____
18. SALARIES - LABOR BURDEN _____ ____ _____ ____
19. SMALL TOOLS & SUPPLIES _____ ____ _____ ____
20. TAXES - MILL/ASSET TAX _____ ____ _____ ____
21. TELEPHONE _____ ____ _____ ____
22. TRAINING/EDUCATION _____ ____ _____ ____
23. TRAVEL & ENTERTAINMENT _____ ____ _____ ____
24. UNIFORMS & SAFETY EQUIPMENT _____ ____ _____ ____
25. UTILITIES _____ ____ _____ ____
26. VECHILES, OVERHEAD _____ ____ _____ ____
27. YARD EXPENSE & LEASEHOLD IMPROV. _____ ____ _____ ____
28. MISCELLANEOUS _____ ____ _____ ____

   _____

   TOTAL OVERHEAD--------------------> _____ ____ _____ ____
   ============= =============

V. NET PROFIT MARGIN (NPM) (III-IV)---> _____ ____ _____ ____

VI. NPM ADJUST. (BONUS, ESOP, ETC.)----> _____ ____ _____ ____
   ============= =============

VII. REVISED NET PROFIT MARGIN (V-VI)---> _____ ____ _____ ____

EXHIBIT 2

2. FIELD-LABOR HOUR (FLH) BUDGET WORKSHEET

	(a)	(b)	(c)	(d)	(e)	(f)

1. AVERAGE WAGE (AW) CALCULATION:

	NO.	HRS/MO	MOS/YR	TOT HRS	HOURLY RATE	TOTAL PAYROLL

A. LANDSCAPE CREW:

(1). FOREMEN	____	_____	_____	_____	$____.__	$_____
(2). FOREMEN	____	_____	_____	_____	$____.__	$_____
(3). LEADMEN	____	_____	_____	_____	$____.__	$_____
(4). LEADMEN	____	_____	_____	_____	$____.__	$_____
(5). LABORERS	____	_____	_____	_____	$____.__	$_____
(6). LABORERS	____	_____	_____	_____	$____.__	$_____
(7). TOTALS:				*_____		** $_____

(8). AW = $\dfrac{\text{TOTAL PAYROLL}}{\text{TOTAL HOURS}}$ = ---------------- = $_____.___

(9). AW MULTIPLIED BY (1+OTF) = $_____.___ X (1.____) = $_____.___

B. IRRIGATION CREW:

(1). FOREMEN	____	_____	_____	_____	$____.__	$_____
(2). FOREMEN	____	_____	_____	_____	$____.__	$_____
(3). LEADMEN	____	_____	_____	_____	$____.__	$_____
(4). LEADMEN	____	_____	_____	_____	$____.__	$_____
(5). LABORERS	____	_____	_____	_____	$____.__	$_____
(6). LABORERS	____	_____	_____	_____	$____.__	$_____
(7). TOTALS:				*_____		** $_____

(8). AW = $\dfrac{\text{TOTAL PAYROLL}}{\text{TOTAL HOURS}}$ = ---------------- = $_____.___

(9). AW MULTIPLIED BY (1+OTF) = $_____.___ X (1.____) = $_____.___

2. FIELD-LABOR HOUR (FLH) BUDGET WORKSHEET

C. MAINTENANCE CREW:

(1). FOREMEN ____ _____ _____ _____ $_____.__ _____

(2). FOREMEN ____ _____ _____ _____ $_____.__ $_____

(3). LEADMEN ____ _____ _____ _____ $_____.__ $_____

(4). LEADMEN ____ _____ _____ _____ $_____.__ $_____

(5). LABORERS ____ _____ _____ _____ $_____.__ $_____

(6). LABORERS ____ _____ _____ _____ $_____.__ $_____

(7). TOTALS: *_____   ** $_____

(8). AW = $\dfrac{\text{TOTAL PAYROLL}}{\text{TOTAL HOURS}}$ = _____ = $_____.___

(9). AW MULTIPLIED BY (1+OTF) = $_____.___ X (1.____) = $_____.___

2. OVERTIME FACTOR (OTF):

	SAMPLE	YOUR CO
A. AVERAGE NO. HOURS WORKED PER MAN PER WEEK:	50	_____
B. AVERAGE NO. OT HOURS WORKED PER MAN PER WEEK:	10	_____
C. AVERAGE NO. OT HOURS PAID PER MAN PER WEEK:	15	_____

D. OTF = $\dfrac{C - B}{A}$ = $\dfrac{15 - 10}{50}$ = $\dfrac{5}{50}$ = 10%

E. L/S OTF = $\dfrac{-}{}$ = _____ = _____%

F. IRRIG OTF = $\dfrac{-}{}$ = _____ = _____%

G. MAINT OTF = $\dfrac{-}{}$ = _____ = _____%

3. RISK FACTOR: (USED ON SPECIFIC JOBS. DO NOT USE IN FLH BUDGET).

4. TOTAL HOURS AND PAYROLL FOR THE YEAR:

A.   TOTAL LABOR HOURS (SEE * ABOVE): _____

B.   TOTAL PAYROLL (SEE ** ABOVE): $_____

C.   LAST YEAR'S FIELD PAYROLL IF AVAIL: $_____

D.   LAST YEAR'S LABOR HOURS IF AVAIL: _____

EXHIBIT 3

3. CREW AVERAGE WAGE (CAW) WORKSHEET

Date:_____/_____/_____   Prepared by:_____

Job:_____   Crew size:_____

Type Crew:_____   Hours per man per day:_____

Type of Labor	Name	Labor Rate
_____	_____	$_____._____
_____	_____	_____._____
_____	_____	_____._____
_____	_____	_____._____
_____	_____	_____._____
_____	_____	_____._____
_____	_____	_____._____
_____	_____	_____._____
_____	_____	_____._____
_____	_____	_____._____
_____	_____	_____._____
_____	_____	_____._____

Total number in crew _____   Labor total $_____._____

1. Divide labor total $_____ by # in crew _____ = _____

2. Add OTF % (_____%).............................._____

    Subtotal..._____

3. Add RF (10-20%)..............................._____

    Subtotal..._____

4. Round up to nearest $ .10 to obtain CAW............._____

EXHIBIT 4

4. LABOR BURDEN WORKSHEET

	CONST.	IRRIG.	MAINT.	OFFICE
F.I.C.A. (COMPANY PORTION) RATE	._____	._____	._____	._____
F.U.T.A. (FEDERAL UMEMP) RATE	._____	._____	._____	._____
S.U.T.A. (STATE UNEMP) RATE	._____	._____	._____	._____
WORKER'S COMPENSATION RATE	._____	._____	._____	._____
GENERAL LIABILITY INSURANCE RATE (1)	._____	._____	._____	._____ NA
VACATIONS (2)	._____	._____	._____	._____ NA
HOLIDAYS (3)	._____	._____	._____	._____ NA
SICK DAYS (4)	._____	._____	._____	._____ NA
MEDICAL/HEALTH INSURANCE (5)	._____	._____	._____	._____ NA
_____	._____	._____	._____	._____
	========	========	========	========
TOTALS--------------------------------->	._____	._____	._____	._____

CALCULATIONS FOR ABOVE:

(1). GENERAL LIABILITY INSURANCE:

USE POLICY RATE FOR HUNDRED DOLLARS OF FIELD PAYROLL OR
DIVIDE ANNUAL PREMIUM $_____ BY PROJECTED ANNUAL FIELD
PAYROLL $_____ = ._____.

(2). FIELD LABOR VACATIONS:

DIVIDE TOTAL FIELD CREW VACATION WEEKS PAID IN THE YEAR _____ BY THE
TOTAL FIELD WORK FORCE WEEKS IN THE YEAR _____ = ._____.

(3). FIELD LABOR HOLIDAYS:

DIVIDE TOTAL FIELD CREW HOLIDAYS PAID IN THE YEAR _____ BY
THE TOTAL FIELD WORK FORCE MAN DAYS IN THE YEAR _____ = ._____.

(4). FIELD LABOR SICK DAYS:

DIVIDE TOTAL FIELD CREW SICK DAYS EARNED IN THE YEAR _____ BY
THE TOTAL FIELD WORK FORCE MAN DAYS IN THE YEAR _____ = ._____.

(5). FIELD LABOR HEALTH/MEDICAL INSURANCE:

DIVIDE TOTAL ANNUAL HEALTH/MEDICAL INSURANCE PAYMENT (FOR FIELD
PERSONNEL ONLY) $_____ BY PROJECTED ANNUAL FIELD PAYROLL
$_____ = ._____.

5. BUDGET DEFINITIONS

1. ADVERTISING:

A. Yellow pages.
B. Newspaper ads/help wanted ads.
C. Magazines.
D. Bill boards.
E. Door hangers.
F. Bulk mail.
G. Company brochures (pro rate over # of years).
H. Garden, trade and home show.
I. Company resume.

Note: Does not include business cards.

2. BAD DEBTS:

A. Usually one-half of 1% of gross sales.
B. Percent of Accounts Receivables which are written off.

3. COMPUTERS and SOFTWARE:

A. Computer, software, and digitizers are included in this category.
B. Depreciate out over a three-to-five year period.
C. Include costs to have PC consultants train your staff (in-house).
D. Maintenance contracts and repairs.

4. DONATIONS:

A. Cash or time.
B. Only a donation that will put you in a good light in the public eye.
C. Team uniforms (i.e., Little League teams).

Note: Donations from profit to reduce taxes are not included.

5. DOWNTIME LABOR:

Time field crews are still paid but are not working in the field and producing billable hours.

A. Rain/weather - can include non-productive winter crew time.
B. Equipment breakdown (i.e., to and from jobs).
C. Busy work - fixing shovels, cleaning/fixing yard.
D. Safety meetings/misc. crew meeting time.
E. Watering nursery material (only if nursery is not separate division).

Note: Time spent working on/repairing equipment not included.

5. BUDGET DEFINITIONS

6. DOWNTIME LABOR BURDEN:

A percent markup on down time labor to cover FICA, FUTA, SUTA, WCI, etc.

7. DUES and SUBSCRIPTIONS:

A. Dues to national organizations.
B. Local contractors organizations.
C. Chamber of Commerce, Better Business Bureau, etc.
D. Professional magazines.
E. Dodge reports, Plan rooms, Green sheets, etc.
F. Any type of business-related publications.
G. Price Club, discount warehouse dues.

8. INSURANCE:

A. Office contents insurance.
B. Health insurance for overhead people (i.e., owner(s), secretary, accountant, bookkeeper, estimator, sales/field supervisors).
C. Life/Key Man insurance is included as long as the company/corporation is the beneficiary.

9. INTEREST and BANK CHARGES:

A. Interest on charge cards and line of credit.
B. Lost interest on company's retained earnings used for working capital (4-5 weeks payroll).
C. Bank supplier interest.
D. Bank bounced check fees and service fees.

   Note: Does not include:

      (1). Past due receivables owed to you.
      (2). Interest on equipment purchases.
      (3). Interest on mortgage payments.

10. LICENSES and SURETY BONDS:

A. Contractor's license for the state in which you operate.
B. Surety bond on license.
C. City licenses not included in general conditions on bids.

11. OFFICE EQUIPMENT DEPRECIATION, REPAIRS and SERVICE CONTRACTS:

(Divide total cost for each category by its years of useful life)

A. Office equipment.

B. Office furniture.
C. Desks (10 years).
D. File cabinets (10 years).
E. Coffee machine, copy machine, blue print machine.
F. Calculators.
G. Plan Rack.
H. Drafting table (10 years).
I. Typewriter.
J. Any piece of office equipment.
K. Plants, pictures, art.
L. Repairs/maintenance contracts.

Note: Does not include field equipment, overhead vehicles, computers, telephones, radios and digitizers.

12. OFFICE SUPPLIES:

A. Pens, pencils, paper, paper clips, etc.
B. Stationery, invoices, printing of forms, etc.
C. Business cards.
D. Postage.

13. PROFESSIONAL FEES:

A. Outside CPA or bookkeeper.
B. Legal fees (not to include cost of law suits).
C. Consultants (other than computer consultants).
D. Outside payroll services.
E. Cost of Corporate Tax preparation at end of year.
F. Incorporation cost spread over five years.

14. RADIO SYSTEM and CAR PHONE:

A. 2-way radio system (depreciate over 5-10 years/useful lifetime).
B. Beepers, pagers.
C. Car phones (depreciate over 2-3 years).
D. Maintenance of above items (to include anticipated repairs).
E. Repeater charge.
F. Monthly line charges.
G. Monthly toll/use charges.

15. RENT (OFFICE and YARD):

A. Office space (fair market value, not higher/not lower).
B. Storage facilities and yard.

5. BUDGET DEFINITIONS

16. SALARIES - OFFICE:

Bookkeeper, office manager, secretary, estimator, designer, field supervisor (pro-rate if a "working" supervisor).

17. SALARIES - OFFICER (S):

Corporate officers salaries at fair market value.

Note: Total of 16 & 17 should total 8%-12% of gross sales.

18. SALARIES - LABOR BURDEN (OFFICE):

A percent markup of 16 & 17 to cover FICA, FUTA, SUTA, etc.

19. SMALL TOOLS and NON-OFFICE SUPPLIES:

A. Wheelbarrows on down (non-motorized items).
B. Hand tools to repair equipment.
C. Rakes, shovels, hoes, etc.
D. Tarps.
E. Nails and miscellaneous hardware.
F. Paint.

20. TAXES (ASSET or MILL TAX):

A. Business privilege tax/Mill tax.
B. Asset tax.

Note: Does not include field equipment portion of tax.

21. TELEPHONE:

A. Line charges.
B. Long distance charges.
C. Rental (if any).
D. Depreciation of system over 5-10 years.
E. Answering service/machine.
F. Repairs and maintenance contracts.

22. TRAINING and EDUCATION:

A. Books, videos, tapes, etc.
B. Workshops and seminar entrance fees.
C. Community college/continuing education classes.
D. Convention entrance fees/registration fees.

Note: Does not include travel and entertainment expense associated with above.

5. BUDGET DEFINITIONS

23. TRAVEL and ENTERTAINMENT:

A. Travel cost(s) to get to seminars, workshops or conventions.
B. Hotel bills associated with seminars, workshops and conventions.
C. Lunches/dinners for clients.
D. Christmas party for employees.
E. Gifts at Thanksgiving.
F. Employee parties/meals if company pays for them.

24. UNIFORMS and SAFETY EQUIPMENT:

A. T-shirts and ball caps.
B. Uniform rental services.
C. Uniform cleaning services.
D. Safety equipment (i.e., goggles, gloves, ear protectors and hard hats if not included in small tools/supplies).

25. UTILITIES:

A. Water, sewer and electricity for office/yard.
B. Gas and oil for heating office/yard spaces.
C. Trash service/non-job related dumpster/dumpster fees.
D. Ofice cleaning services.

26. VEHICLES, OVERHEAD:

A. Vehicles used by overhead personnel for overhead purposes (e.g., trucks that do not go out on the job).
B. Mileage reimbursement for personal vehicles.

27. YARD EXPENSE/LEASEHOLD IMPROVEMENTS:

A. Cost associated with maintaining a storage yard.
B. Leasehold improvements to your office depreciated over the life of improvements (carpet, painting, tenant improvements, etc.).

28. MISCELLANEOUS:

A catch-all category. Keep under $250.

Equipment Cost Per Hour (CPH) Calculations

Piece of Equipment	Use	Purchase Price (Incl. Interest & Deduct Salvage Value)	+	Life Expectancy (In Hours)	=	Acquisition CPH (1)	Anticipated Lifetime Maintenance Insurance & License Costs	+	Life Expectancy	=	Maintenance CPH (2)	Fuel & Oil CPH (3)	TOTAL CPH (1+2+3)
Pick-up ½ ton	owner/office	12,240		8,320		1.47	9,000		8,320		1.08	.71	3.26

EXHIBIT 6

BID WORKSHEET

PROJECT: _____

LOCATION: _____

P.O.C./G.C.: _____

DUE: _____ / _____ / _____ PH.: _____ FAX: _____

TYPE WORK: _____

REMARKS: _____

PAGE (____) OF (____

DATE PREPARED: _____ / _____ / _____

ESTIMATOR: _____

CREW SIZE: _____

HOURS/DAY: _____ CAW: $_____.___

DESCRIPTION	QTY	UNIT	U/C	MAT'L	LABOR	EQUIP	SUBS

SHI FORM 04-A

EXHIBIT 8

8. BID ANALYSIS REVIEW (BAR) WORKSHEET

DATE: _____/_____/_____

JOB: _____

I. PRODUCTION OF FINISHED PRODUCT/SERVICE:

MATERIAL	LABOR	EQUIPMENT	SUBS
(A).$_____	(B).$_____	(C).$_____	(D).$_____

II. GENERAL CONDITIONS:

MATERIAL	LABOR	EQUIPMENT	SUBS
(A).$_____	(B).$_____	(C).$_____	(D).$_____
(SUBTTL 1)$_____	$_____	$_____	$_____

III. MARKUPS:

A. $_____   (TAX _____% ON MATERIALS)

B. $_____   (LABOR BURDEN %_____)

(SUBTTL 2) $_____ $_____ $_____ $_____

SUBTOTAL 3 (SUM OF ALL ITEMS IN SUBTOTAL # 2 ABOVE) $_____

C. OVERHEAD RECOVERY TOTAL............................ $_____

D. **BEP:** (TOTAL OF DIRECT COSTS PLUS OVERHEAD)....... $_____

E. CONTINGENCY FACTOR (IF APPLICABLE) .... $_____

F. NET PROFIT.. $_____

G. TOTAL PRICE FOR THE JOB......(D+E+F).............. $_____

IV. RATIO / PER HOUR ANALYSIS:

A. SPH:	$_____.__	J. MAT/LAB:	_____: 1
B. DCPH:	$_____.__	K. MPH:	$_____.__
C. OPH:	$_____.__	L. EQ/LAB:	_____: 1
D. PPH:	$_____.__	M. EQPH:	$_____.__
E. BEP:	$_____ _____.__ %	N. GC:	$_____
F. OVERHEAD:	$_____ _____.__ %	O. GCPH:	$_____.__
G. NET PROFIT:	$_____ _____.__ %	P. GCH/TH:	_____.__ %
H. GPM:	$_____ _____.__ %	Q. FACTOR:	_____.__
I. GPMPH:	$_____ . . %	R. UNIT PRICE:$_____

V. MORS MARKUPS (DIVIDE LABOR & BURDEN SUBTTL BY G TO GET H).

A. MATERIAL X 10% =...._____   E. III. C. TOTAL OVHD: $_____

B. EQUIPMENT X 25% =..._____   F. MINUS V.D. SUBTOTAL:-_____

C. SUBS X 5% =........._____   G. TOTAL OVHD ON LABOR:$_____

D. SUBTOTAL............_____   H. MORS % LABOR MARKUP: ____.__%

DAILY JOB SITE LOG

TEMP. _____

WEATHER: CLEAR SNOW WINDY

RAIN OVERCAST

JOB SITE: _____ DATE: _____

SUPERVISOR: _____ JOB SUPERVISOR: _____

1. LABOR/EQUIPMENT: WORK NAME

START TIME: _____

QUIT TIME: _____

WORKMEN							TOTAL	INITIALS
TOTALS								

EQUIPMENT ON JOB

ITEM	HOURS
TOTALS	

2. WORK ACCOMPLISHED: _____
_____
_____
_____
_____
_____

3. SUB-CONTRACTORS:

	TIME-IN	TIME-OUT	TIME-IN	TIME-OUT	TOTAL

4. CHANGES: _____
_____
_____

5. EXTRAS: _____
_____
_____

6. COMMENTS: _____
_____
_____
_____

7. JOB SITE SUPERVISOR SIGNATURE: X _____

10. SAMPLE MAINTENANCE FIELD DAILY REPORT (FDR)

DATE: ____ / ____ /199__

WEATHER CONDS: _____

CREW NO: ____

CREW MEMBERS: _____

JOB SITE	START TIME	FINISH TIME (I.E. 7:30-8:45)	MORNING LOAD MANHRS	DRIVETIME LABOR MANHRS	LOAD/UN-LOAD MANHRS	21" MOW LABOR& EQ HRS	36" MOW LABOR& EQ HRS	52" MOW LABOR& EQ HRS	BLOWER LABOR& EQ HRS	TRIMMER LABOR& EQ HRS	WEED & CLN-UP MANHRS	LABOR MANHRS	LABOR MANHRS	SUPVSR. ON-SITE MANHRS	TOTALS MANHRS
TIME PRIOR TO FIRST JOB		TO													
		TO													
		TO													
		TO													
		TO													
		TO													
		TO													
		TO													
		TO													
		TO													
		TO													
		TO													
		TO													
		TO													
		TO													
		TO													
		TO													
		TO													
		TO													
		TO													

REMARKS / MATERIALS / SPECIAL NOTES: _____

EXHIBIT 10

94-C001 HENDERSON RESIDENCE CONSTRUCTION BID   01-Mar-94   page 1

09:27 PM

UNIT PRICE REPORT :   TOTAL BID PRICE 30,906.04

TOTALS -- 530.59 7,032.00 909.82 12,138.96 873.00 20,964.28

BID ITEM	QTY UNIT	LUMP SUM	/ UNIT	SET $$	LABOR HRS	$ LABOR	$ EQUIP	$ MAT	$ SUBS	TOTAL COST	SPH (SALES/ MANHR)	GPMPH (GPM/ MANHR)	PPH (PROFIT/ MANHR)	MAT/ LABOR RATIO
TOTAL IRRIGATION---(IRR)------------>		13,121.82	2.083 $/SF		182.09	2,366.23	156.80	4,705.34		7,228.37	72.06	32.37	22.37	1.99
TOTAL LANDSCAPE----(LSC)------------>		16,002.57	2.540 $/SF		184.47	2,397.09	160.92	7,067.82		9,625.83	86.75	34.57	24.57	2.95
TOTAL CONSTRUCTION-(CON)------------>		754.60	0.255 $/SF		12.00	155.94	36.00	190.80		382.74	62.88	30.99	20.99	1.22
TOTAL MAINTENANCE--(MNT)------------>		1,027.06	0.307 $/SF						873.00	873.00	NA	NA	NA	0.08
TOTAL GENERAL CONDITIONS------>------>------>------>					152.03	2,112.73	556.10	175.00		2,854.33	NA	NA	NA	
POC IRR	1.0 LS	81.30	81.296		2.00	25.99				25.99	40.65	27.65	17.65	
CONTROLER IRR	1.0 EA	542.71	542.708		2.00	25.99		392.20		418.19	271.35	62.26	52.26	15.09
WIRE IRR	2,000 LF	262.36	0.131		4.00	51.98		84.80		136.78	65.59	31.39	21.39	1.63
BFP IRR	1.0 EA	595.32	595.322		4.00	51.98		367.82		419.80	148.83	43.88	33.88	7.08
QCV RB33 IRR	4.0 EA	332.19	83.048		4.00	51.98		144.16		196.14	83.05	34.01	24.01	2.77
ML SCH 40 1.25" IRR	240.0 LF	568.74	2.370		10.00	129.95	19.20	118.72		267.87	56.87	30.09	20.09	0.91
SLEEVE SCH 40 2.5" IRR	120.0 LF	382.19	3.185		5.00	64.97		152.11		217.08	76.44	33.02	23.02	2.34
VALVES TORO 289 IRR	14 EA	2,578.50	184.179		28.00	363.85		1,224.30		1,588.15	92.09	35.37	25.37	3.36
LATS CL200 IRR	520.0 LF	978.58	1.882		21.09	274.07	41.60	61.48		377.15	46.40	28.52	18.52	0.22
LATS CL315 .5" IRR	1,200 LF	2,166.93	1.806		48.00	623.75	96.00	87.45		807.20	45.14	28.33	18.33	0.14
RB 1806 HEADS IRR	170.0 EA	3,184.04	18.730		34.00	441.82		1,531.70		1,973.52	93.65	35.60	25.60	3.47
RB 1804 HEADS IRR	60.0 EA	1,123.78	18.730		12.00	155.94		540.60		696.54	93.65	35.60	25.60	3.47
FLUSH IRR	1 LS	325.18	325.185		8.00	103.96				103.96	40.65	27.65	17.65	

EXHIBIT 11

09:27 PM

UNIT PRICE REPORT :   TOTAL BID PRICE 30,906.04

TOTALS --   530.59   7,032.00   909.82   12,138.96   873.00   20,964.28

BID ITEM	QTY UNIT	LUMP SUM	/ UNIT	SET $$ LABOR HRS	$ LABOR	$ EQUIP	$ MAT	$ SUBS	TOTAL COST	SPH (SALES/ MANHR)	GPMPH (GPM/ MANHR)	PPH (PROFIT/ MANHR)	MAT/ LABOR RATIO
SOIL PREP LSC	6300 SF	1,456.59	0.231	14.28	185.57	115.92	628.79		930.28	102.00	36.86	26.86	3.39
24" BOX TREES LSC	6 EA	1,555.82	259.304	16.00	207.92	27.00	742.64		977.55	97.24	36.14	26.14	3.57
15 GL TREES LSC	15 EA	1,593.93	106.262	21.25	276.14		620.63		896.77	75.01	32.81	22.81	2.25
WASH ROBUSTA LSC	3 EA	1,578.43	526.142	13.00	168.93	18.00	874.50		1,061.43	121.42	39.77	29.77	5.18
GRADE LSC	6300 SF	256.08	0.041	6.30	81.87				81.87	40.65	27.65	17.65	
5 GL SHRUBS LSC	215 EA	3,437.81	15.990	43.00	558.78		1,436.45		1,995.23	79.95	33.55	23.55	2.57
5 GL ESP SHRUBS LSC	3 EA	107.12	35.706	0.75	9.75		65.14		74.88	142.82	42.98	32.98	6.68
1 GL SHRUBS LSC	283 EA	2,070.66	7.317	23.58	306.46		945.24		1,251.70	87.80	34.73	24.73	3.08
1 GL VINES LSC	24 EA	191.47	7.978	2.00	25.99		93.64		119.63	95.73	35.92	25.92	3.60
GROUND COVER 8" OC LSC	23 FLT	725.29	31.535	11.50	149.38		219.33		368.71	63.10	31.02	21.02	1.47
GROUND COVER 12" OC LSC	36 FLT	1,131.93	31.442	17.94	233.13		342.30		575.42	63.10	31.02	21.02	1.47
SOD LSC	3345 SF	1,897.44	0.567	14.87	193.19		1,099.17		1,292.36	127.63	40.70	30.70	5.69
MULCH CON	9 CY	754.60	83.845	12.00	155.94	36.00	190.80		382.74	62.88	30.99	20.99	1.22
90 DAY MAINT MNT	3 MO	1,027.06	342.353					873.00	873.00				

BID RECAP & SETUP REPORTS

BID RECAP REPORT

I. PRODUCTION COSTS:		PRODUCTION DIRECT COST $$ AMOUNT	GEN. COND. DIRECT COST $$ AMOUNT	TOTAL DIRECT COSTS $$ AMOUNT	OH RECOVERY ON DIRECT JOB COSTS $$ AMOUNT
A. MATERIALS COST (w/salestax)-->	38.7%	11,963.96	185.50	12,149.46	1,214.95
B. EQUIPMENT COST-->	1.1%	353.72	556.10	909.82	227.46
C. SUBCONTRACTORS COST-->	2.8%	873.00		873.00	43.65
D. LABOR COST (w/burden,OTF,RF)-->	15.9%	4,919.26	2,112.73	7,032.00	3,819.81
SUBTOTAL (PH. I) DIRECT COSTS-->	58.6%	18,109.94		20,964.28	
II. GENERAL CONDITION COSTS--->	9.2%	2,854.33	2,854.33		
SUBTOTAL (PH. I&II) DIRECT COSTS>	67.8%	20,964.28			
III. MARKUPS:					
A. OVERHEAD RECOVERY--->	17.2%	5,305.86		5,305.86	
(BEP=OH+DIR COSTS)= $26,270	85.0%				
B. PROFIT + SALES COMMISSION--->	15.0%	4,635.91	4,635.91		
SUBTOTAL MARKUPS (GPM)------>	32.2%	9,941.77			
TOTAL BID PRICE------------>	100.0%	30,906.04	30,906.04		

UNIT PRICE SUMMARY

	DIFFERENCE
	0.00
4,635.91	% PROFIT 15.00%
	% MARKUP GPM 32.17%

0.00

:: BID SETUP REPORT ::
:: BID SETUP INPUT AREA

ENTER AMOUNTS / % IN THIS COLUMN
< --------- >

1. BID FILE NAME------------------>94-C001
   (up to 8 characters)
2. BID NAME (Name,GC,Ph #,Bid date)-->HENDERSON RESIDENCE CONSTR
   (up to 25 characters)

OFFICE OVERHEAD RECOVERY TABLE

3. FIELD-LABOR HOUR ANNUAL BUDGET--> 10,000
4. OVERHEAD TO RECOVER/YEAR--------> $100,000
   (5 & 6 are calculated do not adjust)
5. OVHD COST PER FIELD-LABOR HR----> $10.00
6. OPH CALCULATED BY PPH METHOD----> $10.00

7. SALES TAX ON MATERIALS--------> 6.00%
8. SALES COMMISION FOR THIS BID-->
9. LABOR OVERTIME FACTOR % (OTF)--> 20.00%
10. LABOR RISK FACTOR (RF)--------> 10.00%
11. NET PROFIT MARGIN (NPM) TARGET--> 15.00%
12. NET PROFIT/HR (PPH) OVERRIDE--> $.
13. FIELD LABOR BURDEN %---------> 30.00%
14. LABOR TABLE #---------------> 2
15. SQUARE FOOTAGE TABLE:--------> INPUT "SF"
   A. IRRIGATION COVERAGE AREA---> 6,300
   B. LANDSCAPE PLANTING AREA----> 6,300
   C. CONSTRUCTION AREA---------> 2,955
   D. MAINTENANCE AREA----------> 3,345
   E. TOTAL PROJECT COVERAGE----> 6,300

ANALYSIS REPORT

FIVE (5) ESTIMATING METHODS
MARKET COMPARISON ANALYSIS

BID PRICE $30,906.04

AMOUNTS :: RATIO/PER HOUR ANALYSIS:

1. FACTORING (Price divided by materials
   or materials X ?)----------> 2.55

2. UNIT PRICES (See UNIT PRICE RPT):

	SF	$/SF
A. IRRIGATION------->	6,300	$2.083
B. LANDSCAPE-------->	6,300	$2.540
C. CONSTRUCTION----->	2,955	$0.255
D. MAINTENANCE------>	3,345	$0.307
E. TOTAL PROJECT---->	6,300	$4.906

3. GROSS PROFIT MARGIN (GPM): 32.17%

4. MULTIPLE OVERHEAD RECOVERY SYSTEM (MORS):
   A. MATERIALS-----------SET AT -> 10.00%
   B. EQUIPMENT-----------SET AT -> 25.00%
   C. SUBCONTRACTOR-------SET AT -> 5.00%
   D. LABOR & BURDEN------CALC AT-> 54.32%
   E. LABOR OVERRIDE % ----------> %

5. OVERHEAD & PROFIT PER HOUR (OPPH) METHOD:
   A. TOTAL FIELD-LABOR HOURS IN BID--> 530.59
   B. OVERHEAD PER HOUR (OPH)--------> $10.00
   C. NET PROFIT PER HOUR (PPH)------> $8.74
   D. GROSS PROFIT PER HOUR (GPMPH)--> $18.74
   E. SALES PER HOUR (SPH)-----------> $58.25

11. SALES PER HOUR (SPH):

	HOURS	SPH
A. TOTAL JOB---------->	530.6	$58.25
B. IRRIGATION--------->	182.1	$72.06
C. LANDSCAPE---------->	184.5	$86.75
D. CONSTRUCTION------->	12.0	$62.88
E. MAINTENANCE-------->		ERR
F. GENERAL CONDITIONS->	152.0	N/A

12. MATERIALS TOTAL (WITH TAX)------>$12,138.96
    A. MATERIAL TO LABOR (M/L) RATIO--> 1.73 :1
    B. MATERIAL PER HOUR (MPH)-------> $22.88

13. EQUIP. TOTAL (PROD.+ GEN. CONDS)-----> $909.82
    A. EQUIP. TO LABOR (EQ/L) RATIO----> 0.13 :1
    B. EQUIPMENT PER HOUR (EQPH)-------> $1.71

14. GENERAL CONDITIONS TOTAL----------> $2,854.33
    A. AS A PERCENT OF PRICE----------> 9.24%
    B. PER FIELD-LABOR HOUR (GCPH)----> $5.38
    C. GC HRS/TOTAL HRS--------------> 28.65%

15. OVERHEAD (G&A or INDIRECT COST)----> $5,305.86
    A. AS A PERCENT OF PRICE----------> 17.17%
    B. PER FIELD-LABOR HOUR (OPH)-----> $10.00

16. NET PROFIT MARGIN (NPM) IN BID-----> $4,635.91
    A. AS A PERCENT OF PRICE----------> 15.00%
    B. PER FIELD-LABOR HOUR (PPH)-----> $8.74

17. GROSS PROFIT MARGIN (OVHD+NPM)-----> $9,941.77
    A. AS A PERCENT OF PRICE----------> 32.17%
    B. PER FIELD-LABOR HOUR (GPMPH)---> $18.74

18. SUBCONTRACTOR COSTS---------------> $873.00
    A. AS A PERCENT OF PRICE----------> 2.82%

9. CREW AVE WAGE (w/burden, OTF & RF)
   A. TOTAL JOB HOURS---------> $13.25
   B. IRRIGATION-------------> $12.99
   C. LANDSCAPE--------------> $12.99
   D. CONSTRUCTION-----------> $12.99
   E. MAINTENANCE------------> ERR
   F. GENERAL CONDITIONS-----> $13.90

10. SALES COMMISSION----------->
    A. DOLLARS----------------> $30,906
    B. AS A % OF PRICE-------->

WORKSHEET

BID ITEM ID	UNIT QTY UNIT	DATABASE TASKCODE	TASK DESCRIPTION	QUANTITY UNIT	MATL. UNIT $COST	LABOR/EQUIP TYPE	QUANT / 1 HR	TASK HOURS	SUB UNIT $$ COST	TASK/BID UNIT $$ COST	BID UNIT $$ PRICE	TOTAL $$ PRICE
BID ITEM ID POC IRR	1 LS											
			LABOR	2 HR		$II	1	2.0		12.995		
								2.00		25.990	81.296 I	81.30
BID ITEM ID CONTROLER IRR	1.00 EA											
			CONTROLLER & HARDWARE	1 EA	370	$II	0.5	2.0		418.190		
					370			2.00		418.190	542.708 I	542.71
BID ITEM ID WIRE IRR	2,000.00 LF											
			WIRE 14 GAUGE	2000 LF	0.04	$II	500	4.0		0.068		
					0.04			4.00		0.068	0.131 I	262.36
BID ITEM ID BFP IRR	1.00 EA											
			BACKFLOW PREVENTER	1 EA	347	$II	0.25	4.0		419.799		
					347			4.00		419.799	595.322 I	595.32
BID ITEM ID QCV RB33 IRR	4.00 EA	QCV	RB33									
			QUICK COUPLER VALVE	4 EA	34	$II	1.0	4.0		49.035		
					34			4.00		49.035	83.048 I	332.19
BID ITEM ID ML SCH 40 1.2	240 LF	SCH40125	ML									
			SCH 40 PVC SCH 40 1.25"	280 LF	0.32	$II	28	10.0		0.803		
			TRENCHER	240 LF		$120	100	2.4		0.080		
			FITTINGS (25% OF PVC)	280 LF	0.08	$				0.085		
					0.46666			10.00		1.116	2.370 I	568.74
BID ITEM ID SLEEVE SCH 40	120 LF	SCH4025	SLEEVE									
			PVC SCH 40 2.5"	140 LF	0.82	$II	28	5.0		1.333		
			FITTINGS (25% OF PVC)		0.205	$				0.217		
					1.19583			5.00		1.809	3.185 I	382.19

WORKSHEET

| | BID | | LIBRARY | | | MATL. UNIT | LABOR/EQUIP | QUANT | TASK | SUB UNIT | TASK/BID UNIT | BID UNIT | TOTAL |
BID ITEM ID	UNIT QTY UNIT	DATABASE TASKCODE	TASK DESCRIPTION	QUANTITY UNIT	$COST	:TYPE	/ 1 HR	HOURS	$$ COST	$$ COST	$$ PRICE	$$ PRICE
VALVES TORO 2	14 EA VALVES	T-289	TORO 289 VALVES	14 EA	82.5	$II	0.5	28.0		113.440	184.179 I	2,578.50
					82.5			28.00		113.440	184.179 I	2,578.50
BID ITEM ID												
LATS CL200 IR	520 LF LATS	CL200-1	PVC CLASS 200 1"	580 LF	0.08	$II	27.5	21.1		0.557		
			TRENCHER	520 LF		$120	100	5.2		0.080		
			FITTINGS (25% OF PVC)	580 LF	0.02	$				0.021		
					0.11153			21.09		0.725	1.882 I	978.58
BID ITEM ID												
LATS CL315 .5	1200 LF LATS	CL315-5	CL 315 .5" PVC	1320 LF	0.05	$II	27.5	48.0		0.526		
			TRENCHER	1200 LF		$120	100	12.0		0.080		
			FITTINGS (25% OF PVC)	1320 LF	0.0125	$				0.013		
					0.06875			48.00		0.673	1.806 I	2,166.93
BID ITEM ID												
RB 1806 HEADS	170 EA HEADS	RB1806	RB 1806 HEADS	170 EA	8.5	$II	5	34.0		11.609		
					8.5			34.00		11.609	18.730 I	3,184.04
BID ITEM ID												
RB 1804 HEADS	60.00 EA HEADS	RB1804	RB 1804 HEADS	60 EA	8.5	$II	5	12.0		11.609		
					8.5			12.00		11.609	18.730 I	1,123.78
BID ITEM ID												
FLUSH IRR	1 LS		LABOR	8 HR		$II	1	8.0		12.995		
								8.00		103.958	325.185 I	325.18
BID ITEM ID												
SOIL PREP LSC	6300 SF		LABOR	6300 SF		$LL	1500	4.2		0.009		
			ROTOTILLER			$130	1500	4.2		0.004		
			NITRO SAWDUST (4CY/M)	25.2 CY	16			16.960				
			TRACTOR			$115	5	5.0		3.600		

WORKSHEET

BID ITEM ID	UNIT QTY UNIT	DATABASE TASKCODE	TASK DESCRIPTION	QUANTITY UNIT	MATL. UNIT $COST	LABOR/EQUIP TYPE	QUANT / 1 HR	TASK HOURS	SUB UNIT $$ COST	TASK/BID UNIT $$ COST	BID UNIT $$ PRICE	TOTAL $$ PRICE
			OPERATOR	#		#LL	5	5.0		2.599		
			HAND LABOR	#		#LL	5	5.0		2.599		
			FERTILIZER	#	16 BG 7.5 $					7.950		
			GYPSUM	#	10 BG 7 $					7.420		
			============		0.09415			14.28		0.148	0.231 L	1,456.59
24" BOX TREES	6 EA TREES	24-BOX	SPOT/DIG/PLANT	#	6 EA 104.5 #LL		0.5	12.0		136.760		
			PLANT TABS	#	36 EA 0.1 $					0.106		
			STAKE KIT (2 STAKES)	#	6 KIT 9 #LL		3	2.0		13.872		
			AMENDMENTS	#	24 CF 0.67 $					0.707		
			WATER-IN	#	6 EA #LL		12	0.5		1.083		
			TRACTOR	#	1.5 HR $115		1	1.5		18.000		
			OPERATOR	#	#LL		1	1.5		12.995		
			============		116.766			16.00		162.925	259.304 L	1,555.82
15 GL TREES L	15 EA TREES	15-GL	SPOT/DIG/PLANT	#	15 EA 27.3 #LL		1	15.0		41.933		
			PLANT TABS	#	60 EA 0.1 $					0.106		
			STAKE KIT (2 STAKES)	#	15 KIT 10 #LL		3	5.0		14.932		
			AMENDMENTS	#	30 CF 0.67 $					0.707		
			WATER-IN	#	15 EA #LL		12	1.3		1.083		
			============		39.0333			21.25		59.785	106.262 L	1,593.93
WASH ROBUSTA	3 EA TREES	WASHROB12 WASH. ROBUSTA 10'BTF		#	3 EA 275 $					291.500		
			LABOR			#LL	0.25	12.0		51.979		
			TRACTOR			$115	3	1.0		6.000		
			OPERATOR			#LL	3	1.0		4.332		
			============		275			13.00		353.811	526.142 L	1,578.43
GRADE LSC	6300 SF GRADE	FINE	HAND LABOR	#	6300 SF	#LL	1000	6.3		0.013		
			============					6.30		0.013	0.041 l	256.08

WORKSHEET

BID ITEM ID	UNIT	QTY	UNIT	DATABASE TASKCODE	TASK DESCRIPTION	QUANTITY UNIT	MATL. UNIT $COST	LABOR/EQUIP QUANT / 1 HR	TYPE	TASK HOURS	SUB UNIT $$ COST	TASK/BID UNIT $$ COST	BID UNIT $$ PRICE	TOTAL $$ PRICE
BID ITEM ID														
5 GL SHRUBS L	215 EA	SHRUBS	5-GL		SPOT/DIG/PLANT	215 EA	6.3	5	LL	43.0		9.277		
					PLANT TABS	645 EA	0.0		$			0.001		
							6.303			43.00		9.280	15.990 L	3,437.81
BID ITEM ID														
5 GL ESP SHRU	3 EA	SHRUBS	5GALESP		5 GAL ESPALIER	3 EA	20.48	4	LL	0.8		24.958		
					PLANT TABS	9 EA	0.0		$			0.001		
							20.483			0.75		24.961	35.706 L	107.12
BID ITEM ID														
1 GL SHRUBS L	283 EA	SHRUBS	1-GL		SPOT/DIG/PLANT	283 EA	3.15	12	LL	23.6		4.422		
					PLANT TABS	283 EA	0.0		$			0.001		
							3.151			23.58		4.423	7.317 L	2,070.66
BID ITEM ID														
1 GL VINES LS	24 EA	SHRUBS	1GALVINE		1 GAL VINE	24 EA	3.68	12	LL	2.0		4.984		
					PLANT TABS	24 EA	0.0		$			0.001		
							3.681			2.00		4.985	7.978 L	191.47
BID ITEM ID														
GROUND COVER	23 FLT	GRNDCOVR	64-8OC		GC 8" OC (3.51 FLT/H)	655 SF			$					
					FLAT CONVERSION	23 FLT	9	2	LL	11.5		16.037		
							8.99628			11.50		16.031	31.535 L	725.29
BID ITEM ID														
GROUND COVER	36 FLT	GRNDCOVR	64-12OC		GC 12" OC (1.56 FLT/H)	2,300 SF			$					
					FLAT CONVERSION	36 FLT	9	2	LL	17.9		16.037		
							8.97			17.94		15.984	31.442 L	1,131.93
BID ITEM ID														
SOD LSC	3345 SF	SOD	MAR-2		MARATHON II	3345 SF	0.31	225	LL	14.9		0.386		

WORKSHEET

BID ITEM ID	BID UNIT	QTY	UNIT	LIBRARY DATABASE TASKCODE	TASK DESCRIPTION	QUANTITY	UNIT	MATL. $COST	UNIT	LABOR/EQUIP TYPE	QUANT / 1 HR	TASK HOURS	SUB UNIT $$ COST	TASK/BID UNIT $$ COST	BID UNIT $$ PRICE	TOTAL $$ PRICE
								0.31				14.87		0.386	0.567 L	1,897.44

BID ITEM ID
MULCH CON   9 CY

					TASK DESCRIPTION	QUANTITY	UNIT									
					MULCH 1" DEEP	10 CY		18	LL		1	10.0		32.075		
					TRACTOR				LL	115	5	2.0		3.600		
					OPERATOR				LL		5	2.0		2.599		
								20				12.00		42.526	83.845 C	754.60

BID ITEM ID
90 DAY MAINT   3 MO

					TASK DESCRIPTION	QUANTITY	UNIT									
					90 DAY MAINT PERIOD	3 MO							291.00	291.000		
													291.000	291.000	342.353 M	1,027.06

GEN CONDS   1 LS   GENCOND

					TASK DESCRIPTION	QUANTITY	UNIT									
				GC	SUPERINTENDENT (1HR/DAY)	15.0 HR			S		1	15.0		19.500		
				GC	SUPERINTENDENT'S TRUCK	15.0 HR			102		1	15.0		4.000		
				GC	FOREMAN ADMIN (1HR/DAY)	15.0 HR			F		1	15.0		15.600		
				GC	MOB/DE-MOB EQUIP (L)	2.0 HR			C		1	2.0		12.995		
				GC	MOB/DE-MOB EQUIP (EQ)	2.0 HR			105		1	2.0		10.000		
				GC	CREW PICKUP TRUCK	120.0 HR			101		1	120.0		3.250		
				GC	CREW LOAD/UNLOAD TIME	15.0 HR			CC		1	15.0		12.999		
				GC	HAUL MATL-(L)	6.0 HR			C		1	6.0		12.995		
				GC	HAUL MATL-(EQ)	6.0 HR			105		1	6.0		10.000		
				GC	CREW DRIVETIME TO/FM JOB	60.0 HR			CC		1	60.0		12.999		
				GC	WARRANTY-(L)	16.0 HR			C		1	16.0		12.995		
				GC	WARRANTY-(EQ)	8.0 HR			101		1	8.0		3.250		
				GC	TRAILER/TOILET/FENCES	0.0 WK		0.0						0.001		
				GC	PERMITS/LICENSES/FEES	0.0 LS		0.0						0.001		
				GC	CLEAN-UP AT END OF DAY	15.0 HR			CC		1	15.0		12.999		
				GC	DUMP FEES	0.0 LD		0.0						0.001		
				GC	DUMPSTERS	0.0 EA		0.0						0.001		
				GC	STORAGE CONTAINERS	0.0 MO		0.0						0.001		
				GC	HAUL DEBRIS (L)	0.0 HR			D		1	0.0		12.995		
				GC	HAUL DEBRIS (EQ)	0.0 HR			105		1	0.0		10.000		
				GC	SOIL TESTS	1.0 LS		75						79.500		
				GC	CREW WATERING TIME	0.0 HR			CC		1	0.0		12.999		
				GC	COMMISSIONS	0.0 LS		0.0						0.001		

WORKSHEET

BID ITEM ID	DATABASE TASKCODE	TASK DESCRIPTION	QUANTITY UNIT	MATL. UNIT $COST	LABOR/EQUIP TYPE	QUANT / 1 HR	TASK HOURS	SUB UNIT $$ COST	TASK/BID UNIT $$ COST	BID UNIT $$ PRICE	TOTAL $$ PRICE
	GC	DESIGN FEES	0.0 LS	$ 0.0					0.001		
	GC	ESTIMATOR TIME	0.0 HR	$E	1		0.0		18.720		
	GC	AS-BUILTS	1.0 LS	100 $LL	1				106.000		
	GC	DETAIL JOB	8.0 HR	$LL	1		8.0		12.995		
			175.000				152.03		2,854.334		

94-C001 HENDERSON RESIDENCE CONSTRUCTION BID   01-Mar-94   page 1

RF + OTF = 20.00%

LABOR RATE TABLE/SCHEDULE   HOURLY RATES

TYPE CODE	DESCRIPTION	PUT HRLY RATE HERE 1	2	3 PW RATE W/BENEFITS	CREW SIZE	CREW DAYS at B HRS/DAY	PRODUCTION CREW HRS	TOTAL MANHRS	GEN COND CREW HRS	TOTAL MANHRS	TOTALS CREW HRS	TOTAL MANHRS
A	AVERAGE WAGE (GROSS PAY PER HOUR)	8.33	12.99	NA	1	0.0	94.6	378.6	60.5	152.0	155.2	530.6
AA		NA	NA	NA	1	0.0	0.0	0.0	0.0	0.0	0.0	0.0
B	BASIC ENTRY LEVEL LABOR RATE	5.00	7.80	NA	1	0.0	0.0	0.0	0.0	0.0	0.0	0.0
BB		NA	NA	NA	1	0.0	0.0	0.0	0.0	0.0	0.0	0.0
C	CONSTRUCTION CREW (2 MEN)	8.33	12.99	NA	4	0.8	0.0	0.0	6.0	24.0	6.0	24.0
CC	CONSTRUCTION CREW (4 MEN)	8.33	13.00	NA	4	2.8	(0.0)	(0.0)	22.5	90.0	22.5	90.0
D	DRIVER	8.33	12.99	NA	1	0.0	0.0	0.0	0.0	0.0	0.0	0.0
DD		NA	NA	NA	1	0.0	0.0	0.0	0.0	0.0	0.0	0.0
E	ESTIMATOR	12.00	18.72	NA	1	0.0	0.0	0.0	0.0	0.0	0.0	0.0
EE		NA	NA	NA	1	0.0	0.0	0.0	0.0	0.0	0.0	0.0
F	FOREMAN STEP #1	10.00	15.60	NA	1	1.9	0.0	0.0	15.0	15.0	15.0	15.0
FF	FOREMAN STEP #2	15.00	23.40	NA	1	0.0	0.0	0.0	0.0	0.0	0.0	0.0
G		NA	NA	NA	1	0.0	0.0	0.0	0.0	0.0	0.0	0.0
GG		NA	NA	NA	1	0.0	0.0	0.0	0.0	0.0	0.0	0.0
H		NA	NA	NA	1	0.0	0.0	0.0	0.0	0.0	0.0	0.0
HH		NA	NA	NA	1	0.0	0.0	0.0	0.0	0.0	0.0	0.0
I	IRRIGATION CREW (2 MEN)	8.33	12.99	NA	2	0.0	0.0	0.0	0.0	0.0	0.0	0.0
II	IRRIGATION CREW (4 MEN)	8.33	12.99	NA	4	5.7	45.5	182.1	0.0	0.0	45.5	182.1
J		NA	NA	NA	1	0.0	0.0	0.0	0.0	0.0	0.0	0.0
JJ		NA	NA	NA	1	0.0	0.0	0.0	0.0	0.0	0.0	0.0
K	LEADMAN STEP #1	8.00	12.48	NA	1	0.0	0.0	0.0	0.0	0.0	0.0	0.0
KK	LEADMAN STEP #2	9.00	14.04	NA	1	0.0	0.0	0.0	0.0	0.0	0.0	0.0
L	LANDSCAPE CREW (2 MEN)	9.00	14.04	NA	2	0.0	0.0	0.0	0.0	0.0	0.0	0.0
LL	LANDSCAPE CREW (4 MEN)	8.33	12.99	NA	4	6.4	49.1	196.5	2.0	8.0	51.1	204.5
M	MAINT CREW (2 MEN)	9.00	14.04	NA	2	0.0	0.0	0.0	0.0	0.0	0.0	0.0
MM	MAINT CREW (4 MEN)	7.25	11.31	NA	4	0.0	0.0	0.0	0.0	0.0	0.0	0.0
N		NA	NA	NA	1	0.0	0.0	0.0	0.0	0.0	0.0	0.0
NN		NA	NA	NA	1	0.0	0.0	0.0	0.0	0.0	0.0	0.0
O	OPERATOR STEP #1	12.00	18.72	NA	1	0.0	0.0	0.0	0.0	0.0	0.0	0.0
OO	OPERATOR STEP #1	15.00	23.40	NA	1	0.0	0.0	0.0	0.0	0.0	0.0	0.0
P	PLANTING CREW (2 MEN)	9.00	14.04	NA	2	0.0	0.0	0.0	0.0	0.0	0.0	0.0
PP	PLANTING CREW (4 MEN)	7.25	11.31	NA	4	0.0	0.0	0.0	0.0	0.0	0.0	0.0
Q		NA	NA	NA	1	0.0	0.0	0.0	0.0	0.0	0.0	0.0
QQ		NA	NA	NA	1	0.0	0.0	0.0	0.0	0.0	0.0	0.0
R		NA	NA	NA	1	0.0	0.0	0.0	0.0	0.0	0.0	0.0
RR		NA	NA	NA	1	0.0	0.0	0.0	0.0	0.0	0.0	0.0
S	SUPERVISOR STEP #1	12.50	19.50	NA	1	1.9	0.0	0.0	15.0	15.0	15.0	15.0
SS	SUPERVISOR STEP #2	20.00	31.20	NA	1	0.0	0.0	0.0	0.0	0.0	0.0	0.0

RF + OTF =   20.00%

LABOR RATE TABLE/SCHEDULE   HOURLY RATES

		TABLE REFERENCE NO------>	1	2	3	CREW	CREW DAYS	at	---PRODUCTION ---:		:---GEN COND----:		:-----TOTALS------	
			PUT HRLY		PW RATE	SIZE		8	CREW HRS	TOTAL MANHRS	CREW HRS	TOTAL MANHRS	TOTAL CREW HRS	TOTAL MANHRS
TYPE CODE	DESCRIPTION		RATE HERE		W/BENEFITS			HRS/DAY						
T			NA	NA	NA	1		94.6	378.6	60.5	152.0	155.2	530.6
TT			NA	NA	NA	1		0.0	0.0	0.0	0.0	0.0	0.0
U			NA	NA	NA	1		0.0	0.0	0.0	0.0	0.0	0.0
UU			NA	NA	NA	1		0.0	0.0	0.0	0.0	0.0	0.0
V			NA	NA	NA	1		0.0	0.0	0.0	0.0	0.0	0.0
VV			NA	NA	NA	1		0.0	0.0	0.0	0.0	0.0	0.0
W			NA	NA	NA	1		0.0	0.0	0.0	0.0	0.0	0.0
WW			NA	NA	NA	1		0.0	0.0	0.0	0.0	0.0	0.0
X			NA	NA	NA	1		0.0	0.0	0.0	0.0	0.0	0.0
XX			NA	NA	NA	1		0.0	0.0	0.0	0.0	0.0	0.0
Y			NA	NA	NA	1		0.0	0.0	0.0	0.0	0.0	0.0
YY			NA	NA	NA	1		0.0	0.0	0.0	0.0	0.0	0.0
Z			NA	NA	NA	1		0.0	0.0	0.0	0.0	0.0	0.0
ZZ			NA	NA	NA	1		0.0	0.0	0.0	0.0	0.0	0.0

EQUIPMENT RATE TABLE

$

TYPE CODE DESCRIPTION	COST/HR or RATE	PRODUCTION		GEN COND		TOTALS	
		HOURS	COST	HOURS	COST	HOURS	COST
		33.3	$354	151.0	$556	184.4	$910
100 TRUCK, MINI PICKUP	2.75	0.0	0.00	0.0	0.00	0.00	0.00
101 TRUCK, PICKUP-1/2 TON	3.25	0.0	0.00	128.0	416.00	128.00	416.00
102 TRUCK, PICKUP-3/4 TON	4.00	0.0	0.00	15.0	60.00	15.00	60.00
103 TRUCK, PICKUP-1 TON	5.00	0.0	0.00	0.0	0.00	0.00	0.00
104 TRUCK, PICKUP-1.5 TON	8.00	0.0	0.00	0.0	0.00	0.00	0.00
105 TRUCK, 2 TON	10.00	0.0	0.00	8.0	80.10	8.01	80.10
106 DUMPTRUCK, 5 CY	20.00	0.0	0.00	0.0	0.00	0.00	0.00
107 TRUCK W/500 GL HYDRO MULCHER	25.00	0.0	0.00	0.0	0.00	0.00	0.00
108	NA	0.0		0.0		0.00	0.00
109	NA	0.0		0.0		0.00	0.00
110 TRAILER, 1 AXLE	1.00	0.0	0.00	0.0	0.00	0.00	0.00
111 TRAILER, 2 AXLE	1.75	0.0	0.00	0.0	0.00	0.00	0.00
112 TRAILER, TRI-AXLE	2.50	0.0	0.00	0.0	0.00	0.00	0.00
113	NA	0.0		0.0		0.00	0.00
114	NA	0.0		0.0		0.00	0.00
115 TRACTOR, 1/4 CY BUCKET	18.00	9.5	171.72	0.0	0.00	9.54	171.72
116 TRACTOR, 1/2 CY BUCKET	16.00	0.0	0.00	0.0	0.00	0.00	0.00
117 TRACTOR, 1 CY BUCKET	18.00	0.0	0.00	0.0	0.00	0.00	0.00
118 BACKHOE	25.00	0.0	0.00	0.0	0.00	0.00	0.00
119 DOZER, SMALL	25.00	0.0	0.00	0.0	0.00	0.00	0.00
120 TRENCHER, WALK-BEHIND	8.00	19.6	156.80	0.0	0.00	19.60	156.80
121 TRENCHER, SMALL RIDER	15.00	0.0	0.00	0.0	0.00	0.00	0.00
122 TRENCHER, MEDIUM RIDER	17.00	0.0	0.00	0.0	0.00	0.00	0.00
123 TRENCHER, LARGE RIDER	22.00	0.0	0.00	0.0	0.00	0.00	0.00
124	NA	0.0		0.0		0.00	0.00
125 PIPE PULLER, SMALL	15.00	0.0	0.00	0.0	0.00	0.00	0.00
126 PIPE PULLER, MEDIUM	17.00	0.0	0.00	0.0	0.00	0.00	0.00
127 PIPE PULLER, LARGE	20.00	0.0	0.00	0.0	0.00	0.00	0.00
128	NA	0.0		0.0		0.00	0.00
129	NA	0.0		0.0		0.00	0.00
130 ROTOTILLER, SMALL	6.00	4.2	25.20	0.0	0.00	4.20	25.20
131 ROTOTILLER, LARGE	5.00	0.0	0.00	0.0	0.00	0.00	0.00
132	NA	0.0		0.0		0.00	0.00
133	NA	0.0		0.0		0.00	0.00
134 COMPACTOR	5.00	0.0	0.00	0.0	0.00	0.00	0.00
135 SOD CUTTER	4.00	0.0	0.00	0.0	0.00	0.00	0.00
136	NA	0.0		0.0		0.00	0.00
137 CHAIN SAW	3.00	0.0	0.00	0.0	0.00	0.00	0.00

EQUIPMENT RATE TABLE

TYPE CODE	DESCRIPTION	COST/HR or RATE	;--PRODUCTION ---;		;----GEN COND----;		;----TOTALS------;	
			HOURS	COST	HOURS	COST	HOURS	COST
138	CHOP SAW	5.00	33.3	$354	151.0	$556	184.4	$910
139		NA	0.0	0.00	0.0	0.00	0.00	0.00
140	MOWER 18-21"	1.00	0.0	0.00	0.0	0.00	0.00	0.00
141	MOWER 36"	2.00	0.0	0.00	0.0	0.00	0.00	0.00
142	MOWER 48"	2.50	0.0	0.00	0.0	0.00	0.00	0.00
143	MOWER 60"	NA	0.0		0.0		0.00	0.00
144	MOWER 72"	NA	0.0		0.0		0.00	0.00
145		NA	0.0		0.0		0.00	0.00
146		NA	0.0		0.0		0.00	0.00
147		NA	0.0		0.0		0.00	0.00
148	WEEDEATER	2.00	0.0	0.00	0.0	0.00	0.00	0.00
149	EDGER	2.00	0.0	0.00	0.0	0.00	0.00	0.00
150	BACKPACK BLOWER	1.50	0.0	0.00	0.0	0.00	0.00	0.00

94-M002 COMMERCIAL MAINTENANCE BIDDING EXAMPLE   01-Mar-94   page 1

08:52 PM

UNIT PRICE REPORT :   TOTAL BID PRICE $$ 214,552.02

TOTALS --11,935.8 $116,013 $35,969 $9,455 $161,437

BID ITEM	QTY UNIT	LUMP SUM	/ UNIT	SET $$	LABOR HRS	$ LABOR	$ EQUIP	$ MAT	$ SUBS	TOTAL COST	SPH (SALES/ MANHR)	GPMPH (GPM/ MANHR)	PPH (PROFIT MANHR)	MAT/ LABOR RATIO
TOTAL IRRIGATION---(IRR)------>			$/SF											
TOTAL LANDSCAPE----(LSC)------>		14,533.56	14,533.562 $/SF		185.2	1,739	1,082	9,455		12,276	78.47	12.19	9.54	5.44
TOTAL CONSTRUCTION-(CON)------>			$/SF											
TOTAL MAINTENANCE--(MNT)------>		200,018.46	0.00175 $/SF		11,225.6	105,396	25,853			131,248	17.82	6.13	3.47	
TOTAL GENERAL CONDITIONS------>					525.0	8,878	9,035			17,913	NA	NA	NA	
60" MOWER MNT	91520000 SF	38,099.36	0.00042		1,830.4	17,185.42	9,152.00			26,337.42	20.81	6.43	3.77	
60" MOWER-2 MNT	2080 AC	38,099.36	18.317		1,830.4	17,185.42	9,152.00			26,337.42	20.81	6.43	3.77	
48" MOWER MNT	22880000 SF	12,517.39	0.00055		653.7	6,137.65	2,288.00			8,425.65	19.15	6.26	3.61	
48" MOWER-2 MNT	520 AC	12,517.39	24.072		653.7	6,137.65	2,288.00			8,425.65	19.15	6.26	3.61	
21" MOWER MNT	8320000 SF	18,468.30	0.00222		1,109.3	10,415.41	1,386.67			11,802.07	16.65	6.01	3.36	
EDGING MNT	104000 LF	3,636.14	0.03496		208.0	1,952.89	416.00			2,368.89	17.48	6.09	3.44	
BACK-PACK BLOWER MNT	780 HR	13,202.19	16.926		780.0	7,323.33	1,170.00			8,493.33	16.93	6.04	3.38	
WEEDING MNT	2080 HR	31,739.17	15.259		2,080.0	19,528.89				19,528.89	15.26	5.87	3.22	
CLEANUP MNT	2080 HR	31,739.17	15.259		2,080.0	19,528.89				19,528.89	15.26	5.87	3.22	
MULCH LSC	300.00 CY	9,537.77	31.793		150.0	1,408.33	800.00	5,724.00		7,932.33	63.59	10.70	8.05	4.06
FERTILIZE LSC	17,600.00 LB	4,995.79	0.284		35.2	330.49	281.60	3,731.20		4,343.29	141.93	15.88	11.29	

BID RECAP & SETUP REPORTS

BID RECAP REPORT

	PRODUCTION DIRECT COST $$ AMOUNT	GEN. COND. DIRECT COST $$ AMOUNT	TOTAL DIRECT COSTS $$ AMOUNT	OH RECOVERY ON DIRECT JOB COSTS $$ AMOUNT
I. PRODUCTION COSTS:				
A. MATERIALS COST (w/salestax)-->	4.4%	9,455.20	9,455.20	945.52
B. EQUIPMENT COST-->	12.6%	26,934.27	9,035.00	35,969.27	8,992.32
C. SUBCONTRACTORS COST--->				
D. LABOR COST (w/burden,OTF,RF)->	49.9%	107,134.38	8,878.13	116,012.51	21,722.01
SUBTOTAL (PH. I) DIRECT COSTS---	66.9%	143,523.84		161,436.98
II. GENERAL CONDITION COSTS------>	8.3%	17,913.13	17,913.13	
SUBTOTAL (PH. I&II) DIRECT COSTS>	75.2%	161,436.98	<-<-<-<-<-<	161,436.98
III. MARKUPS:				
A. OVERHEAD RECOVERY-------->	14.8%	31,659.85	<-<-<-<-<-<-<-<	31,659.85
(BEP=OH+DIR COSTS)= $193,097	90.0%			
B. PROFIT + SALES COMMISSION--->	10.0%	21,455.20		
SUBTOTAL MARKUPS (GPM)------->	24.8%	53,115.05		
TOTAL BID PRICE-------->	100.0%	214,552.02	214,552.02	24.76%

UNIT PRICE SUMMARY

	DIFFERENCE
	0.00
21,455.20	% PROFIT
	10.00%
	% MARKUP GPM

BID SETUP REPORT

BID SETUP INPUT AREA — ENTER AMOUNTS / % IN THIS COLUMN

1. BID FILE NAME------------->94-M002
    (up to 8 characters)
2. BID NAME (Name,GC,Ph #,Bid date)-->COMMERCIAL MAINTENANCE BID
    (up to 25 characters)

    OFFICE OVERHEAD RECOVERY TABLE

3. FIELD-LABOR HOUR ANNUAL BUDGET---> : 188,500
4. OVERHEAD TO RECOVER/YEAR------> : $500,000
    (5 & 6 are calculated do not adjust)
5. OVHD COST PER FIELD-LABOR HR----> : $2.65
6. OPH CALCULATED BY PPH METHOD----> : $2.65

7. SALES TAX ON MATERIALS--------> : 6.00%

8. SALES COMMISION FOR THIS BID---->

9. LABOR OVERTIME FACTOR % (OTF)---> : 11.11%
10. LABOR RISK FACTOR (RF)--------> : 10.00%

11. NET PROFIT MARGIN (NPM) TARGET---> : 10.00%
12. NET PROFIT/HR (PPH) OVERRIDE----> : $_____

13. FIELD LABOR BURDEN %----------> : 30.00%
14. LABOR TABLE #--------> : 2

15. SQUARE FOOTAGE TABLE: INPUT "SF"
    A. IRRIGATION COVERAGE AREA-----> : 1
    B. LANDSCAPE PLANTING AREA------> : 1
    C. CONSTRUCTION AREA----------> : 1
    D. MAINTENANCE AREA----------> : 114,400,000
    E. TOTAL PROJECT COVERAGE------> : 114,400,000

ANALYSIS REPORT

FIVE (5) ESTIMATING METHODS
MARKET COMPARISON ANALYSIS

```
                                                          BID PRICE
                                                          $214,552
                                 AMOUNTS ::RATIO/PER HOUR ANALYSIS:                          HOURS        SPH
                                         ::11. SALES PER HOUR (SPH):
1. FACTORING (Price divided by materials ::    A. TOTAL JOB---------->11,935.8      $17.98
   or materials X ?)---------->   22.69  ::    B. IRRIGATION----->
2. UNIT PRICES (See UNIT PRICE RPT):     ::    C. LANDSCAPE----->            185.2       $78.47
                           SF     $/SF   ::    D. CONSTRUCTION----->
   A. IRRIGATION----->      1             ::    E. MAINTENANCE----->       11,225.6      $17.82
   B. LANDSCAPE----->       1  ::::::::   ::    F. GENERAL CONDITIONS--->   525.0        N/A
   C. CONSTRUCTION----->     1            ::12. MATERIALS TOTAL (WITH TAX)---->$9,455.20
   D. MAINTENANCE----->114,400,000 $0.00175::    A. MATERIAL TO LABOR (M/L) RATIO-->   0.08 :1
   E. TOTAL PROJECT---->114,400,000 $0.00188::    B. MATERIAL PER HOUR (MPH)----->  $0.79
3. GROSS PROFIT MARGIN (GPM):   24.76%   ::13. EQUIP. TOTAL (PROD.+ GEN. CONDS)---->$35,969.27
                                         ::    A. EQUIP. TO LABOR (EQ/L) RATIO-->   0.31 :1
4. MULTIPLE OVERHEAD RECOVERY SYSTEM (MORS):::    B. EQUIPMENT PER HOUR (EQPH)----->  $3.01
                                         ::14. GENERAL CONDITIONS TOTAL---->$17,913.13
   A. MATERIALS------SET AT ->   10.00%  ::    A. AS A PERCENT OF PRICE----->   8.35%
   B. EQUIPMENT------SET AT ->   25.00%  ::    B. PER FIELD-LABOR HOUR (GCPH)----->  $1.50
   C. SURCONTRACTOR------SET AT ->  5.00%::    C. GC HRS/TOTAL HRS----->   4.40%
   D. LABOR & BURDEN------CALC AT->18.72%::15. OVERHEAD (G&A or INDIRECT COST)---->$31,659.85
   E. LABOR OVERRIDE %------>      %     ::    A. AS A PERCENT OF PRICE----->   14.76%
5. OVERHEAD & PROFIT PER HOUR (OPPH) METHOD:::    B. PER FIELD-LABOR HOUR (OPH)----->  $2.65
   A. TOTAL FIELD-LABOR HOURS IN BID-->11,935.8::16. NET PROFIT MARGIN (NPM) IN BID---->$21,455.20
   B. OVERHEAD PER HOUR (OPH)----->  $2.65::    A. AS A PERCENT OF PRICE----->   10.00%
   C. NET PROFIT PER HOUR (PPH)----->  $1.80::    B. PER FIELD-LABOR HOUR (PPH)----->  $1.80
   D. GROSS PROFIT PER HOUR (GPMPH)--->  $4.45::17. GROSS PROFIT MARGIN (OVHD+NPM)---->$53,115.05
   E. SALES PER HOUR (SPH)----->  $17.98 ::    A. AS A PERCENT OF PRICE----->   24.76%
                                         ::    B. PER FIELD-LABOR HOUR (GPMPH)---->  $4.45
                                         ::18. SUBCONTRACTOR COSTS---->
                                         ::    A. AS A PERCENT OF PRICE----->
```

```
9. CREW AVE WAGE (w/burden, OTF & RF)
   A. TOTAL JOB HOURS----->
   B. IRRIGATION----->
   C. LANDSCAPE----->               $9.72
   D. CONSTRUCTION----->            $9.39
   E. MAINTENANCE----->             $9.39
   F. GENERAL CONDITIONS---->      $16.91
10. SALES COMMISSION---->
   A. DOLLARS---->
   B. AS A % OF PRICE---->
```

WORKSHEET

BID ITEM ID	UNIT QTY	UNIT	LIBRARY DATABASE	TASKCODE	TASK DESCRIPTION	QUANTITY	UNIT	MATL. UNIT $COST	LABOR/EQUIP TYPE	QUANT /1 HR	TASK HOURS	SUB UNIT $$ COST	TASK/BID UNIT $$ COST	BID UNIT $$ PRICE	TOTAL $$ PRICE
BID ITEM ID 60" MOWER MNT	91520000	SF	MOW	60-T-SF	MOWING-OPERATOR	$ 91520000	SF	$M	$143	50000	1,830.4		0.000		
					60" MOWER	$				50000	1,830.4		0.000		
								#######		#####	1,830.40	#######	0.000	0.000 M	38,099.36
BID ITEM ID 60" MOWER-2 M	2080	AC	MOW	60-T-AC	MOWING-OPERATOR CONVERSION TO SQ. FT.	$ 2080	AC	$	$M	50000	1,830.4		0.000		
					60" MOWER	$ 91520000	SF		$143	50000	1,830.4		0.000		
						$		#######		#####	1,830.40	#######	12.662	18.317 M	38,099.36
BID ITEM ID 48" MOWER MNT	22880000	SF	MOW	48-WB-SF	MOWING-OPERATOR	$ 22880000	SF	$M	$142	35000	653.7		0.000		
					48" MOWER	$				35000	653.7		0.000		
								#######		#####	653.71	#######	0.000	0.001 M	12,517.39
BID ITEM ID 48" MOWER-2 M	520	AC	MOW	48-WB-AC	MOWING-OPERATOR CONVERSION TO SQ. FT.	$ 520	AC	$	$M	35000	653.7		0.000		
					48" MOWER	$ 22880000	SF		$142	35000	653.7		0.000		
						$		#######		#####	653.71	#######	16.203	24.072 M	12,517.39
BID ITEM ID 21" MOWER MNT	8320000	SF	MOW	21-P	MOWING-OPERATOR	$ 8320000	SF	$M	$140	7500	1,109.3		0.001		
					21" MOWER	$				7500	1,109.3		0.000		
								#######		#####	1,109.33	#######	0.001	0.002 M	18,468.30
BID ITEM ID EDGING MNT	104000	LF	MOW	EDGE	LABOR	$ 104000	LF	$M	$149	500	208.0		0.019		
					POWER EDGER	$				500	208.0		0.004		
								#######		#####	208.00	#######	0.023	0.035 M	3,636.14

-295-

WORKSHEET

	BID			LIBRARY				MATL.	LABOR/EQUIP					TASK/BID	BID	
BID ITEM ID	UNIT	QTY	UNIT	DATABASE	TASKCODE	TASK DESCRIPTION	QUANTITY	UNIT	UNIT $COST	TYPE	QUANT / 1 HR	TASK HOURS	SUB UNIT $$ COST	UNIT $$ COST	UNIT $$ PRICE	TOTAL $$ PRICE
BACK-PACK BLO	780	HR	MOW	BP-BLOWR		LABOR	780	HR	$	M	$150	1	780.0		9.389		
						BACK-PACK BLOWER		$					780.0		1.500		
												780.00		10.889	16.926 M	13,202.19
BID ITEM ID																
WEEDING MNT	2080	HR	MOW	LABOR		HAND LABOR	2080	HR	$	M		1	2,080.0		9.389		
												2,080.00		9.389	15.259 M	31,739.17
BID ITEM ID																
CLEANUP MNT	2080	HR	MOW	LABOR		HAND LABOR	2080	HR	$	M		1	2,080.0		9.389		
												2,080.00		9.389	15.259 M	31,739.17
BID ITEM ID																
MULCH LSC	300.00	CY	MULCH	RM-2-IN		REDWOOD MULCH 2" DEEP		$	18	$					19.080		
						CONVERSION TO CY	300	CY	$								
						TRACTOR		$			$116	6	50.0		2.667		
						OPERATOR		$			M	6	50.0		1.565		
						HAND LABOR (2 MEN)		$			M	3	100.0		3.130		
									18				150.00		26.441	31.793 L	9,537.77
BID ITEM ID																
FERTILIZE LSC	17,600.00	LB				FERTILIZER	17600	LB	0.2	$					0.212		
						TRACTOR		$			$116	1000	17.6		0.016		
						OPERATOR		$			M	1000	17.6		0.009		
						HAND LABOR (1 MAN)		$			M	1000	17.6		0.009		
									0.2				35.20		0.247	0.284 L	4,995.79
BID ITEM ID																
GEN CONDS	1	LS	GEN-COND	MOWING		SUPERINTENDENT (5 HR/WK)	260	HR	$	S	$100	1	260.0		18.778		
						SUPERINTENDENT'S TRUCK	260	HR	$		$100	1	260.0		2.750		
						FOREMAN ADMIN (1HR/DAY)	260	HR	$	F		1	260.0		15.022		
						CREW PICKUP TRUCK	2080	HR	$		$100	1	2,080.0		2.750		
						ESTIMATOR TIME	5	HR	$	E		1	5.0		18.027		
						GOLF CART	2080	HR	$		$146	1	2,080.0		1.250		

94-M001 SMALL COMM/RESID MAINT BID EXAMPLE   01-Mar-94   page 1

09:13 PM

UNIT PRICE REPORT :   TOTAL BID PRICE   3,742.32

TOTALS --   144.00   1,913.34   522.75   212.00   2,648.09

BID ITEM	QTY UNIT	LUMP SUM	/ UNIT	SET $$	LABOR HRS	$ LABOR	$ EQUIP	$ MAT	$ SUBS	TOTAL COST	SPH (SALES/ MANHR)	GPMPH (GPM/ MANHR)	PPH (PROFIT MANHR)	MAT/ LABOR RATIO
TOTAL IRRIGATION---(IRR)------->			$/SF											
TOTAL CLEANUP------(LSC)------->		1,024.71	0.0013 $/SF		34.00	454.74	85.50	212.00		752.24	30.14	8.01	3.01	0.47
TOTAL CONSTRUCTION-(CON)------->			$/SF											
TOTAL MAINTENANCE--(MNT)------->		2,717.61	0.0035 $/SF		110.00	1,458.60	437.25			1,895.85	24.71	7.47	2.47	
TOTAL GENERAL CONDITIONS------->------>------>------>------>------>------>											NA	NA	NA	
SPRING CLEANUP LSC	1.0 LS	453.47	453.467		17.00	227.37	42.75	53.00		323.12	26.67	7.67	2.67	0.23
MAINT VISITS MNT	28.0 EA	2,717.61	97.058		110.00	1,458.60	437.25			1,895.85	24.71	7.47	2.47	
FALL CLEANUP LSC	1.0 LS	571.24	571.244		17.00	227.37	42.75	159.00		429.12	33.60	8.36	3.36	0.70

BID RECAP & SETUP REPORTS

BID RECAP REPORT

	PRODUCTION DIRECT COST $$ AMOUNT	GEN. COND. DIRECT COST $$ AMOUNT	TOTAL DIRECT COSTS $$ AMOUNT	OH RECOVERY ON DIRECT JOB COSTS $$ AMOUNT	BID SETUP REPORT	ENTER AMOUNTS / % IN THIS COLUMN <---->
					BID SETUP INPUT AREA	
I. PRODUCTION COSTS:						
A. MATERIALS COST (w/salestax)-->	5.7%	212.00	212.00	21.20	1. BID FILE NAME-------------------->94-M001	
					(up to 8 characters)	
B. EQUIPMENT COST--------------->	14.0%	522.75	522.75	130.69	2. BID NAME (Name,GC,Ph #,Bid date)-->SMALL COMM/RESID MAINT BID	
C. SUBCONTRACTORS COST---------->					(up to 25 characters)	
					OFFICE OVERHEAD RECOVERY TABLE	
D. LABOR COST (w/burden,OTF,RF)->	51.1%	1,913.34	1,913.34	568.11	3. FIELD-LABOR HOUR ANNUAL BUDGET-->	20,000
					4. OVERHEAD TO RECOVER/YEAR-------->	$100,000
					(5 & 6 are calculated do not adjust)	
					5. OVHD COST PER FIELD-LABOR HR---->	$5.00
					6. OPH CALCULATED BY PPH METHOD---->	$5.00
SUBTOTAL (PH. I) DIRECT COSTS--->	70.8%	2,648.09	2,648.09			
II. GENERAL CONDITION COSTS---------->					7. SALES TAX ON MATERIALS---------->	6.00%
SUBTOTAL (PH. I&II) DIRECT COSTS>	70.8%	2,648.09	<-<-<-<-<-<	8. SALES COMMISION FOR THIS BID---->		
III. MARKUPS:						
A. OVERHEAD RECOVERY------------>	19.2%	720.00	<-<-<-<-<-<	720.00	9. LABOR OVERTIME FACTOR % (OTF)---->	10.00%
(REP=OH+DIR COSTS)= 3,368.09	90.0%				10. LABOR RISK FACTOR (RF)---------->	10.00%
		UNIT PRICE SUMMARY			11. NET PROFIT MARGIN (NPM) TARGET--->	10.00%
B. PROFIT + SALES COMMISSION---->	10.0%	374.23	374.23		12. NET PROFIT/HR (PPH) OVERRIDE----> $ ____	
			DIFFERENCE (0.00)		13. FIELD LABOR BURDEN %------------>	30.00%
			% PROFIT 10.00%		14. LABOR TABLE #------------------>	2
SUBTOTAL MARKUPS (GPM)--------->	29.2%	1,094.23	% MARKUP GPM		15. SQUARE FOOTAGE TABLE:-----------> INPUT "SF"	
					A. IRRIGATION COVERAGE AREA-------->	1
TOTAL BID PRICE----------------->	100.0%	3,742.32	3,742.32	29.24%	B. LANDSCAPE CLEANUP AREA---------->	784,000
					C. CONSTRUCTION AREA--------------->	1
	0.00				D. MAINTENANCE AREA---------------->	784,000
					E. TOTAL PROJECT COVERAGE---------->	784,000

ANALYSIS REPORT

FIVE (5) ESTIMATING METHODS
MARKET COMPARISON ANALYSIS

                                                          BID PRICE
                                 AMOUNTS :: RATIO/PER HOUR ANALYSIS: $3,742.32

                                          :: 11. SALES PER HOUR (SPH):   HOURS   SPH
1. FACTORING (Price divided by materials ::   A. TOTAL JOB----------------> 144.0 $25.99
   or materials X ?)----------> 17.65   ::   B. IRRIGATION--------------->
2. UNIT PRICES (See UNIT PRICE RPT):   ::   C. CLEANUP------------------> 34.0 $30.14
                      SF $/SF   ::   D. CONSTRUCTION------------->
   A. IRRIGATION--------> 1   ::   E. MAINTENANCE--------------> 110.0 $24.71
   B. CLEANUP----------> 784,000 $0.0013   ::   F. GENERAL CONDITIONS------->   N/A
   C. CONSTRUCTION-----> 1   :: 12. MATERIALS TOTAL (WITH TAX)-------> $212.00
   D. MAINTENANCE------> 784,000 $0.0035   ::   A. MATERIAL TO LABOR (M/L) RATIO--> 0.11 :1
   E. TOTAL PROJECT----> 784,000 $0.0048   ::   B. MATERIAL PER HOUR (MPH)--------> $1.47
                                            ::
3. GROSS PROFIT MARGIN (GPM):   29.24%   :: 13. EQUIP. TOTAL (PROD.+ GEN. CONDS)---> $522.75
                                            ::   A. EQUIP. TO LABOR (EQ/L) RATIO---> 0.27 :1
4. MULTIPLE OVERHEAD RECOVERY SYSTEM (MORS): ::   B. EQUIPMENT PER HOUR (EQPH)------> $3.63
                                            :: 14. GENERAL CONDITIONS TOTAL--------->
   A. MATERIALS--------SET AT -> 10.00%   ::   A. AS A PERCENT OF PRICE----->
   B. EQUIPMENT--------SET AT -> 25.00%   ::   B. PER FIELD-LABOR HOUR (GCPH)----->
   C. SUBCONTRACTOR----SET AT -> 5.00%   ::   C. GC HRS/TOTAL HRS----->
   D. LABOR & BURDEN---CALC AT-> 29.69%   :: 15. OVERHEAD (G&A or INDIRECT COST)-----> $720.00
   E. LABOR OVERRIDE %----------->   %   ::   A. AS A PERCENT OF PRICE-----> 19.24%
                                            ::   B. PER FIELD-LABOR HOUR (OPH)-----> $5.00
5. OVERHEAD & PROFIT PER HOUR (OPPH) METHOD: :: 16. NET PROFIT MARGIN (NPM) IN BID-----> $374.23
                                            ::   A. AS A PERCENT OF PRICE-----> 10.00%
   A. TOTAL FIELD-LABOR HOURS IN BID--> 144.00 ::   B. PER FIELD-LABOR HOUR (PPH)-----> $2.60
   B. OVERHEAD PER HOUR (OPH)--------> $5.00 :: 17. GROSS PROFIT MARGIN (OVHD+NPM)-----> $1,094.23
   C. NET PROFIT PER HOUR (PPH)------> $2.60 ::   A. AS A PERCENT OF PRICE-----> 29.24%
   D. GROSS PROFIT PER HOUR (GPMPH)--> $7.60 ::   B. PER FIELD-LABOR HOUR (GPMPH)-----> $7.60
   E. SALES PER HOUR (SPH)----------> $25.99 :: 18. SUBCONTRACTOR COSTS--------->
                                            ::   A. AS A PERCENT OF PRICE----->

9. CREW AVE WAGE (w/burden, OTF & RF)
   A. TOTAL JOB HOURS---------------> $13.29
   B. IRRIGATION------------------->
   C. LANDSCAPE-------------------> $13.37
   D. CONSTRUCTION--------------->
   E. MAINTENANCE----------------> $13.26
   F. GENERAL CONDITIONS--------->

10. SALES COMMISSION--------------->
    A. DOLLARS--------------------> $3,742
    B. AS A % OF PRICE----------->

WORKSHEET

	BID		LIBRARY			MATL.	LABOR/EQUIP				TASK/BID	BID	TOTAL
BID ITEM ID	UNIT QTY UNIT	DATABASE TASKCODE	TASK DESCRIPTION	QUANTITY UNIT	UNIT $COST	TYPE	QUANT / 1 HR	TASK HOURS	SUB UNIT $$ COST	UNIT $$ COST	UNIT $$ PRICE	$$ PRICE

BID ITEM ID

SPRING CLEANU	1 LS CLEANUP SPRING											
LABOR-2 MAN CREW	14 HR	$M		1	14.0		12.870		
CREW TRUCK	8 HR	$103		1	8.0		5.000		
DUMP FEES	1 LS	$	50				53.000		
LOAD/UNLOAD	1 HR	$M		1	1.0		12.870		
DRIVE TIME	1 HR	$M		1	1.0		12.870		
SUPERVISOR	1 HR	$S		1	1.0		21.450		
SUPERVISOR'S TRUCK	1 HR	$100		1	1.0		2.750		
			50		17.00		323.120	453.467 L	453.47

BID ITEM ID

MAINT VISITS   28.00 EA

Task	QUANTITY UNIT	TYPE	COST	QUANT	TASK HOURS	SUB UNIT COST	TASK/BID UNIT COST	BID UNIT PRICE	TOTAL PRICE
21" MOWER (LABOR)	224000 SF	$M		8000	28.0		0.002		
21" MOWER (EQUIP)	224000 SF	$140		8000	28.0		0.000		
36" MOWER (LABOR)	560000 SF	$M		20000	28.0		0.001		
36" MOWER (EQUIP)	560000 SF	$141		20000	28.0		0.000		
STRING TRIMMER (LABOR)	4900 LF	$M		350	14.0		0.037		
STRING TRIMMER (EQUIP)	4900 LF	$148		350	14.0		0.004		
BACKPACK BLOWER (LABOR)	7 HR	$M		1	7.0		12.870		
BACKPACK BLOWER (EQUIP)	7 HR	$150		1	7.0		1.500		
WEED & CLEAN UP	14 HR	$M		1	14.0		12.870		
DRIVE TIME (LABOR)	7 HR	$M		1	7.0		12.870		
DRIVE TIME (TRUCK)	3.5 HR	$103		1	3.5		5.000		
LOAD/UNLOAD	7 HR	$M		1	7.0		12.870		
CREW TRUCK (ON-SITE)	52.5 HR	$103		1	52.5		5.000		
SUPERVISOR	5 HR	$S		1	5.0		21.450		
SUPERVISOR'S TRUCK	5 HR	$100		1	5.0		2.750		
					110.00		67.709	97.058 M	2,717.61

BID ITEM ID

FALL CLEANUP	1 LS CLEANUP FALL											
LABOR-2 MAN CREW	14 HR	$M		1	14.0		12.870		
CREW TRUCK	8 HR	$103		1	8.0		5.000		
DUMP FEES	1 LS	$	150				159.000		
LOAD/UNLOAD	1 HR	$M		1	1.0		12.870		
DRIVE TIME	1 HR	$M		1	1.0		12.870		
SUPERVISOR	1 HR	$S		1	1.0		21.450		
SUPERVISOR'S TRUCK	1 HR	$100		1	1.0		2.750		

ILLUSTRATIONS

FOR FURTHER READING/REFERENCE LIST

Title	Author(s)	Publisher
In Search of Excellence	Peters & Waterman	Warner Books
Thriving on Chaos	Peters	Knopf
Quality is Free	Crosby	Mentor
Further Up the Organization	Townsend	Harper & Row
The "E" Myth	Gerber	Ballinger
Finding a Job You Can Love	Mattson & Miller	Nelson
Hand Me Another Brick	Swindoll	Nelson
Landscape Data Manual	Griffin	CLCA, Inc. (1)
Quality and the Landscape Contractor	Crystal Ball Committee	ALCA, Inc. (2)
Kerr's Cost Data Manual for Landscape Contractors	Dietrich	Van Nostrand Reinhold (3)
The Eternally Successful Organization	Crosby	Plume
Strategic Planning for L&I Contractors	Huston	Smith Huston, Inc.
Preparing for & Responding to a Down Economy	Huston	Smith Huston, Inc.
Means Site Work & Landscape Cost Data Manual	Hollman Fee	R.S. Means Co., Inc. (4)
Operating Cost Study	HRI, Inc.	HRI, Inc. (5)
The Gardening of America	Cetron/Davies	Chilton (6)
Legal Practices Manual for Construction Subcontractors	Sklar	American Subcontractors Assn. (7)
Labor & Equipment Production Times for Landscape Const.	Vander Kooi	Vander Kooi & Assoc., Inc. (8)

AGC's Partnering: A Concept   AGC   (9)
   for Success (video)

Winning Management Strategies Peters &   Nightingale-Conant
   for the Real World (audio)   Townsend   (10)

Road Construction Ahead   Focus Video   (11)
   (video)

Contractor's Equipment Cost   Dataquest   (12)

---

(1). California Landscape Contractor's Association   916 448-2522
     2021 N Street, Suite 300; Sacramento CA 95814

(2). Associated Landscape Contractors of America   703 620-6363
     12200 Sunrise Valley Dr., Suite 150; Reston VA
     22091

(3). 115 Fifth Avenue, New York NY 10003   800 842-3636

(4). Construction Consultants & Publishers   617 747-1270
     100 Construction Plaza, P.O. Box 800; Kingston
     MA 02364-0800

(5). Horticultural Research Institute, Inc.   202 789-2900
     1250 I Street, NW, Suite 500; Washington DC
     20005

(6). Dickson Felix, Inc. 1441 Que Street, NW;   202 328-1540
     Washington DC 20009

(7). American Subcontractors Assoc., 1004 Duke Street   703 684-3450
     Alexandria VA 22314-3512

(8). Vander Kooi & Assoc., Inc., P.O. Box 621759,   303 697-6467
     Littleton CO 80162

(9). Associated General Contractors of America,   202 393-2040
     1957 E. Street, NW, Washington DC 20006

(10).Nightingale-Conant Corp., 7300 N. Lehigh Ave.,   800 323-5552
     Chicago IL 60648

(11).Focus Video Productions, RD #5, Box 2108, Mont-   800 843-3686
     pelier VT, 05602

(12).Dataquest, A Company of the Dun & Bradstreet Corp.800 669-3282

ASSOCIATIONS

American Association of Nurserymen (AAN)
1250 I Street NW, Suite 500
Washington, DC 20005
202 789-2900/1893 FAX

American Council of Turfgrass Soil and Crop Science Center
Texas A&M University
College Station, TX 77848
409 845-4826

American Society of Golf Course Architects
221 North LaSalle Street, Suite 3500
Chicago, IL 60601
312 372-7090/6160 FAX

American Society of Irrigation Consultants (ASIC)
P.O. Box 426
Byron CA 94514
510 516-1124/1301 FAX

American Society of Landscape Architects (ASLA)
4401 Connecticut Avenue, 5th Floor
Washington DC 20008
202 686-2752/1001 FAX

American Society of Professional Estimators (ASPE)
11141 Georgia Avenue, Suite 412
Wheaton MD 20902
301 929-8848/0231 FAX

American Sod Producers Association
1855-A Hicks Road
Rolling Meadows IL 60008
708 705-9898/8347 FAX

American Subcontractors Association
1004 Duke Street
Alexandria VA 22314-3512
703 684-3450/836-3482 FAX

Associated Builders and Contractors (ABC)
1300 N 17th Street
Rosslyn VA 22209
703 812-2000/8235 FAX

Associated General Contractors of America (AGC)
1957 E Street NW
Washington DC 20006
202 393-2040/347-4004 FAX

Associated Landscape Contractors of America (ALCA)
12200 Sunrise Valley Drive, Suite 150
Reston, VA 22091
800 395-2522, 703 620-6363
703 620-6365 FAX

California Landscape Contractors Association
2021 North Street, Suite 300
Sacramento CA 95814
800 448-2522, 916 448-2522
916 446-7692 FAX

Canadian Nursery Trades Association
1293 Matheson Blvd. East
Mississauga Ontario Canada L4W1R1
416 629-1367/4438 FAX

The Council of Tree & Landscape Appraisers
1250 I Street NW, Suite 500
Washington DC 20005
202 789-2900/1893 FAX

The Garden Centers of America
1250 I Street NW, Suite 500
Washington DC 20005
202 789-2900/1893 FAX

The Garden Council
10210 Bald Hill Road
Mitchellville MD 20721
301 577-4073/459-6533 FAX

Golf Course Superintendents Asociation of America
1421 Research Park Dr.
Lawrence KS 66049-3859
913 841-2240/832-4455 FAX

Horticultural Research Institute
1250 I Street NW, Suite 500
Washington DC 20005
202 789-2900/1893 FAX

International Society of Arboriculture
P.O. Box GG
Savoy IL 61874
217 355-9411/9516 FAX

International Turfgrass Society
Crop & Soil Environmental Sciences, VPI-SU
Blacksburg VA 24061-0403
703 231-9736

Irrigation Association
1911 North Fort Myer Drive, Suite 1009
Arlington VA 22209
703 524-1200/9544 FAX

Landscape Materials Information Service
P.O. Box 216
Callicoon NY 12723
914 887-4401/4401 FAX

Landscape Ontario Horticultural Trade Association
1293 Matheson Blvd. East
Mississauga Ontario Canada L4W1R1
416 629-1184/4438 FAX

The Lawn Institute
1509 Johnson Ferry Road NE, Suite 190
Marietta GA 30062
404 977-5492/8205 FAX

National Arborist Association
P.O. Box 1094
Amherst NH 03031-1094
603 673-3311/672-2613 FAX

National Association of Plumbing-Heating-Cooling Contractors
P.O. Box 6808
Falls Church VA 22040
703 237-8100/237-7442 FAX

National Golf Foundation
1150 South US Highway One
Jupiter FL 33477
407 744-6006/744-6107 FAX

National Landscape Association (NLA)
1250 I Street NW, Suite 500
Washington DC 20005
202 789-2900/1893 FAX

National Lighting Bureau
2101 L Street NW, Suite 300
Washington DC 20037
202 457-8437

Nursery Marketing Council
1250 I Street NW, Suite 500
Washington DC 20005
202 789-2900/1893 FAX

Plants for Clean Air Council
10210 Bald Hill Road
Mitchellville MD 20721
301 459-9625/6533 FAX

Professional Lawn Care Association of America (PLCAA)
1000 Johnson Ferry Road, Suite C-135
Marietta GA 30068-2112
800 458-3466, 404 977-5222
404 578-6071 FAX

Professional Grounds Management Society (PGMS)
120 Cockeysville Road, Suite 104
Hunt Valley MD 21031
410 584-9754/9756 FAX

Responsible Industry for a Sound Environment (RISE)
1156 15th Street NW, Suite 400
Washington DC 20005
202 872-3860/463-0474 FAX

Sports Turf Managers Association
401 N. Michigan Avenue
Chicago IL 60611-4267
312 644-6610/245-1084 FAX

SURVEY

Name: _____ Position: _____
Company: _____
Address: _____
City: _____State: _____ Zip: _____
Phone #: (_____) _____-_____ Fax: (_____) _____-_____

We at Smith Huston, Inc. greatly appreciate your providing us your feedback regarding our book, *Estimating for Landscape & Irrigation Contractors (ELIC)*. Your comments about our products and services are indispensable and are an integral part of the never-ending process to improve and to add to these products and services. Please circle the number which corresponds to your answer. The numbers represent the following: 1 = no, not at all, or unsatisfactory; 2 = poor; 3 = fair; 4 = good; 5 = yes, very much, or excellent.

	NO/ UNSAT	POOR	FAIR	GOOD	YES/ EXCEL
1. Is the cover and format of the book pleasing to the eye?	1	2	3	4	5
2. Is the Table of Contents laid out well and does it facilitate locating key topics?	1	2	3	4	5
3. Is the book easy to read and understand?	1	2	3	4	5
4. Will you recommend to other contractors that they read *ELIC*?	1	2	3	4	5

5. What did you find most helpful in the book?_____
_____
_____
_____
_____
_____

6. What could have been added to or improved in *ELIC*?_____
_____
_____
_____
_____
_____

7. What immediate changes will you make in your company as a result of reading *ELIC*? __

_____
_____
_____
_____
_____
_____

8. Any additional comments that you would like to make or that you think would be helpful?

_____
_____
_____
_____
_____
_____
_____
_____
_____
_____
_____

9. Have you read our other books?

Strategic Planning for L&I Contractors   _____ Yes _____ No

Preparing for & Responding to a Down Economy _____ Yes _____ No

Thank you for your participation in this survey. Please mail it to:

SMITH HUSTON, INC.
P.O. Box 1244
Englewood, CO 80150-1244

ORDER FORM
(Return Entire Page)

Please send me the following books:

Qty	Titles	Unit Price	Total
	Strategic Planning for Landscape & Irrigation Contractors	$49.00	
	Preparing for and Responding to a Down Economy	$49.00	
	Estimating for Landscape & Irrigation Contractors	$75.00	

Totals _____

Colorado Add 6% Tax _____

Shipping & Handling* _____

* Shipping (Allow 4 to 6 weeks for delivery):
    Book Rate: $3.00 for the first book and $1.00 for each additional book.

Name: _____

Company: _____

Address: _____

City: _____ State: _____ Zip: _____-_____

Phone No: (_____)_____ Best Time To Call: _____

FAX No: (_____)_____

Mail Check To: **Smith Huston, Inc.**
           P.O. Box 1244
           Englewood, CO 80150-1244

E-mail: shi@smith-huston.com

Any Questions Call   800-451-5588
Or Fax   800-451-5494

ORDER FORM
(Return Entire Page)

Please send me the following books:

Qty	Titles	Unit Price	Total
	Strategic Planning for Landscape & Irrigation Contractors	$49.00	
	Preparing for and Responding to a Down Economy	$49.00	
	Estimating for Landscape & Irrigation Contractors	$75.00	

Totals _____ _____

Colorado Add 6% Tax _____

Shipping & Handling* _____

* Shipping (Allow 4 to 6 weeks for delivery):
    Book Rate: $3.00 for the first book and $1.00 for each additional book.

Name: _____

Company: _____

Address: _____

City: _____ State: _____ Zip: _____-___

Phone No: (_____)_____ Best Time To Call: _____

FAX No: (_____)_____

Mail Check To: **Smith Huston, Inc.**
P.O. Box 1244
Englewood, CO 80150-1244

E-mail: shi@smith-huston.com

Any Questions Call   **800-451-5588**
Or Fax   **800-451-5494**

INDEX